# Global Perspectives on Health Geography

**Series editor**

Valorie Crooks, Department of Geography, Simon Fraser University,
Burnaby, BC, Canada

Global Perspectives on Health Geography showcases cutting-edge health geography research that addresses pressing, contemporary aspects of the health-place interface. The bi-directional influence between health and place has been acknowledged for centuries, and understanding traditional and contemporary aspects of this connection is at the core of the discipline of health geography. Health geographers, for example, have: shown the complex ways in which places influence and directly impact our health; documented how and why we seek specific spaces to improve our wellbeing; and revealed how policies and practices across multiple scales affect health care delivery and receipt.

The series publishes a comprehensive portfolio of monographs and edited volumes that document the latest research in this important discipline. Proposals are accepted across a broad and ever-developing swath of topics as diverse as the discipline of health geography itself, including transnational health mobilities, experiential accounts of health and wellbeing, global-local health policies and practices, mHealth, environmental health (in)equity, theoretical approaches, and emerging spatial technologies as they relate to health and health services. Volumes in this series draw forth new methods, ways of thinking, and approaches to examining spatial and place-based aspects of health and health care across scales. They also weave together connections between health geography and other health and social science disciplines, and in doing so highlight the importance of spatial thinking.

Dr. Valorie Crooks (Simon Fraser University, crooks@sfu.ca) is the Series Editor of Global Perspectives on Health Geography. An author/editor questionnaire and book proposal form can be obtained from Publishing Editor Zachary Romano (zachary.romano@springer.com).

More information about this series at http://www.springer.com/series/15801

Yongmei Lu • Eric Delmelle
Editors

# Geospatial Technologies for Urban Health

Springer

*Editors*
Yongmei Lu
Department of Geography
Texas State University
San Marcos, TX, USA

Eric Delmelle
Department of Geography and Earth Sciences
University of North Carolina at Charlotte
Charlotte, NC, USA

ISSN 2522-8005 ISSN 2522-8013 (electronic)
Global Perspectives on Health Geography
ISBN 978-3-030-19572-4 ISBN 978-3-030-19573-1 (eBook)
https://doi.org/10.1007/978-3-030-19573-1



This Springer imprint is published by the registered company Springer Nature Switzerland AG
The registered company address is: Gewerbestrasse 11, 6330 Cham, Switzerland

# Acknowledgments

This book would not be possible without the strong support we received from our colleagues, friends, and family members. First, the editors would like to thank the reviewers for the manuscripts included in this book. Each chapter went through at least two rounds of rigorous reviews. Through investing their time and sharing their valuable suggestions, these scholars (in alphabetical order) have helped improve the book significantly: Angela Antipova, Department of Earth Sciences, University of Memphis; Luke Bergman, Department of Geography, University of British Columbia; Ryan Burns, Department of Geography, University of Calgary; Irene Casas, School of History and Social Science, Louisiana Tech University; Xiang (Peter) Chen, Department of Emergency Management, Arkansas Tech University; Serena Coetzee, Department Geography, Geoinformatics and Meteorology, University of Pretoria; Dajun Dai, Department of Geosciences, Georgia State University; Michael Desjardins, Department of Geography and Earth Sciences, University of North Carolina, Charlotte; Coline Dony, American Association of Geographers; Fazlay Faruque, Department of Preventive Medicine, John D. Bower, School of Population Health, University of Mississippi; David Hondula, School of Geographical Sciences and Urban Planning, Arizona State University; Karen Kemp, Spatial Sciences Institute, University of Southern California, Dornsife; Wen Lin, School of Geography, Politics and Sociology, Newcastle University; Yingru Li, Department of Sociology, University of Central Florida; Sara McLafferty, Department of Geography and Geographic Information Science, University of Illinois; Lan Mu, Department of Geography, University of Georgia; Alan Murray, Department of Geography, University of California, Santa Barbara; Tonny Oyana, Department of Preventive Medicine, University of Tennessee Health Science Center; Molly Richardson, Department of Population Health Sciences, Virginia Polytechnic Institute and State University; Rick Sadler, Department of Family Medicine, Michigan State University; Alexander (Sasha) Savelyev, Department of Geography, Texas State University; Jerry Shannon, Department of Geography, University of Georgia; Michael Widener, Department of Geography and Planning, University of Toronto; and Benjamin Zhan, Department of Geography, Texas State University.

We wish to express our gratitude to our friends at Springer Sciences. Special thanks go to Zachary Romano, Associate Editor, Earth Sciences, Geography and Environment. Without Zachary's initiation for a discussion on such a book project and his support throughout the whole process from project proposal to approval, this book would never be conceived, let alone come into being. We would like to also thank Aaron Schiller, Editorial Assistant, Earth Sciences, Geography and Environment, for his consistent assistance throughout this project. Further thanks go to our book project coordinators, Dinesh Shanmugam (until March 2018) and Krishnan Sathyamurthy (since March 2018), both of whom are Production Editors at Springer Sciences.

Both editors of this book work in academia, and we always highly appreciate the freedom for intellectual exploration and the support for such. Our book project would not be possible without the support from our respective universities. Yongmei Lu would like to express her special appreciation to the faculty and staff at the Department of Geography, Texas State University, especially for the level of support she received during the transition period to uptake administrative duties while working on this book project. Eric Delmelle wishes to thank his current and former graduate students for their support with this book project, particularly Michael Desjardins, Claudio Owusu, Yu Lan, Alexander Hohl, and Coline Dony.

Last but not the least, we are indebted to our families and loved ones. No support is stronger than a morning kiss after the many evening-into-early-morning hours of working on the book project. No understanding is more touching than a pizza dinner without complaint when dinner cooking time is donated to this book project. Yongmei Lu's deepest thanks go to her husband, James, and her children, Kati and Jeffrey, for their love and support during and beyond this book project. Eric Delmelle is grateful to his family for the continued support they have provided over the years.

# Contents

# Contributors

**Chris I. Ardern** School of Kinesiology and Health Science, York University, Toronto, ON, Canada

**Lisa Azu-Popow** Community Services/External Affairs, Northwestern Memorial HealthCare, Chicago, IL, USA

**Dongmei Chen** Department of Geography and Planning, Queen's University, Kingston, ON, Canada

**Sunghwan Cho** Korea Land and Geospatial Informatrix Corporation, Deokjin-gu, Jeonju-si, Jeollabuk-do, South Korea

**Anrong Dang** School of Architecture, Tsinghua University, Beijing, China

**Eric Delmelle** Department of Geography and Earth Sciences, The University of North Carolina at Charlotte, Charlotte, NC, USA

**Coline C. Dony** American Association of Geographers, Washington, DC, USA

**Qinjin Fan** Department of Geography, University of Georgia, Athens, GA, USA

**Emily Fekete** American Association of Geographers, Washington, DC, USA

**Christopher M. Fuhrmann** Department of Geosciences, Mississippi State University, Starkville, MS, USA

**Kathryn Grace** Department of Geography, Environment and Society, University of Minnesota, Twin Cities, MN, USA

**Megan Hinchy** Consortium to Lower Obesity in Chicago's Children, Ann and Robert H. Lurie Children's Hospital, Chicago, IL, USA

**Lauren Jewett** Department of Geography & Planning, University of Toronto, Toronto, ON, Canada

**Xiangyu Jiang** Department of Geography, State University of New York at Buffalo, Buffalo, NY, USA

**Dohyeong Kim** University of Texas at Dallas, Richardson, TX, USA

**Marynia Kolak** Center for Spatial Data Science, University of Chicago, Chicago, IL, USA

**Yongmei Lu** Department of Geography, Texas State University, San Marcos, TX, USA

**Ross Maciejewski** School of Computing, Informatics, and Decision Systems Engineering, Arizona State University, Tempe, AZ, USA

**Geraldine Malana** Erie Humboldt Park Health Center, Chicago, IL, USA

**Holly Manprisio** Community Services/External Affairs, Northwestern Memorial HealthCare, Chicago, IL, USA

**Lan Mu** Department of Geography, University of Georgia, Athens, GA, USA

**Alan T. Murray** Department of Geography, University of California, Santa Barbara, CA, USA

**Reilley Plue** Department of Geography & Planning, University of Toronto, Toronto, ON, Canada

**Jennifer D. Runkle** North Carolina Institute for Climate Studies, North Carolina State University, Asheville, NC, USA

**Michael Steptoe** School of Computing, Informatics, and Decision Systems Engineering, Arizona State University, Tempe, AZ, USA

**Margaret M. Sugg** Department of Geography and Planning, Appalachian State University, Boone, NC, USA

**Michael Vingiello** The Water Institute of the Gulf, Baton Rouge, LA, USA

**Fahui Wang** Department of Geography and Anthropology, Louisiana State University, Baton Rouge, LA, USA

**Lei Wang** Institute of Remote Sensing and Digital Earth, Chinese Academy of Sciences, Beijing, China

Department of Geography and Planning, Queen's University, Kingston, ON, Canada

**Ran Wei** School of Public Policy and Center for Geospatial Sciences, University of California, Riverside, CA, USA

**Michael J. Widener** Department of Geography & Planning, University of Toronto, Toronto, ON, Canada

Dalla Lana School of Public Health, University of Toronto, Toronto, ON, Canada

**Imam M. Xierali** Department of Family and Community Medicine, University of Texas Southwestern Medical Center, Dallas, TX, USA

**Xiaobai A. Yao** Department of Geography, University of Georgia, Athens, GA, USA

**Eun-Hye Yoo** Department of Geography, State University of New York at Buffalo, Buffalo, NY, USA

**Xuan Zhang** Department of Geography, University of Georgia, Athens, GA, USA

# Introduction

**Yongmei Lu and Eric Delmelle**

**Abstract** This chapter provides an overview of the background and content of this book. Starting with a discussion on the recent edited volumes on or closely related to urban health, this chapter highlights the need for a book on geospatial technologies for the study of urban health. The uniqueness of geospatial approaches to investigate urban health issues can be attributed to the spatial perspective and the lens of place. This chapter further argues that the continuous development in geospatial technologies, coupled with recent advances in communication and information technologies, portable sensor technologies, and the various social media and open data, has played an essential role for the modelling of environment exposure and health risk. However, there still exist challenges for urban health studies. These challenges maybe rooted in, among the multiple causes, a lack of understanding of the micro-level health decisions and the methodological limitation to address the Uncertain Geospatial Contextual Problem. This chapter finishes with a section-by-section and chapter-by-chapter overview of the empirical studies included in this book volume. This overview is provided to illustrate the organization of this book and to serve as a guide for a reader to navigate through the book chapters.

## 1 Overview

With 55% of the world's population living in urban areas and an expectation that the proportion of urban population worldwide will increase to 68% by 2050 (UN DESA 2018), urban health is among the top agenda items for governments, researchers, and the public. This book is an edited volume of research papers to showcase how

Y. Lu (✉)
Department of Geography, Texas State University, San Marcos, TX, USA
e-mail: YL10@txstate.edu

E. Delmelle
Department of Geography & Earth Sciences, The University of North Carolina at Charlotte, Charlotte, NC, USA



Y. Lu, E. Delmelle (eds.), *Geospatial Technologies for Urban Health*, Global Perspectives on Health Geography, https://doi.org/10.1007/978-3-030-19573-1_1

geospatial technologies are used to empower our understanding of urban health. *Urban health* refers to not only disease burdens and the related disparities in urban areas, but also health services and access to such, health behaviour and lifestyle, and the impact of health policies and practices in urban areas. *Geospatial technologies* include both the traditional Geographic Information System (GIS) and Remote Sensing (RS) technologies, and more importantly, the continuous development in Global Positioning System (GPS) and tracking/locational technologies, location-enabled online services and social media, volunteered geographic Information (VGI), and portable sensors, as well as the advances of such technologies in urban health applications with support from big and open data.

A number of edited books have investigated urban health and the related issues from different perspectives. Among them are some well-received volumes that were published since the turn of this century. The book by Galea and Vlahov (2006) examines how cities and city lifestyle may affect the overall population health. The volume by Corburn (2009) and the one by Sarkar et al. (2014) adopt the lens of urban planning and urban management to investigate urban health. Some books promote interdisciplinary approach for understanding the impact of urban settings on health (e.g. Freudenberg et al. 2009). Other books underscore the need for empowering local data to examine health disparity (e.g. Whitman et al. 2011). When putting into a broad context and a large spatial scale, some other books emphasize the importance of global change for urban health by connecting to demographic, climate, and globalization dynamics (e.g. Vlahov et al. 2010). The edited volume by Hynes and Lopez (2009) is one of the few books that address the role of space and geography for urban health; this book discusses the impact of urban environment on the health situation of U.S. cities through examining the social, built, and physical environments. However, there is a lack of recognition by the existing books of the potential of geographical approach and geospatial technologies for the study of urban health.

A geographical approach allows the examination of urban health from a spatial perspective and through a place lens. The spatial perspective emphasizes how and why health risks and disease burdens are spatially distributed and connected the way they are. The place lens supports the investigation of how the social, cultural, economic, and physical environments interact with people within a specific urban environment to shape the health of its population. With its holistic worldview that supports an integrated examination of the multiple aspects of man-environment interaction and of the cross-scale dynamics of risk factors and disease patterns, geography brings to health studies the unique geospatial approach. Health geography, as a subdiscipline of human geography (Dummer 2008), has been leading the examination of health issues within a geospatial context, which can be defined by the physical, socioeconomic, cultural, political, and policy aspects of a place. Applying the geospatial approach to urban health research is a subfield of health geography that highlights the unique opportunities and challenges for understanding health issues in an urban environment, including the highly concentrated population and resources, as well as the urban pollution and other human impacts. Moreover, the geospatial approach to urban health provides a channel for geography to offer important methodologic contributions to urban health studies. The potentials of geographic information system (GIS) for health research have been well recognized in literature (e.g. McLafferty 2003; Nykiforuk and Flaman 2011; Kirby et al. 2017). GIS is commonly used for urban health research to analyse and visualize disease and risk patterns, the spatiotemporal association of such with the selected socioeconomic, environmental and policy context, and the distribution of and access to health services.

The recent development in geospatial technologies has created new opportunities for urban health studies to further the understanding of health challenges and to develop appropriate health management strategies. With the continuous development in location acquisition and communication technologies and in sensor data and technologies, the studies of urban health are able to examine environment exposure and health risk at a much finer scale and (near) real-time. Dummer (2008) is among the earliest to recognize that GIS can be aligned with global positioning system (GPS) to monitor and analyse the movement of people and their interaction with environment for health studies. Integrating GIS and GPS provides a promising solution to the Uncertain Geospatial Contextual Problem, a challenge for geographical studies in general (Kwan 2012). With a focus on health research, Fang and Lu (2012) conducted a comprehensive review of the different approaches to integrate GIS, GPS, and portable sensors for individual-level environmental exposure assessment. Lu and Fang (2015) reported one of the earliest experimental studies that integrates GIS, GPS, and air quality sensor in an urban area for real-time individual-level air pollution exposure and health risk modelling and visualization. Park and Kwan (2017) took this approach further to evaluate the environment exposure injustice at a multi-contextual scale. Chapter 2 in this book is yet another example of modelling individual environment exposure by integrating GIS, GPS, and portable sensor technologies.

Geospatial technologies can play a pivotal role in augmenting the traditional data with the various new and large real-time datasets for urban health studies. Geotagged social media data can be incorporated as an important data source for health studies. Empirical studies have successfully embraced such data into the geospatial analyses of health issues to detect the spatial patterns of depression among population (Yang and Mu 2015), examine the neighbourhood happiness, diet, and physical activity patterns (e.g. Nguyen et al. 2016), and evaluate urban dwellers' access to and utilization of physical activity facilities (Lu and Lu 2018), just to name a few. In addition, crowdsourcing data, data from smart phone and wearable devices, data collected through Internet of Things (IoT), and the various open data in combination with the traditional large geospatial data can be integrated and big data technologies be used for research and practice to improve urban health and citizen's well-being (Miller and Tolle 2016; Wang and Moriaty 2018). This may be extended to the various virtual reality technologies and applications as well. Researchers (e.g. Althoff et al. 2016) have confirmed a substantial short-term change in physical activity behaviour as a result of people engaging in playing mobile apps game such as Pokémon Go. Further, as argued by Boulos et al. (2017), the applications of virtual reality GIS (VGIS) and augmented reality GIS (ARGIS) may be incorporated into urban planning and emergency training to develop better urban health management and public health response.

Nevertheless, challenges still exist, some of which are due to the gaps in understanding urban health and the related issues while others are rooted in the current limitations of geospatial technologies and methods. One of the long-lasting challenges is to model micro-level human health behaviour, including both spatial decision and activity /lifestyle choice. While geospatial technologies can serve as the backbone to model the socioeconomic, cultural, and physical environments, there is limited means to incorporate the behaviour decision at sub-neighbourhood level (let alone individual level) into a health behaviour or lifestyle model. As discussed in Chap. 6 of this book, modelling the food environment based on activity space is not hard; the challenge is to discern if an individual is "passively exposed to a space or actively seek it out" when making food choice decision. This aligns with the difficulty in explaining the discrepancies between individuals' utilization of health services or physical activity facilities when their accessibilities are the same and the related sociodemographic variables are controlled. Some of the new data sources, such as geotagged social media data, may potentially help improve our understanding of such individual spatial decision through sentiment analysis and /or semantic analysis of fine-scale data (e.g. Lu and Lu 2018; Chaps. 8 and 9 of this book), but the accuracy of such analyses and their scalability need further examination.

Another challenge is related to the Uncertain Geospatial Contextual Problem (Kwan 2012), an inherited problem to the current geospatial approaches when environmental exposure is of concern. With the rapid development in data technologies, data for urban health studies have been growing in both volumes and types. While this provides great potentials for better capturing individual-level data, the challenge exists when linking these individual-level data with the environmental context data in order to model environmental exposure and to assess individual-level health risk. As pointed out by Robertson and Feick (2018), the uncertainties generated when linking the individual-level data with contextual information may lead to alternative findings. Fang and Lu (2011) proposed a framework using space–time cube to estimate the environmental exposure for a spatiotemporally located point or trajectory. Further studies are needed to evaluate the efficacy and scalability of such approach.

With the background discussed above, we are excited to present this book with the intention to illustrate the many potentials of geospatial technologies for urban health studies. Although there is a plethora of conference papers and journal articles that apply geospatial technologies to examine the aspects of urban health issues, there remains a lack of an edited volume that showcases the current status of research on the theme of geospatial technologies for urban research. The chapters included in this book each reports a unique application of geospatial technologies in tackling an urban health challenge. This edited volume collectively provides a snapshot of the current status in the field of applying geospatial technologies for urban studies. However, it is by no means our claim to capture a complete picture of all the promises geospatial technologies may offer for urban health studies. That would be an extremely challenging job given the constant and rapid development in geospatial technologies, data, and modelling.

## 2 Parts of This Book

The themes throughout this book reflect the advancement at the unique juxtaposition of urban health studies and geospatial technologies. This edited volume is articulated around four parts: (1) Urban Health Risk and Disease, (2) Urban Health Service Access, (3) Healthy Behaviour and Urban Lifestyle, and (4) Health Policies and Urban Health Management. These four parts are organized to reflect four of the most recognized aspects for urban health issues, with no intention of disclaiming the importance of other urban health themes. The health risk and disease patterns aspect is about what health problems occur where in an urban environment. Access to health service in an urban area reflects how the relevant resources and the locating and management of such are responsive, or not, to urban health challenges. Research on healthy behaviour and lifestyle examines how people interact with the living environment in urban areas through adopting certain lifestyles or behaviour preferences or patterns as related to the health outcomes. The theme on health policy and management addresses how geographical perspective and geospatial technologies can contribute to informed decisions at policy-making and health management levels. These parts together reflect the holistic perspective of health geography in general (Dummer 2008) and that of urban health studies supported by the contemporary geospatial technologies in particular.

The first part, *Urban Health Risk and Disease* contains three chapters that address an urban health risk or disease of broad concern. In Chap. 2, Sugg, Furhmann and Runkle provide a review of geospatial technologies to monitor extreme heat and the associated correlation with individual vulnerability in urban settings. Recent and projected changes in temperature extremes, including the intensification of heat waves, present a persistent health threat for urban residents. The authors argue that rapid advancements in low-cost wearable sensors and other mobile technologies can be leveraged to capture geo-referenced environmental exposure and health data to better understand and quantify the impacts of variations in individual microclimates. The chapter suggests that the emergence of new technologies and rich spatial datasets requires multi-disciplinary collaboration to advance the science on place-based exposure to thermal extremes and the associated health impacts for at-risk populations in urban environments. The authors advocate for the use of wearable, GPS-enabled sensors to enhance current exposure assessment methods by enabling researchers to continuously monitor time-activity patterns over extended time frames and construct dynamic and individualized spatial units for heat-health analysis in urban settings.

Chapter 3 by Wang, Arden and Chen reports on an empirical study that utilizes GIS and spatial analysis to enhance Cardiovascular Disease (CVD) surveillance through identifying the disease patterns and the relationships between CVD mortality and the risk factors. Ordinary Least Squares Regression (OLS) and Geographically Weighted Regression (GWR) techniques were applied to reveal the geospatial clustering of CVD in a mixed rural-suburban setting in Ontario, Canada. Built environment and immigrant time were found to be significantly associated with the CVD mortality. Moreover, this pilot work suggests that the integration of geospatial information with routinely collected surveillance data is a feasible means within the structure and resources of local public health units to assist in the identification of regional variation in CVD burden.

The association between particulate matters (PM2.5) exposure and adverse health effects has been well documented in the literature. However, many of these epidemiological studies rely primarily on data collected from sparse monitoring sites that operated only every so often. In Chap. 4, Jiang and Yoo present an approach that evaluates the effect of domain size on Community Multiscale Air Quality (CMAQ) modelling performance. CMAQ is a three-dimensional air quality model designed to describe chemical and physical processes in the atmosphere at multiple spatial scales over varying time periods. Increasingly, CMAQ model has been used in urban health studies to estimate spatially varying air pollution exposure.

The second part of this book, *Urban Health Service Access* contains three chapters that address accessibility issue to health services in urban environment through spatiotemporal analysis. These chapters demonstrate applications of both classical and new spatial technologies in modelling and depicting how different segments of urban population are facing varied challenges of health service accessibility. In Chap. 5, Wang, Vingiello, and Xierali examine spatial accessibility of primary care in Baton Rouge, Louisiana. The authors apply two popular accessibility measures (a proximity metric using travel time from the nearest facility, and the two-step floating catchment area -2SFCA). The authors demonstrate that the residents in urban areas generally enjoy shorter travel time from their nearest service providers as well as higher accessibility scores than the rural residents. Overall, disproportionally higher percentages of African Americans are in areas with shorter travel time to the nearest primary care providers and higher accessibility scores, so do the residents in areas of high poverty rates. However, the authors argue that this "reversed racial advantage" in spatial accessibility does not capture the nonspatial obstacles related to financial and other socioeconomic factors for African Americans (and population in poverty).

The topic of food access (and food deserts) has received a tremendous attention in the literature. Advancements in geospatial technologies including GIS and GPS have provided insights on how the retail food environment might be contributing to the ongoing obesity epidemic. Caution has been raised, however, around the potential for research that uses GPS-captured activity spaces to overestimate the impact that exposure to food retailers has on food choices and behaviour. It may become difficult to discern whether an individual is passively exposed to a space or actively seeks it out, and this phenomenon is generally referred to as a 'selective (daily) mobility bias'. In Chap. 6, Plue, Jewett and Widener review recent literature to identify and critique the methods proposed for handling this bias and offer recommendations to consider as the use of GPS-activity space studies continues to grow.

Rapid emergency response is critically important in the context of urban health. Previous research has suggested that providing prompt access to emergency medical services (EMS) may greatly improve the health outcomes of patients with urgent conditions. It is in this context that in Chap. 7, Cho and Kim apply a dynamic maximal covering location model to optimally locate the dispatch services of medical service to respond to emergency calls in the Gyeongnam Province (Korea) in 2014. The authors use Long Short Term Memory (LSTM) method (a machine learning approach) to forecast EMS demands based on historical data. Their results indicate that machine learning algorithms have the potential to support more efficient allocation of medical and health service resources, especially when the resources are limited.

The chapters in the third part, *Healthy Behaviour and Urban Lifestyle*, focus on incorporating geospatial technologies for the studies of health behaviour and urban lifestyle. These studies demonstrate how geospatial technologies can enable us to investigate the interaction of human beings with the built environment at both collective and individual levels. This in turn helps us understand how different health behaviour and lifestyle may have been developed and sometimes sustained/confined by certain population or society segments. The findings contribute to building a health culture that promotes active lifestyle and facilitates positive human and built environment interaction.

Existing walkability measurements have not considered some important components of the built environment, pedestrians' preferences, or walking purposes. As area-based measurements, they may also overlook some detailed walkability changes. In Chap. 8, Zhang and Mu propose the Perceived importance and Objective measure of Walkability in the built Environment Rating (POWER), considering both the perception of pedestrians and subjective characterizing of the urban built environment. Their approach incorporates online surveys and social media data; the survey is efficient in customizing for the specific urban environment and capturing the preferences of a local population, while the social media component aims at obtaining the general opinions from a broader audience. Using social media and survey can bring two scales together to provide a more complete understanding of walkability.

In Chap. 9, Dony and Fekete use data extracted from different social media platforms and apply sentiment analysis and maps to quantify and visualize aggregated opinions about public parks. This approach is particularly useful for city governments to leverage these publicly available data to complement the assessments they already perform about their park system, such as satisfaction surveys or quality assessments. The authors use public parks in Mecklenburg County, North Carolina (which encompasses the City of Charlotte) as a case study. Social media data are generated by urban residents continuously and in real-time; they capture citizen's needs, suggestions, and satisfaction of public spaces. Leveraging social media is not only a cost-effective complement to already existing data collection methods, but it also offers cities new ways to engage with their residents.

Part IV, *Health Policies and Urban Health Management* addresses urban health issue from the perspective of policy and management. The contributions are from those who conduct research in urban health management and policy development. In Chap. 10, Fan and Yao use spatiotemporal analysis and data mining to examine the 2014–2016 Ebola Virus Disease (EVD) outbreak in West Africa. Specifically, the authors mine spatial associations between disease patterns and other geographically distributed factors. The authors use fine-grained population data obtained through a population interpolation method to conduct healthcare accessibility analysis. Their results suggest that (1) poor accessibility to healthcare facilities and EVD clusters are identified in many urban areas as well as some remote areas and (2) EVD cases were more likely to be found in border areas of these countries. The findings suggest that planners and practitioners in this region should pay special attention to the border areas and cities of high population density when fighting to reduce the morbidity and mortality rates of EVD in the future.

Community asset mapping is an essential step in public health practice for identifying community strengths, needs, and ultimately health intervention strategies. In Chap. 11, Kolak, and colleagues advocate that new systems are needed to extend existing Volunteered Geographic Information (VGI) concepts to bridge community groups and health systems in collaboration. The authors demonstrate the usefulness of an open participatory asset mapping infrastructure developed with a Chicago community using VGI concepts, participatory design principles, and geospatial Software as a Service (SaaS) in an open software environment. Open infrastructures using decentralized system architecture can link data and mapping services, transforming siloed datasets to integrated systems managed and shared across multiple organizations.

In Chap. 12, Grace, Murray, and Wei develop and apply quantitative models that rely on remotely sensed data and health survey data to highlight the importance of different aspects of demand for food aid in urban spaces. Chronic food insecurity significantly constrains short- and long-term health, as well as the development of individuals and households, ultimately impacting economic progress in some of the poorest and fastest growing communities on the planet. Ensuring that food aid reaches the neediest people, however, is an ongoing challenge. In their chapter, the authors explore the use of geospatial technologies as part of a framework for improving food aid targeting in Bamako, Mali. The results highlight the usefulness of this approach for food aid planning in urban areas where food need is unevenly distributed over a densely populated area.

In summary, the papers in this book form a timely collection reporting on the progress, opportunities, and challenges regarding how urban health studies may benefit from the advancements of geospatial technologies. Meanwhile, this volume contributes to the conversation of how geospatial technologies and the related GIScience research may be enhanced through continuously addressing and responding to the data, modelling, and analytical challenges in urban health studies. This book targets audience with a background or interest in health and medical geography (including spatial epidemiology), social epidemiology, urban health management, health behaviour and lifestyle research, and healthcare delivery and access assessment. The book can also help experts in geospatial technologies and sciences broaden their application studies to urban health issues and challenges. The book is suitable for readers from both academic background and practical walks in urban health management and policy-making.

## References


Althoff, T., White, R. W., & Horvitz, E. (2016). Influence of Pokémon Go on physical activity: study and implications. *Journal of Medical Internet Research., 18*(12), e315.

Boulos, M. N. K., Lu, Z., Guerrero, P., Jennett, C., & Steed, A. (2017). From urban planning and emergency training to Pokémon Go: Applications of virtual reality GIS (VRGIS) and augmented reality GIS (ARGIS) in personal, public and environmental health. *International Journal of Health Geographics, 16*(7), 1–11.

Corburn, J. (2009). *Towards the healthy city: People, places, and the politics of urban planning*. Cambridge, MA: The MIT Press.

Dummer, T. J. (2008). Health geography: Supporting public health policy and planning. *CMAJ: Canadian Medical Association journal = journal de l'Association medicale canadienne, 178*(9), 1177–1180.

Fang, B. T., & Lu, Y. (2011). Constructing near real-time space-time cube to depict urban ambient air pollution scenario. *Transactions in GIS, 15*(5), 635–649.

Fang, T. B., & Lu, Y. (2012). Personal real-time air pollution exposure assessment methods promoted by information technological advances. *Annals of GIS, 18*(4), 279–288.

Freudenberg, N., Klitzman, S., & Saegert, S. (2009). *Urban health and society: Interdisciplinary approaches to research and practice*. San Francisco: Jpssey-Bass.

Galea, S., & Vlahov, D. (2006). *Handbook of urban health: Populations, methods, and practice*. New York: Springer-Verlag.

Hynes, H. P., & Lopez, R. (2009). *Urban health: Readings in the social, built, and physical environments of U.S. Cities*. Sudbury, MA: Jones and Bartlett Publishers.

Kirby, R. S., Delmelle, E., & Eberth, J. M. (2017). Advances in spatial epidemiology and geographic information systems. *Annals of Epidemiology, 27*(1), 1–9.

Kwan, M.-P. (2012). The uncertain geographic context problem. *Annals of the Association of American Geographers, 102*(5), 958–968.

Lu, Y., & Fang, T. B. (2015). Examining personal air pollution exposure, intake, and health danger zone using time geography and 3d geovisualization. *ISPRS International Journal of Geo-Information., 4*(1), 32–46.

Lu, Y., & Lu, F. (2018). *Physical activities, BMI, and accessibility to and utilization of facilities*. Paper presented at the *Annual Meeting of American Association of Geographers*. New Orleans, LA. April 10–14, 2018.

McLafferty, S. L. (2003). GIS and health care. *Annual Review of Public Health, 24*, 25–42.

Miller, H. J., & Tolle, K. (2016). Big data for healthy cities: Using location-aware technologies, open data and 3D urban models to design healthier built environments. *Built Environment, 42*(3), 441–456.

Nykiforuk, C. I., & Flaman, L. M. (2011). Geographic information systems (GIS) for health promotion and public health: A review. *Health Promotion Practice, 12*, 63–73.

Nguyen, Q. C., Kath, S., Meng, H. W., Li, D., Smith, K. R., VanDerslice, J. A., Wen, M., & Li, F. (2016). Leveraging geotagged Twitter data to examine neighborhood happiness, diet, and physical activity. *Applied geography (Sevenoaks, England), 73*, 77–88.

Park, Y. M., & Kwan, M.-P. (2017). Multi-contextual segregation and environmental justice research: Toward fine-scale spatiotemporal approaches. *International Journal of Environmental Research and Public Health, 14*, 1205.

Robertson, C., & Feick, R. (2018). Inference and analysis across spatial supports in the big data era: Uncertain point observations and geographic context. *Transactions in GIS, 22*, 455–476. https://doi.org/10.1111/tgis.12321.

Sarkar, C., Webster, C., & Gallacher, J. (2014). *Healthy cities: Public health through urban planning*. Cheltenham: Edward Elgar.

United Nations, Department of Economic and Social Affairs (UN DESA). (2018). *World Urbanization Prospects*. https://population.un.org/wup/. Last accessed on 23 Feb 2019.

Vlahov, D. J., Boufford, I., Pearson, C., & Norris, L. (2010). *Urban health: Global perspective*. San Francisco: John Wilson & Sons.

Wang, S., & Moriarty, P. (2018). Big data for urban health and Well-being. In S. J. Wang & P. Moriarty (Eds.), *Big Data for Urban Sustainability* (pp. 119–140). Cham: Springer International Publishing AG.

Whitman, S., Shah, A., & Benjamins, M. (2011). *Urban health: Combating disparities with local data*. New York: Oxford University Press.

Yang, W., & Mu, L. (2015). GIS analysis of depression among Twitter users. *Applied Geography, 60*, 217–223. https://doi.org/10.1016/j.apgeog.2014.10.016.

**Yongmei Lu** is a Professor and Chair of the Department of Geography, Texas State University. Dr. Lu's teaching and research interests fall under the broad umbrella of GIS and its application on human–environment interaction studies, particularly health and environmental issues, disease and crime patterns, access to services, and disparities. Dr. Lu's research has been supported by federal, state, and university funding.

**Eric M. Delmelle** is an Associate Professor of Geography and Earth Sciences at the University of North Carolina at Charlotte where he teaches undergraduate and graduate courses in GIScience, spatial optimization, geovisualization, GIS programming, and medical geography. Dr. Delmelle's research interests lie in GIScience, spatial analysis, epidemiology, and uncertainty.

# Part I
# Urban Health Risk and Disease

# Geospatial Approaches to Measuring Personal Heat Exposure and Related Health Effects in Urban Settings

**Margaret M. Sugg, Christopher M. Fuhrmann, and Jennifer D. Runkle**

**Abstract** Recent and projected changes in temperature extremes, including the intensification of heat waves, present a persistent health threat for urban residents. Due to limitations in data availability and the spatial representativeness of fixed-site temperature observations, there exists a notable gap in the geospatial sciences on the multi-scale characterization of geographic patterns of extreme heat and the associated correlation with individual vulnerability in urban settings. Studies employing individual-level exposure assessment methodologies are sparse. Yet rapid advancements in low-cost wearable sensors and other mobile technologies can be leveraged to capture geo-referenced environmental exposure (e.g., temperature) and health data (e.g., physiologic strain) to better understand and quantify the impacts of variations in individual microclimates. The emergence of new technologies and rich spatial datasets requires multi-disciplinary collaboration to advance the science on place-based exposure to thermal extremes and the associated health impacts for at-risk populations in urban environments.

M. M. Sugg (✉)
Department of Geography and Planning, Appalachian State University, Boone, NC, USA
e-mail: kovachmm@appstate.edu

C. M. Fuhrmann
Department of Geosciences, Mississippi State University, Starkville, MS, USA
e-mail: cmf396@msstate.edu

J. D. Runkle
North Carolina Institute for Climate Studies, North Carolina State University, Asheville, NC, USA
e-mail: jrrunkle@ncsu.edu

 
Y. Lu, E. Delmelle (eds.), *Geospatial Technologies for Urban Health*,
Global Perspectives on Health Geography, https://doi.org/10.1007/978-3-030-19573-1_2

# 1 Introduction

Heat is one of the leading causes of weather-related death in the USA (NWS 2019), and two thousand temperature-related deaths are estimated to occur annually (Berko et al. 2014). Average temperatures across the USA increased by 1–2 °F over the past century, and climate change models project an increase in average temperatures ranging from 2 to 10 °F by the turn of the twenty-first century (NCA 2018). Recent evidence suggests that there is a limit to human adaptive capacity and our ability to adapt may likely be exceeded if climate change continues unmitigated (Sherwood and Huber 2010a; b).

Climate change-related increases in the intensity and frequency of hotter ambient temperatures will continue to negatively impact public health, particularly in densely populated urban areas where extreme temperatures are amplified by the urban heat island effect (Macintyre et al. 2018; Friel et al. 2011; Heaviside et al. 2017). In urban centers, prolonged exposure to high ambient temperatures and small seasonal deviations from average temperatures during the warmer months have been linked to increased risk of heat-related illness, exacerbation of chronic conditions like asthma or cardiovascular disease, and in severe cases, heat-related mortality (Sarofim et al. 2016). Yet, limited examples exist of the public health efforts in establishing real-time urban surveillance networks or deriving early warning systems targeting vulnerable segments of the population (Ebi et al. 2004).

The adverse health impacts of exposure to thermal extremes vary geographically and across vulnerable segments of the population, making it difficult to apply universal temperature-health thresholds across a range of urban environments. Large spatio-temporal variations exist in heat exposure due to individual-level differences in mobility patterns and microenvironments. Traditionally, thermal exposure has been estimated using temperature observations from fixed-site (in situ) weather stations or spatially and temporally coarse remotely sensed imagery, which is often limited by cloud cover and the timing of satellite orbits. However, the spatial distribution of these data is not sufficient to assess the fine-scale spatial patterns of temperature needed to provide the necessary context behind temperature-health associations. Indeed, a major limitation in the study of temperature exposure is the paucity of individual-level data, resulting in potential exposure misclassification and biased estimates of heat-related health effects. In recent years, a variety of low-cost environmental sensors have been used in crowd-sourced participatory sensing projects with a particular focus on real-time and continuous monitoring of personal exposure to air pollution (e.g., De Nazelle et al. 2013; Steinle et al. 2015; Castell et al. 2017; Schneider et al. 2017; Heimann et al. 2015; Gao et al. 2015; Dewulf et al. 2016).

This chapter reviews contemporary themes for exposure assessment in the context of heat-health and personal heat exposure in urban areas. In Sect. 2, we address the need for advances in personal heat exposure assessment studies by discussing the spatial variations in heat risk within cities and the differential vulnerability across urban populations. Contemporary studies and current methods for measuring personal exposures are discussed in Sect. 3. In Sect. 4, we provide examples of the theoretical implications of personal monitoring devices and how such methodologies address previous limitations of public health and geographic research. We conclude this chapter by discussing the future implications and research needs to further advance geospatial analysis and monitoring of personal heat exposure in an urban environment:

## 2 Spatial Variation in Urban Heat Exposure and Individual Health Risk

The adverse health impacts of exposure to thermal extremes vary within and between urban communities and across vulnerable subgroups, including the young and elderly, the chronically ill, outdoor workers, athletes, and low-income persons (Sarofim et al. 2016), making it challenging to identify universal temperature-health warning thresholds within an urban environment. Certain social and physical features of the urban environment are associated with increased risk of adverse heat-health effects, including recent increases in population growth and density, population age, housing type, preexisting conditions, and location within the urban heat island (Macintyre et al. 2018; Vlahov and Galea 2002). In fact, research has demonstrated a social gradient in heat-related health risks whereby the urban poor, characterized by lower socioeconomic status, and minority racial and ethnic groups are more likely to live in warmer neighborhoods lacking green space and work in hotter and more humid environments, including poorly ventilated buildings (Friel et al. 2011).

Urban populations may be disproportionately vulnerable to hotter ambient temperatures due to both increased greenhouse gas concentrations and the urban heat island (UHI) effect (Hondula et al. 2017), which involves areas where vegetative surfaces or natural covering that typically reflect heat have been replaced with impervious surfaces that retain heat and are thereby associated with elevated daytime and nighttime temperatures compared to less urban or more rural landscapes (Wong et al. 2011; Heaviside et al. 2017). For example, densely populated urban communities that lack green space experience maximum daytime temperatures that are on average up to 4 °F hotter than urban communities with parks and greenscapes (Friel et al. 2011; Wong et al. 2011). Moreover, these urban-rural temperatures differences are maximized in the nighttime hours, a time when many individuals require cooler temperatures to mitigate their cumulative daily heat exposure (Fischer et al. 2012). As a result, heat exposure for urban populations exhibits significant variation across urban surfaces due to inherent spatial variations in the built and physical environment that is also highly influenced by the UHI. These variations have and will likely continue to be magnified at the scale of the individual by social determinants of health (e.g., poverty, low health literacy, access to care, social isolation, green space, high-crime neighborhoods, and poor housing stock) (Reid et al. 2009; Hondula et al. 2015a, b). As cities continue to grow in physical size and population, so will the potential health burden on urban residents (Hondula et al. 2015a).

The study of climate impacts on urban health presents new scientific and methodological challenges, particularly the assessment of climate-related changes in individual-level temperature exposure and associated health risks. A large body of evidence from the fields of epidemiology and medical geography have demonstrated the significant influence of place on health, even after adjusting for individual factors and behaviors, and research has shown that this relationship is highly dynamic and comprised of a series of spatially and temporally interdependent exposure relationships that are context-specific (e.g., Macintyre et al. 2002; Tunstall et al. 2004; Hondula et al. 2015b). Yet, population health experts have traditionally relied on survey responses, personal observations, or time-activity diaries to reconstruct temperature exposure histories, which are subject to recall bias and may result in exposure misclassification (i.e., dilution or underestimation of the true effect of temperature exposure on a particular health endpoint). On the other hand, geographers routinely rely on publicly available, static datasets for heat-health research, whereby exposure is aggregated to a single spatial unit (e.g., census tract) and point in time, resulting in further misclassification of the context in which individual variation in health status changes in response to fluctuations in temperature exposure. Recent advancements in GPS-tracking technology and low-cost wearable sensors have significant potential to broaden the geographic and time scales of environmental exposure measurement, especially as it pertains to establishing smart city surveillance networks for monitoring climate impacts on vulnerable urban populations (e.g., Muller et al. 2015; Chapman et al. 2015; Meier et al. 2017; Chapman et al. 2017). In the urban context, wearable environmental sensors have already been used to measure a range of toxic and harmful environmental exposures including pesticides, air pollution (e.g., PM2.5, PM10), and carbon monoxide to name a few (Dons et al. 2017; Rainham 2016). There is a growing effort to harness sensor applications in the design of smart cities (Hancke et al. 2012), but very few studies have employed personal monitoring of individually experienced ambient temperatures (Kuras et al. 2015; Bernhard et al. 2015; Basu and Samet 2002; Uejio et al. 2018). These GPS-enabled personal monitoring technologies have the power to transform scientific understanding of how characteristics of geographic location (i.e., "place") and the context of social and environmental exposures interact over time to influence health at the individual level. Wearable sensors can be used to enhance current exposure assessment methods by enabling researchers to continuously monitor time-activity patterns over extended time frames and construct dynamic and individualized spatial units for heat-health analysis in an urban setting. These data can be used to record physiologic response (e.g., heart rate) in real time in response to changing environmental conditions, quantify daily patterns of exposure and corresponding physiologic response that can be harnessed to establish personalized baselines for at-risk individuals, and detect adverse health events or provide early warning systems in advance of an adverse health event. Public health professionals can then rely on these data to provide situational awareness in which detected variations or trends in health can be used to make recommendations on heat reduction strategies and subsequent health risks. The introduction of time-location data provides finer-scale spatial and temporal context to then make inferences on the types of daily activities, duration of exposure, and behavioral modifications that influence heat-health outcomes. Wearable technology will empower underrepresented urban communities to provide high-resolution environmental monitoring data to better understand and creatively address place-based heat-health concerns.

## 3 Measuring Personal Heat Exposure

Most studies in urban climatology and biometeorology have focused on measuring urban-rural temperature differences and their impacts at the city scale (Hondula et al. 2017; Sheridan and Allen 2018). As such, there is considerable information on regional variability in UHI structure and heat-related health risks (Karimi et al. 2017; Sheridan and Allen 2018).

In contrast, there are relatively fewer studies that have examined fine-scale (i.e., intra-city) variability in temperature and associated health effects (Hondula et al. 2017). However, such studies are becoming more common, as it is recognized that not all urban residents are equally vulnerable to extreme heat or experience the same thermal environments (Sheridan and Allen 2018). For example, Hondula et al. (2015b), in a study of seven US cities, found significant increases in mortality during extreme heat events in only about half of the postal codes within each city. Demographic information from each of these postal codes revealed specific risk factors that may have been masked at the broader city scale. A better understanding of the spatial structure of urban temperature and associated health outcomes may result in more targeted intervention strategies focused on specific locations within a city where resources should be allocated (Hondula et al. 2015b).

### *3.1 Methodological Approaches*

There are three general approaches that have been taken to obtain fine-scale measurements of temperature in urban areas (Vant-Hull et al. 2014). The most common approach is the use of fixed-site weather stations, such as those maintained by the US National Weather Service and Federal Aviation Administration. These stations, many of which are automated, provide continuous observations of numerous meteorological variables at high temporal resolution (seconds to hours). Such stations are often restricted to airports and other remote locations, though some instrument packages and data loggers (e.g., HOBO Micro-Stations) may be mounted on lampposts to measure the influence of buildings and trees (e.g., skyview fraction) on the street-level spatial structure of the urban climate (Karimi et al. 2017).

Another approach is the use of remotely sensed data from satellites, such as MODIS, Landsat, and ASTER. While satellite-based measurements of temperature provide better spatial resolution than most fixed-site station networks (10s to 100 s of meters), they are hindered by intermittent temporal coverage and cloud cover.

Detailed satellite observations of the urban environment, particularly at street level, can also be obstructed by buildings. In addition, satellites typically measure surface temperature, such as that on rooftops, treetops, and parking lots, not the overlying air temperature (Karimi et al. 2015; Karimi et al. 2017).

The third approach, which overcomes many of the limitations of fixed-site and satellite approaches, involves the use of mobile instruments (e.g., thermometers) to identify local "hotspots" within the city. Examples include walking campaigns where individuals use handheld devices or sensors attached to their clothing or carried in a backpack to record street-level temperatures (Kuras et al. 2017; Karimi et al. 2015; Karimi et al. 2017; Vant-Hull et al. 2014; Tsin et al. 2016). More sophisticated mobile data packages may include additional instruments to measure radiation, humidity, and wind, which can be used to model the thermal comfort of urban residents (Vant-Hull et al. 2014). When combined with information on building geometry, land cover characteristics, and elevation, these measurements can inform both short-term meteorological forecasts and long-term planning of more efficient and comfortable urban spaces (Karimi et al. 2015).

## 3.2 *Recent Advancements*

While these approaches have helped identify the hottest places in cities, they do not, on their own, reveal how often, how long, and under what circumstances urban residents actually encounter these conditions. Such information may be obtained through personal heat exposure research, which shifts the focus from places and populations to people and individuals. Since fine-scale thermal variability has been well documented in urban areas, this type of research may be particularly beneficial, as urban residents move through several different thermal environments over the course of a day (Dias and Tchepel 2014; Kuras et al. 2017; Dėdelė et al. 2018; Reis et al. 2018).

Recent studies have found substantial variability in personal heat exposure not only within urban areas (Kuras et al. 2015; Basu and Samet 2002; Uejio et al. 2018) but across more rural and heterogeneous land cover types (Bernhard et al. 2015; Sugg et al. 2018). Compared to fixed-site observations, which have traditionally been used to estimate personal heat exposure, individually experienced temperatures (IETs, Kuras et al. 2015) may be warmer or cooler depending on social and behavioral factors, as well as adaptive capacity (e.g., mitigation strategies) (Kuras et al. 2017). In cities, personal exposure is also affected by aspects of the built environment, such as the spatial and temporal structure of the UHI and access to shading and green spaces (Jenerette et al. 2016). Time-activity diaries can provide complementary information on the circumstances surrounding personal heat exposure, such as whether the individual was indoors or outdoors, in transit, or participating in a strenuous activity that might result in heat-related illness or injury (Sugg et al. 2018). By pairing individual temperature observations with location-specific time-activity patterns, researchers can create a citywide "hazard-scape" that paints a more comprehensive image of heat vulnerability at the individual level (Mehdipoor et al. 2017).

## 3.3 Methodological Considerations and Limitations

Gaining a better understanding of vulnerability to extreme heat requires measuring environmental conditions that individuals actually experience. As previously discussed, traditional research methods have involved either direct or indirect measurements of outdoor conditions. However, it has been found that most people spend up to 90% of their day indoors (Klepeis et al. 2001), which typically provides a respite from extreme outdoor conditions by reducing exposure to solar radiation and maintaining comfortable and consistent thermal conditions through the use of air conditioning (Kuras et al. 2017). As such, it is likely that traditional research methods using fixed-site and remotely sensed measurements are misclassifying actual exposure at the individual level (Bernhard et al. 2015).

Data on individual-level variation of indoor conditions within an urban environment is currently limited. Previous studies have attempted to relate outdoor conditions to indoor conditions, but the results are generally inconclusive (Hondula et al. 2017). Some studies have found a strong relationship between indoor and outdoor summer temperatures (Uejio et al. 2016; Quinn et al. 2014; Nguyen et al. 2014) and significantly higher indoor temperatures in a small subset of vulnerable patients who required emergency medical attention (Uejio et al. 2016). The relationship between extreme heat, indoor environments, and personal exposure is particularly complicated in urban areas, as some buildings may not be properly climate-controlled or constructed, and residents may be unwilling to open their windows due to the threat of crime. In these situations, indoor temperatures may exceed outdoor temperatures. In fact, during some severe heat events (e.g., July 1995 in the Midwest USA), most individuals who died of heat-related illness in cities were found in their homes with the windows closed (Klinenberg 2002).

Obtaining data on personal heat exposure in cities is now easier with emerging sensor technologies that are becoming more affordable and convenient, thereby allowing for the generation of large amounts of digital data at resolutions that can better inform public policy on themes such as urban design and environmental health (Mehdipoor et al. 2017). What remains uncertain is the accessibility of these sensors, particularly among low-income and underrepresented urban residents (i.e., the "digitally invisible," Longo et al. 2017).

While personal heat exposure research requires individual participation, most sensors are non-intrusive and do not interfere with daily activities, reducing the burden placed on study subjects (Sugg et al. 2018). However, sensor placement, particularly on clothing, should consider the contributions of body heat and perspiration to the thermal environment and experience of an individual (Kuras et al. 2017). This is particularly the case for those performing high-intensity activities, such as exercise or heavy lifting.

Due to the need for more precise locational information to assess personal heat exposure, it is important to consider the limitations of technologies that rely on satellite-derived information (e.g., GPS instruments and smart watches). In particular, the density and geometry of buildings in urban areas may result in decreased locational accuracy due to signal interference (Sugg et al. 2018). Daily activity diaries may supplement GPS data as well as provide important contextual information on exposure (e.g., time and duration of specific activities). However, such information is largely subjective and documentation may vary in detail from person to person (Kuras et al. 2017).

## 4 Geospatial Theoretical and Methodological Advancements Utilizing Wearable Sensor Technology

Today, recent development and widespread diffusion of geospatial data and technology (e.g., remote sensing, Global Positioning Systems, geographic information systems) are enabling the creation of highly accurate multidimensional spatial datasets that significantly enhance temporally linked health research. These advances warrant new methodological approaches in exposure assessment that couple geo-location with personal monitoring measurements to provide precise time-activity patterns of individuals as they move throughout urban environments. This inclusion of geolocation and personal monitoring measurements has shaped a new field in geography that addresses previous theoretical limitations, such as the modifiable areal unit problem and the uncertain geographic context problem. By addressing theoretical constraints within the field of geography, personal wearable devices are rapidly expanding new geospatial and digital public health methodologies for data collection and analysis, thus creating novel opportunities for public health education and targeted intervention for urban populations.

### *4.1 Theoretical Contributions*

Historically, geographers have been constrained by scale limitations in efforts to monitor finer patterns of environmental exposure and have expressed the need for more conceptual and methodological developments in space-time-geography to characterize environmental exposures, mobility patterns, or behavioral responses at the individual or neighborhood level (Kestens et al. 2017). Hagerstrand (1967, 1970) noted that accounting for the movement of people within their individual time-activity space is a crucial determinant of personal exposure assessment and provides the necessary context needed to characterize patterns of individual variations in heat-health responses. Despite this understanding, few studies have assessed personal exposure, particularly in the context of temperature. In this section, we address how personal wearable sensors provide solutions to common geographic problems, including the modifiable areal unit problem, and most recently, the uncertain geographic context problem.

### 4.1.1 Modifiable Areal Unit Problem

The modifiable areal unit problem (MAUP) was brought forth by Openshaw (1984) and describes the problems that arise from the analysis of zone-based data or delineating areal boundaries. Both urban and health geographers are often restricted by the MAUP as data are available only at aggregate units, such as administrative units, and restricted at the individual level due to privacy issues (Kwan 2012). For health and medical geographers, the MAUP problem is further compounded, as many studies use residential addresses as a proxy for temperature exposure and therefore fail to account for an individual's complex daily time-activity patterns. Researchers often use multilevel models to examine correlations between individual and area-based ambient temperature exposures on health outcomes to reduce biased inference originating from the MAUP (Diez-Roux 2000).

Despite this methodological progress, temperature exposure estimates derived from local weather stations are typically homogenously aggregated across a well-defined geographic unit (e.g., county, zip code, census tract), and multilevel models use these geographically aggregate units, which are not intended for health or environmental exposure research. Wearable sensor technologies enable the measurement of exposure to account for the "true spatial configuration" of an individual's exposure by recording their temperature as they move throughout their daily environment, subsequently addressing MAUP and accurately identifying temperature exposure (Kwan 2009).

### 4.1.2 Uncertain Geographic Context Problem

Recently, Kwan (2012) presented a new geographic theoretical limitation to health and mobility research that also applies directly to exposure assessment research. Unlike the MAUP, the Uncertain Geographic Context Problem (UGCoP), "arises because of the spatial uncertainty in the actual areas that exert contextual influences on the individuals being studied and the temporal uncertainty in the timing and duration in which individuals experienced these contextual influences" (Kwan 2012, p. 959). Thus, the UGCoP describes problems that arise in exposure assessment when the *exact* location and *timing* of the exposure are unknown. Many studies of environmental exposure have been designed in static spatial terms and, therefore, have largely ignored the roles of time and mobility that contribute to exposure (Kwan 2012, 2013). This can lead the underestimation or overestimation of the true exposure response in health studies (Kwan 2013). The emergence of wearable sensors enables researchers to conduct space-time studies and account for spatio-temporal patterns that address *where* exposure is occurring and the circumstances that result in adverse health outcomes.

In addition to addressing the UGCoP, new research can identify the temporal patterns that result in adverse health outcomes. Exposure can occur over multiple time periods and cumulative exposure, rather than intermittent exposure, may potentially result in health outcomes of varying severity. By examining the cascade of potential health outcomes using personal wearable devices, ranging from a slight state change in which an individual's physiologic response starts to deviate outside the "normal" range to a more severe response that includes heat strain, researchers can begin to identify the "temporal etiology" of certain temperature-related conditions (i.e., variations in health outcomes in response to intermittent and/or cumulative exposure), thereby providing new insights into the spatial and contextual processes that link changes in an individual's environment with corresponding changes in mobility, behavior, and health response.

## 4.2 *Methodological Needs and Examples*

### 4.2.1 GPS Tracking Technologies

Although travel and activity diaries have been used extensively to describe mobility patterns across various micro-environments, their utilization is time consuming, accuracy is limited by participant recall, and is burdensome for research participants over extended time periods. Global Position Systems (GPS) provide an objective and automated method to record mobility patterns with limited human effort and high accuracy for larger populations, particularly those in urban areas. Moreover, the inclusion of GPS with time and activity diaries provides quantitative positioning to the contextual details of participants' mobility patterns (i.e., activity type, participant comfort level, behavior modifications, etc.).

The inclusion of GPS coordinates into exposure assessment approaches can provide researchers with the ability to construct high-resolution spatio-temporal simulation models that indirectly calculate a range of exposures across a heterogeneous urban environment. These models have been used extensively in air quality research and have recently been employed in temperature studies (e.g., Steinle et al. 2015; Ryan et al. 2015; Nethery et al. 2014). Although more accurate than studies that disregard time-activity patterns, simulation models are limited by significant uncertainty as model estimation assumes many parameters, ignores contextual factors, and can disregard estimates of indoor exposure (Kuras et al. 2017). Wearable sensors that incorporate temperature data, as well as GPS, allow researchers to reduce uncertainty and provide datasets for model improvement and validation.

The utilization of GPS technology in personal exposure research can be enhanced with the use of smartphone technology. Smartphones provide a convenient, low-cost method to recruit participants for research and passively collect geo-located changes in daily activity levels, behavior, environmental exposures, and clinical characteristics (e.g., Fang and Lu 2012; Chan et al. 2018). An estimated 77% of Americans carry a smartphone, while slightly more, 8 out of 10, urban residents own a smartphone. Smartphone technology adoption has become pervasive in society and is embraced by individuals of all ages, races, education, and income brackets (Pew Research 2018). Moreover, smartphones provide a high-tech platform equipped with in-built sensors that allow for simultaneous sensing of multiple environmental and physiologic parameters, thus reducing participant burden and increasing data collection for researchers (Oliver et al. 2015; Helbich 2018). Future research is needed on the integration of smartphone-enabled passive collection of GPS and temperature studies to provide high-resolution spatio-temporal temperature data for a larger population that adequately characterizes mobility patterns.

### 4.2.2 Integration of Continuous Physiologic Monitoring

Health exposure assessments can also be enhanced with wearable sensors that provide measurements of physiologic well-being (e.g., heart rate, core body temperature, blood pressure). By combining ambient environmental conditions with personal physiologic measures, researchers can identify the precise environmental conditions that result in heat strain or other adverse health outcomes. These data can be used to determine thresholds for early warning systems and inform targeted public health interventions, thereby providing more informed climate change health risk assessments of environmental exposure and their resulting health impacts now and in the future.

### 4.2.3 Visualizing and Analyzing Space-Time Data

Kwan (2000) pioneered the space-time visualizations in the field of geovisual analytics by creating space-time methodological examples. Since then, multiple researchers have created visualization to assess space-time patterns of exposure, including clustering metrics, space-time tests, and path comparison indexes (An et al. 2015; Demšar and Virrantaus 2010). Unlike traditional geospatial outputs, space-time data and visualization still require significant computational resources, and previous work has utilized methods including parallel computing and decomposition algorithms to provide space-time interpolations and visual outputs (Desjardins et al. 2018). Presently, widespread GIS software is required to quickly create high-resolution space-time visualizations for pattern recognition of point data. Newer versions of ESRI products, including ArcPro, provide tools such as 3D space-time cubes and Emerging Hot Spot Analysis (i.e., space-time clustering detection) (ESRI 2018). However, their use is still restricted to point vector data, and these products fail to readily incorporate more dimensions beyond two-dimensional space and one-dimensional time, thus not allowing for the incorporation of other environmental exposure variables or advanced space-time interpolations. Geographers, computer scientists, and biostatisticians should focus on creating space-time models and other methodologies that allow for readily available space-time pattern recognition and the quick inclusion of multiple variables (e.g., temperature, physiographic strain). Until such progress is made, individual space-time behavior will continue to be studied at a relatively coarse spatial scale and discrete time periods (Desjardins et al. 2018). Recent developments in air quality research have been successful at the near-real time creation of an urban ambient air pollution cube, allowing for simultaneous collection of information on where, when, and what. Yet, such methods need to be integrated into sources like a WebGIS, for use among practitioners and interested stakeholders (Fang and Lu 2011).

### 4.2.4 Challenges with Geospatial Wearable Sensor Technologies

Numerous limitations still exist with wearable sensor technologies. First, capturing high-resolution geographic data for dynamic temperature exposure assessment is still data-intensive, requiring collection from large population sizes over extended time periods. Current personal exposure research for temperature is limited to short time spans (i.e., less than 1 week) and small populations (i.e., less than 100 participants) (Sugg et al. 2018; Kuras et al. 2015; Bernhard et al. 2015; Basu and Samet 2002). This research is limited due to short battery life, low memory capacity, high instrument costs, and low compliance, resulting in research studies that utilize a shorter exposure period on a smaller number of participants (Helbich 2018; Fang and Lu 2012). New research designs are required that utilize ubiquitous technologies (i.e., smartphones) that reduce participant burden and allow for long-term, large-sample research that identifies exposure and other factors that result in adverse health outcomes. Other limitations to wearable sensor technologies, particularly those involving the geospatial sciences, include GPS data collection. Gaps can exist in location tracking when the GPS signal is lost due to satellite disruption or malfunction, atmospheric conditions, multipath signal reflection, or signal loss or blocking (e.g., individuals moving into indoor environments) (Yoo et al. 2015). Solutions are needed to address data lapses from GPS, such as utilizing Wi-Fi networks as proxies for location. Until researchers identify best practices to address these limitations, widespread use of wearable technology will remain limited. Lastly, new research shows that potential users of wearable sensor technology may be concerned with privacy issues collected for research purposes. However, the recent Quantified Self movement has ushered in general public acceptance and trust concerning self-tracking or the sharing of user-generated data on health and well-being, as well as productivity, with commercial corporations despite poorly defined data use, ownership, and privacy policies (Ostherr et al. 2017). In order to better understand the contextual factors driving personal exposure on a large scale, participants must be willing to provide GPS coordinates without it being seen as an infringement of their personal rights. Data storage and processing should be done within a secure information technology environment requiring effective protection conditions that respect the privacy of participants. Geographers will need to consider reframing recruitment strategies and materials that address participants' social conception of privacy (e.g., loose federal guidelines governing commercial use of user-generated data in comparison with stringent ethical supervision and approval process imposed upon scientific researchers).

## 5 Future Directions

Moving forward, personal heat exposure research will benefit from further incorporation of GIS, which can help merge and visualize individual-level temperature observations with time-activity patterns. Such information may reveal how personal exposure is linked to various aspects of the urban environment, such as urban form, poverty, housing quality, and adaptive capacity. Therefore, personal heat exposure research can help evaluate and provide guidance on heat mitigation strategies (e.g., tree planting) and the allocation of resources (e.g., cooling centers) to areas of the city with the greatest risk for heat-related impact.

Despite significant declines in heat-related mortality over the past several decades (Sheridan and Allen 2018), most projections of heat-related mortality through the rest of the twenty-first century show dramatic increases, some on the order of multiple orders of magnitude (Hondula et al. 2015a, b). One of the factors that may contribute to increased heat-related mortality is urbanization. Missing from these projections, however, is the effect of adaptation, which could potentially cut the projected mortality estimates in half (Hondula et al. 2015a, b). To date, few epidemiological studies have attempted to measure adaptive behaviors in response to extreme heat. Personal heat exposure research may provide an opportunity to document these adaptive behaviors and link them with individual temperature observations and time-activity patterns. Other forms of adaptation, such as physiologic (e.g., acclimatization) and infrastructure adaptation, may also benefit from this approach by considering seasonal changes in time-activity patterns and exposure and relationships between urban form, building design, and indoor versus outdoor exposure, respectively (Hondula et al. 2015a, b; Karimi et al. 2015, 2017). By emphasizing exposure at the individual level, instead of focusing broadly on exposure at the city level, our understanding of where and why adaptation strategies have succeeded may greatly improve (Sheridan and Allen 2018).

Future research on personal heat exposure should focus on indoor environments, which are largely unaccounted for in most environmental health and exposure studies, particularly in urban areas (Hondula et al. 2017). As the relationships between indoor and outdoor temperatures remain mostly unclear, personal heat exposure research may provide new insights into the connections between indoor exposure and heat-related health outcomes. Lastly, as citizen science becomes more popular and widespread, opportunities to use the latest in affordable and convenient sensor technology will increase significantly, thereby empowering individuals in cities (and elsewhere) to participate in observing their thermal environment and providing policy-makers with the information necessary to develop more targeted and efficient heat mitigation strategies (Mehdipoor et al. 2017).

## 6 Conclusion

Assessing personal heat exposure remains a challenge, as an individual's experienced temperature is driven not only by the spatio-temporal patterns of their thermal environment but also by their mobility patterns. The emergence of new technologies and rich spatial datasets requires multi-disciplinary collaboration to advance the science on place-based exposure to thermal extremes and the associated health impacts for at-risk populations in urban environments. The recent emergence of low-cost, convenient, portable sensors for environmental exposure applications provides a platform for recording data at high spatial and temporal resolution.

Using the novel application of consumer-based "wearable" sensor technology, new research at the intersection of geospatial science and public health will lay the groundwork for translating personalized temperature exposure measures to technology solutions and tailored prevention strategies in urban areas. As mobile technology progresses, real-time monitoring and analysis of environmental conditions and health effects at the individual level will become more feasible and, ultimately, a standard approach in the field.

## References

An, L., Tsou, M. H., Crook, S. E., Chun, Y., Spitzberg, B., Gawron, J. M., & Gupta, D. K. (2015). Space–time analysis: Concepts, quantitative methods, and future directions. *Annals of the Association of American Geographers, 105*(5), 891–914.

Basu, R., & Samet, J. M. (2002). An exposure assessment study of ambient heat exposure in an elderly population in Baltimore, Maryland. *Environmental Health Perspectives, 110*(12), 1219.

Berko, J., Ingram, D. D., Saha, S., & Parker, J. D. (2014). Deaths attributed to heat, cold, and other weather events in the United States, 2006–2010. *National Health Statistics Reports, 30*, 1–15.

Bernhard, M. C., Kent, S. T., Sloan, M. E., Evans, M. B., McClure, L. A., & Gohlke, J. M. (2015). Measuring personal heat exposure in an urban and rural environment. *Environmental Research, 137*, 410–418.

Castell, N., Dauge, F. R., Schneider, P., Vogt, M., Lerner, U., Fishbain, B., et al. (2017). Can commercial low-cost sensor platforms contribute to air quality monitoring and exposure estimates? *Environment International, 99*, 293–302.

Chan, Y. F. Y., Bot, B. M., Zweig, M., Tignor, N., Ma, W., Suver, C., et al. (2018). The asthma mobile health study, smartphone data collected using ResearchKit. *Scientific Data, 5*, 180096.

Chapman, L., Muller, C. L., Young, D. T., Warren, E. L., Grimmond, C. S. B., Cai, X. M., & Ferranti, E. J. (2015). The Birmingham urban climate laboratory: An open meteorological test bed and challenges of the smart city. *Bulletin of the American Meteorological Society, 96*(9), 1545–1560.

Chapman, L., Bell, C., & Bell, S. (2017). Can the crowdsourcing data paradigm take atmospheric science to a new level? A case study of the urban heat island of London quantified using Netatmo weather stations. *International Journal of Climatology, 37*(9), 3597–3605.

Dėdelė, A., Miškinytė, A., Česnakaitė, I., & Gražulevičienė, R. (2018). Effects of individual and environmental factors on GPS-based time allocation in Urban microenvironments using GIS. *Applied Sciences, 8*(10), 2007.

Demšar, U., & Virrantaus, K. (2010). Space-time density of trajectories: Exploring spatiotemporal patterns in movement data. *International Journal of Geographical Information Science, 24*, 1527–1542.

De Nazelle, A., Seto, E., Donaire-Gonzalez, D., Mendez, M., Matamala, J., Nieuwenhuijsen, M. J., & Jerrett, M. (2013). Improving estimates of air pollution exposure through ubiquitous sensing technologies. *Environmental Pollution, 176*, 92–99.

Desjardins, M. R., Hohl, A., Griffith, A., & Delmelle, E. (2018). A space–time parallel framework for fine-scale visualization of pollen levels across the Eastern United States. *Cartography and Geographic Information Science*, 1–13. https://doi.org/10.1080/15230406.2018.1515664

Dewulf, B., Neutens, T., Van Dyck, D., De Bourdeaudhuij, I., Panis, L. I., Beckx, C., & Van de Weghe, N. (2016). Dynamic assessment of inhaled air pollution using GPS and accelerometer data. *Journal of Transport & Health, 3*(1), 114–123.

Dias, D., & Tchepel, O. (2014). Modelling of human exposure to air pollution in the urban environment: A GPS-based approach. *Environmental Science and Pollution Research, 21*(5), 3558–3571.

Diez-Roux, A. V. (2000). Multilevel analysis in public health research. *Annual Review of Public Health, 21*(1), 171–192.

Dons, E., Laeremans, M., Orjuela, J. P., Avila-Palencia, I., Carrasco-Turigas, G., Cole-Hunter, T., et al. (2017). Wearable sensors for personal monitoring and estimation of inhaled traffic-related air pollution: Evaluation of methods. *Environmental Science & Technology, 51*(3), 1859–1867.

Ebi, K. L., Teisberg, T. J., Kalkstein, L. S., Robinson, L., & Weiher, R. F. (2004). Heat watch/warning systems save lives: Estimated costs and benefits for Philadelphia 1995–98. *Bulletin of the American Meteorological Society, 85*(8), 1067–1074.

ESRI. (2018). *ArcPro: Release 2.2.4*. Redlands: Environmental Systems Research Institute.

Fang, T. B., & Lu, Y. (2011). Constructing a near real-time space-time cube to depict urban ambient air pollution scenario. *Transactions in GIS, 15*(5), 635–649.

Fang, T. B., & Lu, Y. (2012). Personal real-time air pollution exposure assessment methods promoted by information technological advances. *Annals of GIS, 18*(4), 279–288.

Fischer, E. M., Oleson, K. W., & Lawrence, D. M. (2012). Contrasting urban and rural heat stress responses to climate change. *Geophysical Research Letters, 39*(3), L03705. https://doi.org/10.1029/2011GL050576

Friel, S., Hancock, T., Kjellstrom, T., McGranahan, G., Monge, P., & Roy, J. (2011). Urban health inequities and the added pressure of climate change: An action-oriented research agenda. *Journal of Urban Health, 88*(5), 886.

Gao, M., Cao, J., & Seto, E. (2015). A distributed network of low-cost continuous reading sensors to measure spatiotemporal variations of PM2. 5 in Xi'an, China. *Environmental Pollution, 199*, 56–65.

Hägerstrand, T. (1967). *Innovation diffusion as a spatial process*. Chicago: The University of Chicago Press.

Hägerstrand, T. (1970). What about people in regional science? *Papers of the Regional Science Association, 24*, 7–21.

Hancke, G. P., Silva Bde, C., & Hancke, G. P., Jr. (2012). The role of advanced sensing in smart cities. *Sensors, 13*(1), 393–425.

Heaviside, C., Macintyre, H., & Vardoulakis, S. (2017). The urban heat island: Implications for health in a changing environment. *Current Environmental Health Reports, 4*(3), 296–305.

Heimann, I., Bright, V. B., McLeod, M. W., Mead, M. I., Popoola, O. A. M., Stewart, G. B., & Jones, R. L. (2015). Source attribution of air pollution by spatial scale separation using high spatial density networks of low cost air quality sensors. *Atmospheric Environment, 113*, 10–19.

Helbich, M. (2018). Toward dynamic urban environmental exposure assessments in mental health research. *Environmental Research, 161*, 129–135.

Hondula, D. M., Balling, R. C., Andrade, R., Krayenhoff, E. S., Middel, A., Urban, A., Georgescu, M., & Sailor, D. J. (2017). Biometeorology for cities. *International Journal of Biometeorology, 61*, S59–S69.

Hondula, D. M., Balling, R. C., Vanos, J. K., & Georgescu, M. (2015a). *Rising temperatures, human health, and the role of adaptation*. Curr Clim Change Rep (Vol. 1, p. 144).

Hondula, D. M., Davis, R. E., Saha, M. V., Wegner, C. R., & Veazey, L. M. (2015b). Geographic dimensions of heat-related mortality in seven U.S. cities. *Environmental Research, 138*, 439–452.

Jenerette, G. D., Harlan, S., Buyanteuv, A., Stefanov, W. L., Declet-Barreto, J., Ruddel, B. L., Wyint, S. W., Kaplan, S., & Li, X. (2016). Micro-scale urban surface temperatures are related to land-cover features and residential heat related health impacts in Phoenix, AZ USA. *Landscape Ecology, 31*(4), 745–760.

Karimi, M., Nazari, R., Vant-Hull, B., & Khanbilvardi, R. (2015). Urban heat island assessment with temperature maps using high resolution datasets measured at street level. *International Journal of the Constructed Environment, 6*, 17–26.

Karimi, M., Vant-Hull, B., Nazari, R., Mittenzwei, M., & Khanbilvardi, R. (2017). Predicting surface temperature variation in urban settings using real-time weather forecasts. *Urban Climate, 20*, 192–201.

Kestens, Y., Wasfi, R., Naud, A., & Chaix, B. (2017). "Contextualizing context": Reconciling environmental exposures, social networks, and location preferences in health research. *Current Environmental Health Reports, 4*(1), 51–60.

Klepeis, N. E., Nelson, W. C., Ott, W. R., Robinson, J. P., Tsang, A. M., Switzer, P., Behar, J. V., Hern, S. C., & Engelmann, W. H. (2001). The National Human Activity Pattern Survey (NHAPS): A resource for assessing exposure to environmental pollutants. *Journal of Exposure Analysis and Environmental Epidemiology, 11*, 231–252.

Klinenberg, E. (2002). *Heat wave: A social autopsy of disaster in Chicago*. Chicago: University of Chicago Press.

Kuras, E. R., Hondula, D. M., & Brown-Saracino, J. (2015). Heterogeneity in individually experienced temperatures (IETs) within an urban neighborhood: Insights from a new approach to measuring heat exposure. *International Journal of Biometeorology, 59*(10), 1363–1372.

Kuras, E., Bernhard, M., Calkins, M., Ebi, K., Hess, J., Kintziger, K., Jagger, M., Middel, A., Scott, A., Spector, J., Uejio, C., Vanos, J., Zaitchik, B., Gohlke, J., & Hondula, D. (2017). Opportunities and challenges for personal heat exposure research. *Environmental Health Perspectives, 125*, 085001.

Kwan, M. P. (2009). From place-based to people-based exposure measures. *Social Science & Medicine, 69*(9), 1311–1313.

Kwan, M. P. (2012). How GIS can help address the uncertain geographic context problem in social science research. *Annals of GIS, 18*(4), 245–255.

Kwan, M. P. (2013). Beyond space (as we knew it): Toward temporally integrated geographies of segregation, health, and accessibility: Space–time integration in geography and GIScience. *Annals of the Association of American Geographers, 103*(5), 1078–1086.

Kwan, M.-P. (2000). Interactive geovisualization of activity travel patterns using three-dimensional geographical information systems: A methodological exploration with a large data set. *Transportation Research Part C, 8*, 185–203.

Longo, J., Kuras, E., Smith, H., Hondula, D. M., & Johnston, E. (2017). Technology use, exposure to natural hazards, and being digitally invisible: Implications for policy analytics. *Policy & Internet, 9*(1), 76–108.

Macintyre, H. L., Heaviside, C., Taylor, J., Picetti, R., Symonds, P., Cai, X. M., & Vardoulakis, S. (2018). Assessing urban population vulnerability and environmental risks across an urban area during heatwaves–Implications for health protection. *Science of the Total Environment, 610*, 678–690.

Macintyre, S., Ellaway, A., & Cummins, S. (2002). Place effects on health: How can we conceptualise, operationalise and measure them? *Social Science & Medicine, 55*(1), 125–139.

Mehdipoor, H., Vanos, J. K., Zurita-Milla, R., & Cao, G. (2017). Emerging technologies for biometeorology. *International Journal of Biometeorology, 61*, S81–S88.

Meier, F., Fenner, D., Grassmann, T., Otto, M., & Scherer, D. (2017). Crowdsourcing air temperature from citizen weather stations for urban climate research. *Urban Climate, 19*, 170–191.

Muller, C. L., Chapman, L., Johnston, S., Kidd, C., Illingworth, S., Foody, G., et al. (2015). Crowdsourcing for climate and atmospheric sciences: Current status and future potential. *International Journal of Climatology, 35*(11), 3185–3203.

National Oceanic and Atmospheric Administration. (2019). Natural hazard statistics. National Weather Service, Office of Climate, Water, and Weather Services. http://www.nws.noaa.gov/om/hazstats.html.

NCA4 Health Ch, Ebi, K. L., Balbus, J. M., Luber, G., Bole, A., Crimmins, A., Glass, G., Saha, S., Shimamoto, M. M., Trtanj, J., & White-Newsome, J. L. (2018). Human Health. In D. R. Reidmiller, C. W. Avery, D. R. Easterling, K. E. Kunkel, K. L. M. Lewis, T. K. Maycock, & B. C. Stewart (Eds.), *Impacts, risks, and adaptation in the United States: Fourth National Climate Assessment, Volume II*. Washington, DC: U.S. Global Change Research Program. https://doi.org/10.7930/NCA4.2018.CH14.

Nethery, E., Mallach, G., Rainham, D., Goldberg, M. S., & Wheeler, A. J. (2014). Using Global Positioning Systems (GPS) and temperature data to generate time-activity classifications for estimating personal exposure in air monitoring studies: An automated method. *Environmental Health, 13*(1), 33.

Nguyen, J. L., Schwartz, J., & Dockery, D. W. (2014). The relationship between indoor and outdoor temperature, apparent temperature, relative humidity, and absolute humidity. *Indoor Air, 24*(1), 103–112.

Oliver, N., Matic, A., & Frias-Martinez, E. (2015). Mobile network data for public health: Opportunities and challenges. *Frontiers in Public Health, 3*, 189.

Openshaw, S. (1984). *The modifiable areal unit problem*. Norwich: Geo Books.

Ostherr, K., Borodina, S., Bracken, R. C., Lotterman, C., Storer, E., & Williams, B. (2017). Trust and privacy in the context of user-generated health data. *Big Data & Society, 4*(1), 2053951717704673.

Quinn, A., Tamerius, J. D., Perzanowski, M., Jacobson, J. S., Goldstein, I., Acosta, L., & Shaman, J. (2014). Predicting indoor heat exposure risk during extreme heat events. *Science of the Total Environment, 490*, 686–693.

Reid, C. E., O'neill, M. S., Gronlund, C. J., Brines, S. J., Brown, D. G., Diez-Roux, A. V., & Schwartz, J. (2009). Mapping community determinants of heat vulnerability. *Environmental Health Perspectives, 117*(11), 1730.

Reis, S., Liška, T., Vieno, M., Carnell, E. J., Beck, R., Clemens, T., et al. (2018). The influence of residential and workday population mobility on exposure to air pollution in the UK. *Environment International, 121*, 803–813.

Rainham, D. (2016). A wireless sensor network for urban environmental health monitoring: UrbanSense. *IOP Conference Series: Earth and Environmental Science, 34*(1), 012028. IOP Publishing.

Ryan, P. H., Son, S. Y., Wolfe, C., Lockey, J., Brokamp, C., & LeMasters, G. (2015). A field application of a personal sensor for ultrafine particle exposure in children. *Science of the Total Environment, 508*, 366–373.

Sarofim, M. C., Saha, S., Hawkins, M. D., Mills, D. M., Hess, J., Horton, R., Kinney, P., Schwartz, J., & Juliana, A. S. (2016). Ch. 2: Temperature-related death and illness. In *The impacts of climate change on human health in the United States: A scientific assessment* (pp. 43–68). Washington, DC: U.S. Global Change Research Program. https://doi.org/10.7930/J0MG7MDX.

Schneider, P., Castell, N., Vogt, M., Dauge, F. R., Lahoz, W. A., & Bartonova, A. (2017). Mapping urban air quality in near real-time using observations from low-cost sensors and model information. *Environment International, 106*, 234–247.

Sheridan, S. C., & Allen, M. J. (2018). Temporal trends in human vulnerability to excessive heat. *Environmental Research Letters, 13*, 043001.

Sherwood, S. C., & Huber, M. (2010a). An adaptability limit to climate change due to heat stress. *Proceedings of the National Academy of Sciences, 107*(21), 9552–9555.

Steinle, S., Reis, S., Sabel, C. E., Semple, S., Twigg, M. M., Braban, C. F., et al. (2015). Personal exposure monitoring of PM2. 5 in indoor and outdoor microenvironments. *Science of the Total Environment, 508*, 383–394.

Sherwood, S. C., & Huber, M. (2010b). An adaptability limit to climate change due to heat stress. *Proceedings of the National Academy of Sciences, 107*(21), 9552–9555. https://doi.org/10.1073/pnas.0913352107.

Sugg, M. M., Fuhrmann, C. M., & Runkle, J. D. (2018). Temporal and spatial variation in personal ambient temperatures for outdoor working populations in the southeastern USA. *International Journal of Biometeorology, 62*, 1521.

Tsin, P. K., Knudby, A., Krayenhoff, E. S., Ho, H. C., Brauer, M., & Henderson, S. B. (2016). Microscale mobile monitoring of urban air temperature. *Urban Climate, 18*, 58–72.

Tunstall, H. V., Shaw, M., & Dorling, D. (2004). Places and health. *Journal of Epidemiology & Community Health, 58*(1), 6–10.

Uejio, C. K., Morano, L. H., Jung, J., Kintziger, K., Jagger, M., Chalmers, J., & Holmes, T. (2018). Occupational heat exposure among municipal workers. *International Archives of Occupational and Environmental Health, 91*, 705–715.

Vant-Hull, B., Karimi, M., Sossa, A., Wisanto, J., Nazari, R., & Khanbilvardi, R. (2014). Fine structure in Manhattan's daytime urban heat island: A new dataset. *Journal of Urban and Environmental Engineering, 8*, 59–74.

Vlahov, D., & Galea, S. (2002). Urbanization, urbanicity, and health. *Journal of Urban Health, 79*(1), S1–S12.

Wong, E., Akbari, H., Bell, R., & Cole, D. (2011). *Reducing urban heat islands: Compendium of strategies*. Environmental Protection Agency. Retrieved 12 May 2011.

Yoo, E., Rudra, C., Glasgow, M., & Mu, L. (2015). Geospatial estimation of individual exposure to air pollutants: Moving from static monitoring to activity-based dynamic exposure assessment. *Annals of the Association of American Geographers, 105*(5), 915–926.

**Margaret M. Sugg** is an Assistant Professor in the Department of Geography and Planning at Appalachian State University. Her research uses innovative geospatial technologies and methodologies to address climate-health interactions. She holds a PhD in Geography from the University of North Carolina at Chapel Hill.

**Dr. Chris Fuhrmann** is an Assistant Professor in the Department of Geosciences at Mississippi State University. He also serves as the Assistant State Climatologist. His research interests are in the fields of applied and synoptic climatology, where he studies the effects of weather and climate on society and the role of large-scale circulation features on the distribution and intensity of surface weather events. He earned a B.A. and Ph.D. in Geography from the University of North Carolina at Chapel Hill, and a M.S. in Geography from the University of Georgia.

**Dr. Jennifer Runkle** is a Research Scholar at the North Carolina Institute for Climate Studies at North Carolina State University. Her research interests include examining the health effects of climate change and variability, with particular interests in characterizing localized impacts for vulnerable populations like pregnant women and outdoor workers. She is interested in advancing the science around how social and environmental factors work independently and jointly to influence climate-health outcome associations and using this information to identify community-level pathways to resilience. She holds a PhD in Environmental Epidemiology from the University of South Carolina Arnold School of Public Health and completed postdoctoral training in environmental and occupational epidemiology at Emory University.

# Geographic Variation in Cardiovascular Disease Mortality: A Study of Linking Risk Factors and Built Environment at a Local Health Unit in Canada

**Lei Wang, Chris I. Ardern, and Dongmei Chen**

**Abstract** Cardiovascular disease (CVD) is one of the leading causes of death in Canada. CVD risk factors and outcome data are used to determine trends of disease risk to inform public health program planning for prevention and control of disease and risk reduction or elimination. Recent efforts to map CVD and its associated risk factors at the health region level have provided further insights into variation in determinants across populations. In this chapter, geographic information system (GIS) and spatial analysis were utilized to enhance CVD surveillance to identify the patterns and relationships between CVD mortality and its potential risk factors. Ordinary Least Squares (OLS) regression and Geographically Weighted Regression (GWR) approaches were used to explore geographical variation in the rate of CVD mortality. After consideration of potential environmental, epidemiological, demographic, and socioeconomic factors, spatial statistics analysis revealed geospatial clustering for CVD mortality and the "hot spots" or "cold spots." Within a mixed rural-suburban setting in Ontario, Canada, there was an evidence of significant built environmental factors and immigrant time associated with the rate of CVD mortality. Moreover, this pilot work suggests that the integration of geospatial information with routinely collected surveillance data appears feasible within the structure and resources of local public health units as a means to assist in the identification of regional variation in the burden of CVD.

---

L. Wang
Institute of Remote Sensing and Digital Earth, Chinese Academy of Sciences, Beijing, China

Department of Geography and Planning, Queen's University, Kingston, ON, Canada

C. I. Ardern
School of Kinesiology and Health Science, York University, Toronto, ON, Canada

D. Chen (✉)
Department of Geography and Planning, Queen's University, Kingston, ON, Canada
e-mail: chendm@queensu.ca



Y. Lu, E. Delmelle (eds.), *Geospatial Technologies for Urban Health*,
Global Perspectives on Health Geography, https://doi.org/10.1007/978-3-030-19573-1_3

# 1 Background

Cardiovascular disease (CVD) is one of the leading causes of death in Canada, representing 22.7% of all deaths in 2009 (Public Health Agency of Canada 2016). Data from Statistics Canada show that the mean 10-year risk of CVD events in the population aged 20–79 was 8.9% during 2007–2011 (Statistics Canada 2017), and data from the Canadian Community Health Survey (CCHS) suggest that four in five of the population between the ages of 20 and 59 years have at least one modifiable risk factor (Heart and Stroke Foundation of Canada 2016). Many modifiable and non-modifiable risk factors can contribute to the high prevalence of CVD, and it is also well known that the burden of CVD is unequally distributed in outcomes, determinants and risk factors across subgroups of the population (Tanuseputro et al. 2003; O'Donnell and Elosua 2008). In broader terms, there is also marked geographic difference in CVD indicators, determinants, and risk factors, as well as mortality (Chow et al. 2005; Filate et al. 2003; Hall and Tu 2003; Lee et al. 2009; Leal and Chaix 2011). Recent efforts to map CVD and its associated risk factors (e.g., smoking, obesity, inactivity, low income, hypertension, and diabetes) at the health region level have provided further insight into variation in determinants across populations (Tu et al. 2006; CDC 2017).

Early studies have shown that more than 70% of global CVD is attributable to modifiable risk factors such as unhealthy lifestyles, policy factors, as well as features of the social and built environment (Ezzati et al. 2003; Sallis et al. 2012; Malambo et al. 2016). The "built environment" comprises urban design, land use, and the transportation system, and encompasses patterns of human activity within the physical environment (Handy et al. 2002; Sallis et al. 2012). Although the importance of individual-level determinants (such as age, gender, income, education) on physical activity and obesity is well described, the influence of environmental determinants of health relating to "place" (i.e., the social experience of the environment) and "space" (i.e., the physical environment) is infrequently integrated into chronic disease surveillance and may offer considerable insight into risk factor clustering of cardiovascular morbidity and mortality through modifiable risk factors such as physical inactivity and obesity (Heath et al. 2006; McCormack et al. 2004; Sallis et al. 2012). The link between the built environment and health has been the focus of an increasing number of studies in recent years (Chum and O'Campo 2015; Malambo et al. 2016). However, the importance of the neighborhood built environment across a range of health outcomes has not been fully explored, and there is currently no consensus as to the relative impact of the built environment and collective community factors on cardiovascular morbidity and mortality (Malambo et al. 2016).

In Canada, CVD risk factor surveillance data sources, including vital statistics, hospitalization records, census and health surveys, are commonly used to inform public health program planning for prevention and control of CVD and risk reduction or elimination. Although existing sources of data for chronic disease surveillance include information that can be geocoded to the municipality, city, or community, application of such frameworks to enhance routine surveillance of CVD at the local level has rarely been implemented (Holowaty et al. 2010; Odoi et al. 2005; Caley 2004). More discrete geographical units with other community-level health determinants should be considered as vital elements to future surveillance strategies, as this would allow for informed public health decision-making and targeted program planning for the areas of highest need. This approach may be particularly informative for the coordination, allocation, and delivery of public health services and interventions within the context of a rapidly growing, geographically and demographically distinct areas. The fast pattern of growth in both residential and employment areas suggests a need to monitor cardiovascular disease risk, morbidity, and mortality risk factors within the public health unit and to explore their relation with the built environment. Furthermore, monitoring various risk factors could provide opportunities to identify areas or regions where disease risk factors are clustered together which could then be investigated to help inform future policy-makers and urban planners how the neighborhood could be altered in future development plans to decrease the overall number of cases. Therefore, it is important to document the methodology and process by which geospatial analysis may be implemented, and to assess whether or not this strategy would help identify clusters of disease determinants that will allow for targeted public health programs and policies to those most at risk.

Application of geographic information system (GIS) and spatial statistics to assess built environment and improve public health, epidemiology, and health planning has been growing in the last two decades (Pickle 2002; Yiannakoulias et al. 2009; Cerin et al. 2009; Thornton et al. 2011). However, there has yet to be a surveillance system that monitors disease outcomes, associated risk factors, and social determinants, using a spatial framework on an ongoing basis to detect temporal and spatial trends. When taken together, the persistence of regional differences in CVD outcomes and risk factors in Canada emphasizes the need for effective surveillance of chronic disease risk factors in addition to patterns of healthcare utilization. The purpose of this chapter is to, therefore, evaluate the use of spatial approaches to analyze the spatial variation of CVD mortality at the local public health unit level in Ontario, considering the potential impact of the "built" physical environment. To date, there is no existing single surveillance system in place that monitors all disease outcomes, associated risk factors and social determinants for CVD.

This pilot study, funded by the Public Health Agency of Canada, brought together urban planners, public health officials, epidemiologists, and policy-makers from The Regional Municipality of York, with academic researchers to explore the relationship between CVD risk factors and built environment. To achieve the project objectives, a combination of respondent-level risk factor data from the Canadian Community Health Survey (CCHS), determinant data from the Census of Canada, and CVD morbidity and mortality outcome data from intelliHEALTH ONTARIO in concert with spatial data was used.

# 2 Methods

## *2.1 Study Area*

The study area was the York region of southern Ontario, Canada (Fig. 1). It belongs to the Greater Toronto Area and is about 1762.17 km$^2$ in area, consists of 155 census tracts (CTs), and had a population of 1,032,524 in the 2011 Census based on Statistics Canada (2016). The population in the 155 CTs ranged from 1970 to 18,959 persons and the population density ranged from 22 to 8580 persons per square kilometer in 2011. During the period of 1996–2001, York Region was one of the fastest-growing census divisions in Canada (Bryan et al. 2006).

Risk factor surveillance in York Region was limited to individual-level survey data provided by routinely collected sources such as the Canadian Community Health Survey and Rapid Risk Factor Surveillance System. In light of the consistent finding of regional (e.g., provincial and rural/urban) and demographic (e.g., ethnicity and time-in-country) variation in traditional CVD risk factors (Tremblay et al. 2005, 2006), critical insight into contributors to inequities in cardiovascular morbidity and mortality may be provided by the integration of geospatial information with

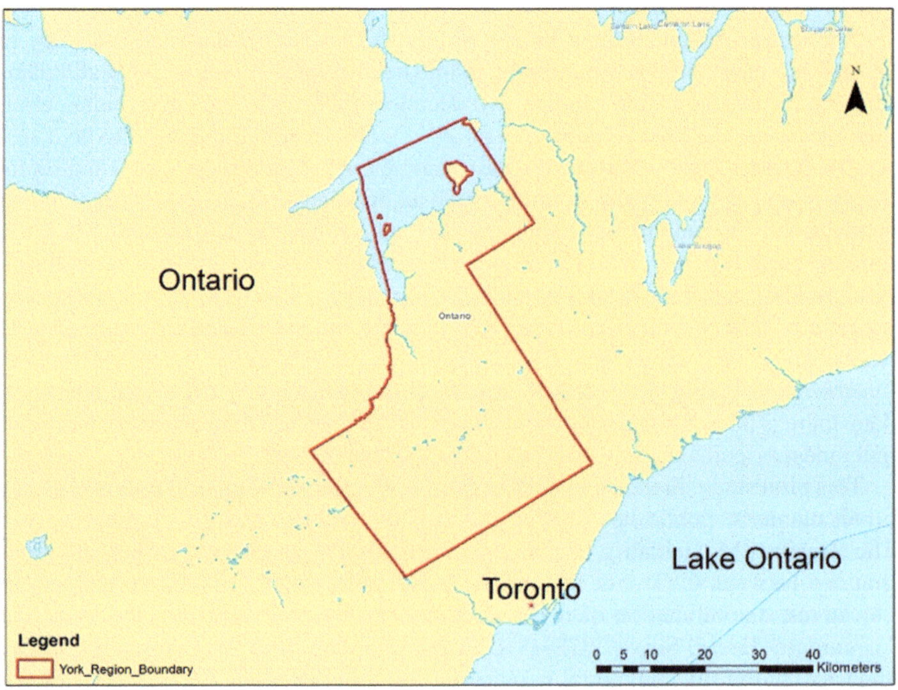


**Fig. 1** The location of the study area

existing risk factors and health event data. However, to date, only limited attempts have been directed to multi-level modeling and surveillance to assess the joint effects. The coordination and integration of multiple sources and levels of data will provide a resource on which to build a system that can integrate individual and community-level determinants and risk factors in an effort to enhance existing primary prevention strategies.

## 2.2 *Data*

Multiple independent variables were captured from the CCHS dataset, census, and GIS data to account for environmental, epidemiological, demographic, and socioeconomic characteristics and risk factor for CVD morbidity and mortality. The values of poorer states of health of each variable (i.e., obesity, hypertension, diabetes, heavy drinking, heavy smoking, and sedentary lifestyle) were included within the models for spatial analysis.

## 2.3 *Canadian Community Health Survey (CCHS)*

CCHS is a nationally representative population-based cross-sectional survey conducted by Statistics Canada. The CCHS collects information on the health status, healthcare use, and health determinants of Canadians aged 12 years or older living in private households. The target population of the CCHS included household residents in all provinces and territories. Residents of indigenous lands, institutions, some remote areas, and military bases were not included. While there was one randomly selected respondent per household, planned over-sampling of youths resulted in a second member of some households being interviewed. Participants provided their demographic, socioeconomic, behavioral, and health-related information. Cycle 3.1 (2005) was used for use in this study to match the available mortality and morbidity data. For Cycle 3.1, interviews were conducted between January and December 2005. The response rate was 79%, yielding a national sample of 132,947 respondents, with a total of 1681 respondents in York Region. Three sampling frames were used to select the sample of households: 49% of the sample of households came from an area frame; 50% from a list frame of telephone numbers; and the remaining 1% from a Random Digit Dialing (RDD) sampling frame. The distribution of the samples in the study area is shown in Fig. 2.

A number of CVD risk factors were identified after conducting a literature review of cardiovascular disease risk factors. Table 1 lists these data and risk factors selected in this study and their description and rationales.

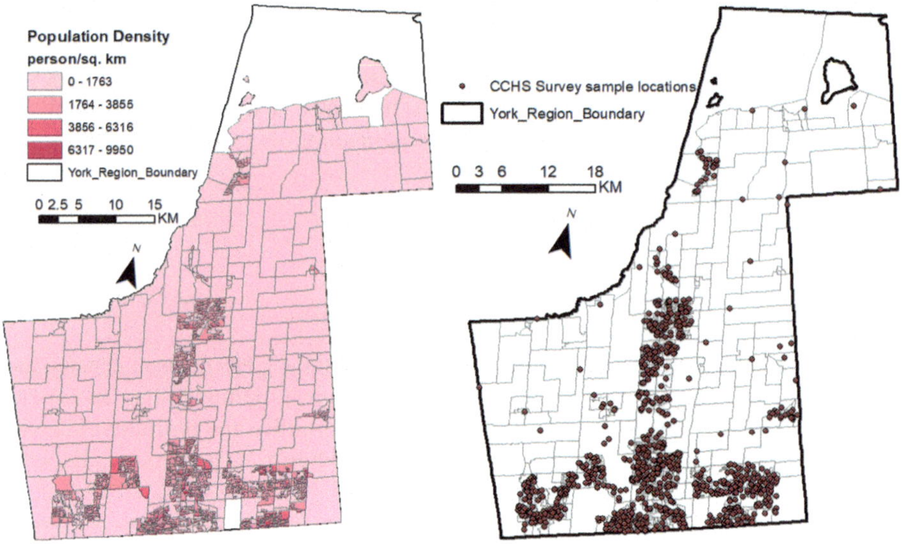


**Fig. 2** The population density (left) and location of CCHS survey samples (right) geocoded based on their six-digit postal codes within the York Region

## 2.4 *Postal Data*

The postal data used in this study was the unique enhanced postal (UEP) codes data produced by DMTI Spatial Inc. (https://www.dmtispatial.com/). The data contains postal code points positioned to the most representative address and allows for a 1:1 relationship wherein one postal code matches to one postal code location. Each postal code is attributed by its spatial coordinates, census population, and other determinant data. In UEP, postal code regions are determined based on their corresponding dissemination area (DA) regions. Where postal codes serve more than one DA (such as in both rural and urban areas of Canada), postal codes are assigned to DAs based on an unbiased population weighted random allocation method. In cases where valid postal codes cannot be used to assign the full range of geographic identifiers, the first two or three characters in the postal code are used to assign partial geography.

A six-digit postal code residential information was captured for each respondent from the share file of the CCHS database. Geocoding was subsequently applied in ArcGIS to retrieve the associated geographic coordinates of each CCHS respondent using UEP codes for the purpose of visualization of patterns and further analysis. Since the analysis unit of this study is census tracts (CT), a spatial join was applied in ArcGIS to assign CCHS respondent into census tract units to get the count number of CCHS respondent in each CT. CVD risk factor rates (obesity, hypertension, diabetes, heavy drinking, heavy smoking, sedentary lifestyle, low income, low

**Table 1** CVD data and its risk factors from CCHS Cycle 3.1

| Category and indicator | Quality of data (has indicator been used in other research?) | Description of data | Is data available for use? | Selected for analysis? |
|---|---|---|---|---|
| *Demographics* | | | | |
| Age | Shigematsu et al. (2009) | Age accurate to single year | Yes | Yes |
| Sex | Bennett et al. (2007) | Female/male | Yes | Yes |
| Education | Berrigan and Troiano (2002) | Based on respondent's highest level of educational attainment | Yes | Yes |
| Income | Gordon-Larsen et al. (2006) | Based on respondent's income level | Yes | Yes |
| Housing | Agreement by senior members | Household size (number of residents) | Yes | Yes |
| Country of birth | Berrigan and Troiano (2002) | Considered white or visible minority | Yes | Yes |
| Recent immigrant status | Agreement by senior members | Average length of time in Canada since immigration | Yes | No |
| *Health indicators (risk factors/ behaviors)* | | | | |
| Leisure time physical activity index | Hoehner et al. (2005) | Based on extensive list of activities with questions relating to frequency and duration | Yes | Yes |
| Smoking | Ross (2000) | Smoking classification for frequency and type | Yes | Yes |
| High blood pressure | Li et al. (2009) | Self-reported physician-diagnosed high blood pressure | Yes | Yes |
| BMI | Evenson et al. (2007) | Based on self-reported height and weight measurements | Yes | Yes |
| Fruit and vegetable consumption | Ball et al. (2009) | Daily consumption of fruits and vegetables | Yes | Yes |
| Diabetes | Agreement by senior members | Based on self-reported response to physician-diagnosed diabetes | Yes | Yes |
| Access to physicians | Agreement by senior members | Access to a medical physician | Yes | Yes |

consumption of fruit and vegetable, and inaccessible to physicians) were calculated by the frequency of each risk factor by the count number of population in each CT. Average age, percentage of males, percentage of rent dwelling, and average length of time in Canada since immigration were also calculated.

## 2.5 Census

Socioeconomic and demographic data were derived from Census of Canada profiles. The census is carried out every 5 years and is a reliable source of social and demographic information for the population of Canada. Socioeconomic information was collected from 20% of the households, surpassing the sample size of any available population-based survey. In urbanized areas of Canada, Statistics Canada classifies Canadian geography using the Statistical Area Classification (SAC) for data dissemination purposes and breaks down areas of Canada into census metropolitan areas (CMAs), census agglomeration areas (CAs), CTs and DAs. CTs are small, relatively stable geographic areas with a population of ~2500 to 8000, whereas DAs are the smallest geographic unit at which Statistics Canada reports complete census information, and typically consist of between 400 and 700 people. Considering the distribution of CCHS cases, after comparing the case maps at CT and DA levels, CT was selected as the unit of analysis for characterization of spatial autocorrelation and regression analysis, as many DAs did not have a sufficient number of cases (Table 2).

**Table 2** Risk factors obtained from 2006 census data

| Category and indicator | Quality of data (has indicator been used in other research?) | Description of data | Is data available for use? | Selected for analysis? |
|---|---|---|---|---|
| *Demographics* | | | | |
| Total number of low education (no certificate, diploma or, degree) | Berrigan and Troiano (2002) | Total number of population with no certificate, diploma, or degree in CT | Yes | Yes |
| Average income | Gordon-Larsen et al. (2006) | Average income in CT | Yes | Yes |
| Total number of occupied private dwellings | Agreement by senior members | Total number of occupied private dwellings in CT | Yes | Yes |
| Total number of owned dwellings | Agreement by senior members | Total number of owned dwellings in CT | Yes | Yes |
| Average value of dwelling | Djietror and Inungu (2007) | Average value of dwelling in CT | Yes | Yes |
| Total visible minority population | Berrigan and Troiano (2002) | Total visible minority population in CT | Yes | Yes |
| Total aboriginal identity population | | Total aboriginal identity population in CT | Yes | Yes |
| Total recent immigrants | Agreement by senior members | Total recent immigrants in CT | Yes | Yes |
| Unemployment rate | | Unemployment rate in CT | Yes | Yes |
| Number of dependents | Agreement by senior members | Average number of children at home per census family in CT | Yes | No |

## 2.6 CVD Mortality

CVD mortality data was obtained from the Ministry of Health and Long-Term Care (2000–2005, $N = 5872$ cases) and used for the present analysis. Causes of death were subsequently classified as: Chronic rheumatic disease (ICD-9 codes: I05-I09), Hypertensive disease (I10-I15), Ischemic heart disease (I20-I25), Pulmonary heart disease and related (I26-I28), Non rheumatic valve disorders (I34-I36), Cardiac arrest (I46), Cardiac arrhythmias (I44-I49), Heat failure and complication, ill-defined heart disease (I50-I51), Cardiomegaly (I51.7), Cerebrovascular diseases (I60-I69), Atherosclerosis (I70), and Aortic aneurysm and dissection (I71-I72). The R96 classification of "Other sudden death, cause unknown" (including "Instantaneous Death" (R96.0) and "Death occurring less than 24 hours from onset of symptoms, not otherwise explained" (R96.1)) were not included, and treated as censored (non-cardiac) events. As such, the mortality data related to CVD death are likely an underestimate of the true total number of mortality cases within the region.

Among these data, 5238 cases had postal code residential information and could be geocoded for spatial analysis. After elimination of postal codes outside of the catchment area, the final analytic sample included 4992 cases. The mortality sample was then spatially linked to the CT boundary file to reveal the total number of CVD-related deaths in each CT. Mortality rates were subsequently calculated by using the total number of deaths divided by total number of population by CT from Statistics Canada. Rates were based on averaged mortality rate for 6 years – 2000 to 2005 – to enable more stable estimates at the CT level. The overall mortality rate of York Region was 74 per 100,000 population (using 2006 census population).

## 2.7 Geospatial Factors

The "built environment" comprises urban design, land use, and the transportation system and encompasses patterns of human activity within the physical environment. There is currently no consensus as to the relative importance of the built environment and community collective factors in influencing cardiovascular morbidity and mortality. Based on the literature review and discussion with the senior offices at York Public Health Unit, a list of geospatial indicator data was used for representing the neighborhood built environment, including:

- Distance-based accessibility index: the average distance (m) for people to the nearest fitness facilities, hospitals, recreation sites, long-term care facilities, bus stops, sidewalk, trails, bike paths, and green spaces.
- Street network connectivity: the number of street connectivity in each CT.
- Building density: the percentage of building areas in each CT.
- Vegetation cover: the vegetation area percentage in each CT. Remote sensing image processing was applied on Landsat Thematic Mapper (TM) Images, Queen's University Library, 2004, to get the vegetation area in each CT, and the vegetation area percentage was got by dividing vegetation area by CT area.

- Average number of opportunities: average number of opportunities such as fast-food restaurants, convenience stores, and grocery store in each CT.

Multiple independent variables were captured from the CCHS dataset, census and GIS data to account for environmental, epidemiological, demographic, and socioeconomic characteristics. The values of poorer states of health of each variable (i.e., obesity, hypertension, diabetes, heavy drinking, heavy smoking, and sedentary lifestyle) were included within the models for spatial analysis. Table 3 lists and describes these variables.

**Table 3** CVD risk factors related to neighborhood built environment extracted from GIS data

| Category and indicator | Quality of data (has indicator been used in other research?) | Description of data | Is data available for use? | Selected for analysis? |
|---|---|---|---|---|
| *Urban design (base information)* | | | | |
| Municipal boundaries | | | Yes | Yes |
| CTs | | Small geographic areas with populations b/w 2500–8000 | Yes | Yes |
| Roads | Saelens et al. (2003) | Files for existing street network in York region | Yes | Yes |
| Water bodies | Humpel et al. (2004) | Bodies of water | Yes | Yes |
| Social housing | Agreement by senior members | Rental and subsidized housing | Yes | Yes |
| *Urban design (density)* | | | | |
| Density of buildings | Handy et al. (2002) | Number of buildings per square km | Yes | Yes |
| *Connectivity* | | | | |
| Number of intersections per square area | Frank et al. (2005) | Measure of street connectivity | Yes | Yes |
| *Transportation systems* | | | | |
| Sidewalks | Hoehner et al. (2005) | Indication of pedestrian walkways and pedestrian traffic | Yes | Yes |
| Roads | Agreement by senior members | Location of motor vehicle routes | Yes | Yes |
| Hiking trails | Hoehner et al. (2005) | Areas designated for leisure-time activity | Yes | Yes |
| Biking trails | Hoehner et al. (2005) | Indication of active transportation for leisure; transport-related commute routes | Yes | Yes |
| Bus stops | Evenson et al. (2009) | Designated bus stops | Yes | Yes |

(continued)

**Table 3** (continued)

| Category and indicator | Quality of data (has indicator been used in other research?) | Description of data | Is data available for use? | Selected for analysis? |
|---|---|---|---|---|
| *Land use designations* | | | | |
| Fast-food locations | Jones et al. (2009) | Restaurants/chains offering high-calorie/nutritionally deficient food | Yes | Yes |
| Fitness facilities | Hoehner et al. (2005) | Fitness/health facilities within region | Yes | Yes |
| Tobacco vendors | Agreement by senior members | List of current establishments licensed to sell tobacco products | Yes | Yes |
| Schools | Saelens et al. (2003) | Location of primary and secondary schools | Yes | Yes |
| Healthcare facilities – hospitals, LTC | Agreement by senior members | Location of hospitals, long-term care facilities, and healthcare centers | Yes | Yes |
| *Air quality* | | | | |
| Modeling data for air quality | Agreement by senior members | The length of the major roads (km) in that CT allows for an approximation of the CVD burden due to traffic | Yes | Yes |
| *Open space* | | | | |
| Percentage of green space | Coombes et al. (2010) | Percent of land zoned as green space | Yes | Yes |
| Park locations | Coombes et al. (2010) | Open/free access to designated parks | Yes | Yes |
| Green fields | Agreement by senior members | Land designated as green space lacking developmental plans | Yes | Yes |

## 2.8 Statistical Analysis

Two different spatial statistical techniques were applied to evaluate individual CVD risk factors or outcomes, including Moran's statistic to measure whether there is a significant spatial variation in the rates of CVD mortality and risk factors throughout York Region based on their locations and attribute values and hot spot analysis to see where significant spatial variation was. Ordinary Least Squares regression and Geographically Weighted Regression (GWR) were subsequently applied to determine the contribution of each geographic, demographic, and lifestyle factors on CVD mortality rate. OLS is a global regression method while GWR is a local, spatial, regression method that allows the relationships being modeled to vary across the study area. GWR subsequently constructs separate equations by incorporating the dependent and explanatory variables of features falling within the bandwidth of each target feature.

CVD mortality rate per 100,000 population was used as dependent variable, and population density; percentages of males and females; low education population; average income; total number of occupied private dwellings; average value of dwelling; total visible minority population; aboriginal identity population; total recent immigrants; air quality index (total length of the major roads (km)); distance-based accessibility index (average distance (m)); building density; number of street network connectivity; obesity rate per 100,000 population; diabetes rate per 100,000 population; hypertension rate per 100,000 population; sedentary lifestyle rate per 100,000 population; low consumption of fruit and vegetable rate per 100,000 population; low income rate per 100,000 population; inaccessible to physicians rate per 100,000 population; heavy smoking rate per 100,000 population; heavy drinking rate per 100,000 population; average age; average value of dwelling; unemployment rate; average household size; percentage of rent dwelling; average number of fast-food restaurants, convenience stores, and grocery stores; and average length of time in Canada since immigration were used as independent variables.

## 3 Results

### *3.1 Prevalence of CVD Risk Factors*

Table 4 describes the prevalence of CVD risk factors by age, sex, education, and location of dwelling (living in urban or rural environment) within the CCHS samples. As expected, younger adults tended to have a better CVD risk profile than older adults, with lower prevalence of hypertension and diabetes. The prevalence of diabetes and hypertension increased with age, and older adults tended to be more inactive and more overweight than younger adults. Indeed, the prevalence of inactivity in 12- to 19-year-olds was 29% but increased to around 50% in 20- to 75-year-olds. Similarly, the overweight rate increased from 9.4% in 12- to 19-year-olds to over 30% after the age of 20 years. These age-related patterns persisted for the prevalence of non-smokers (12–19 years, 88%, vs. 20+ years, <50%) and high consumption of fruits and vegetation (12–19 years, 50.4%, vs. 20–44, <34.8%). Interestingly the heavy drinkers are more popular in young age groups than old groups. The rates of heavy drinkers were 23.9% and 21.6% for the age groups of 12–19 and 20–44, respectively, but this rate has reduced to 3.9% at the age group of 75+.

In general, males had higher rate of physical activity than females. However, over half of males were classified as either overweight or obese, while only one-third of women fell into this category. Compared with males, females had a much higher percentage of non-smokers and non-drinkers with higher consumption of fruits and vegetables. The rate of diabetes was slightly higher in males than in females, while the opposite trend existed in the rate of hypertension.

Overall, the majority of respondents had completed at least their high school degree, were living in an urban setting, and had regular access to a family physician.

**Table 4** Prevalence (%) of demographic characteristics for York Region (weighted samples)

| | | Age group (years) | | | | | Sex | | Education | | Location | |
|---|---|---|---|---|---|---|---|---|---|---|---|---|
| Risk factor | | 12–19 | 20–44 | 45–64 | 65–74 | 75+ | Male | Female | <High school | ≥High school | Urban | Rural |
| Physical activity ($N = 1646$) | Inactive | 29.0 | 47.5 | 51.6 | 46.4 | 64.1 | 40.5 | 51.9 | 41.4 | 47.7 | 47.2 | 40.4 |
| | Moderately active | 20.6 | 28.2 | 27.4 | 29.6 | 27.2 | 26.9 | 26.7 | 21.9 | 28.2 | 26.5 | 29.8 |
| | Sufficiently active | 50.4 | 24.2 | 21.0 | 24.0 | 8.7 | 32.6 | 21.4 | 36.6 | 24.1 | 26.3 | 29.8 |
| BMI category ($N = 1626$) | Normal weight | 88.0 | 59.1 | 45.5 | 37.0 | 50.0 | 48.0 | 66.2 | 67.8 | 54.8 | 58.5 | 48.2 |
| | Overweight | 9.4 | 31.0 | 37.5 | 39.4 | 43.0 | 39.0 | 23.7 | 22.9 | 33.1 | 29.8 | 42.7 |
| | Obese | 2.6 | 9.9 | 16.9 | 23.6 | 7.0 | 13.0 | 10.1 | 9.3 | 12.1 | 11.8 | 9.1 |
| Smoking status ($N = 1619$) | Heavy smokers | 2.9 | 19.0 | 14.7 | 8.5 | 5.3 | 15.6 | 12.0 | 9.3 | 15.0 | 13.0 | 20.2 |
| | Former smokers | 9.1 | 37.2 | 44.0 | 43.4 | 47.4 | 41.1 | 31.4 | 22.4 | 39.8 | 35.7 | 38.0 |
| | Non-smokers | 88.0 | 43.8 | 41.3 | 48.1 | 47.4 | 43.2 | 56.5 | 68.3 | 45.2 | 51.3 | 41.7 |
| Drinking status ($N = 1242$) | Heavy drinkers | 23.9 | 21.6 | 14.0 | 5.5 | 3.9 | 25.8 | 8.9 | 16.5 | 17.6 | 16.8 | 23.3 |
| | Few drinks per week | 27.2 | 29.4 | 17.4 | 7.7 | 5.2 | 26.5 | 19.3 | 20.3 | 23.2 | 23.1 | 21.8 |
| | Non-drinkers | 48.9 | 49.1 | 68.6 | 86.8 | 90.9 | 47.7 | 71.7 | 63.3 | 59.2 | 60.1 | 54.9 |
| Income ($N = 1681$) | Low income | 3.6 | 3.2 | 4.5 | 7.8 | 13.7 | 3.4 | 5.7 | 8.3 | 3.6 | 4.6 | 5.4 |
| | High income | 96.4 | 96.8 | 95.5 | 92.2 | 86.3 | 96.6 | 94.3 | 91.7 | 96.4 | 95.4 | 94.6 |
| Fruit and vegetable consumption ($N = 1598$) | Low consumption | 49.6 | 63.2 | 57.0 | 54.2 | 51.0 | 63.5 | 53.4 | 55.9 | 59.1 | 58.4 | 56.6 |
| | High consumption | 50.4 | 36.8 | 43.0 | 45.8 | 49.0 | 36.5 | 46.6 | 44.1 | 40.9 | 41.6 | 43.4 |
| Family doctor ($N = 1681$) | No family doctor | 4.0 | 9.9 | 4.5 | 4.7 | 2.6 | 8.3 | 5.3 | 7.4 | 6.6 | 6.5 | 8.4 |
| | Has a family doctor | 96.0 | 90.1 | 95.5 | 95.3 | 97.4 | 91.7 | 94.7 | 92.6 | 93.4 | 93.5 | 91.6 |
| Comorbidities | Diabetes ($N = 1679$) | 0.0 | 0.5 | 7.9 | 14.7 | 14.5 | 4.9 | 3.8 | 6.0 | 3.8 | 4.2 | 5.4 |
| | Hypertension ($N = 1674$) | 0.8 | 2.9 | 21.1 | 43.0 | 47.9 | 12.0 | 14.6 | 16.2 | 12.6 | 13.5 | 12.7 |

The group without a high school degree had a slightly higher rate of diabetes and hypertension than those with a high school degree. Here again, higher education was associated with a higher income, but higher rates of inactivity, overweight or obesity, and smoking and drinking. Finally, on average, those who lived in rural areas had higher physical activity, but lower income than those in urban areas. Rural areas also had higher prevalence of self-reported overweight and heavy smokers and drinkers. The rate of diabetes was only marginally higher in rural areas than in urban, where hypertension was slightly higher.

## 3.2 Hot Spot Analysis of CVD Mortality and Risk Factor Rate

Within the entire region, a weak spatial autocorrelation existed for CVD mortality rate (Moran's I Index < 0.1, $p < 0.01$). For CVD risk factors, random dispersal was also observed for diabetes and hypertension, rates of physical inactivity, low income, and respondents without a regular medical doctor. On the other hand, several risk factors showed significant but weak spatial autocorrelation, including obesity (Moran's I Index = 0.2, $p < 0.01$), alcohol consumption (Moran's I Index = 0.2, $p < 0.01$), and regular cigarette smoking (Moran's I Index = 0.2, $p < 0.01$).

For rates of CVD mortality and risk factors (obesity, heavy drinking, and smoking) in which a significant weak spatial autocorrelation was found, hot spot analysis was subsequently applied to identify where these clusters were located. Spatial clusters of high values (hot spots) were identified in the northern regions, while spatial clusters of low values (cold spots) were identified in the southern region, indicating regional differences in risk factors. (See Fig. 3b for an example.)

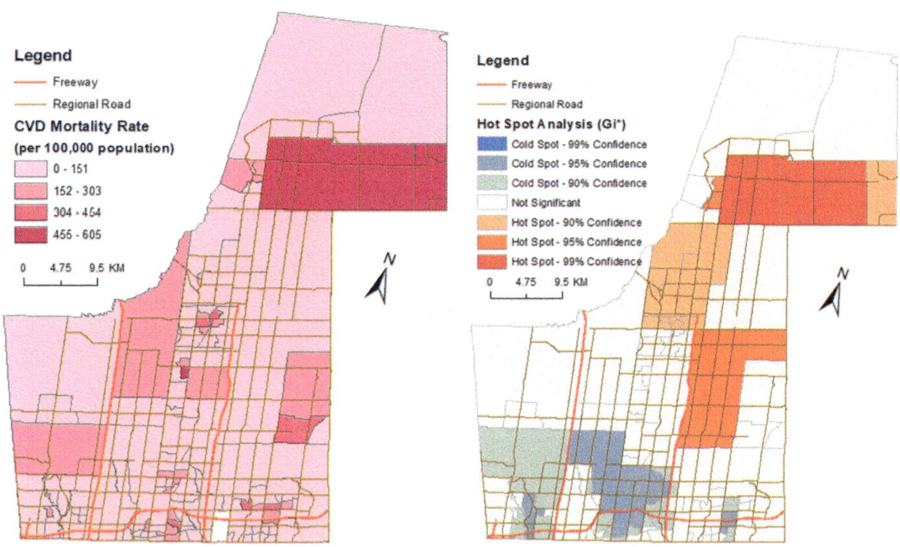


**Fig. 3** CVD mortality rate (left) and its hot spot analysis result (right) at census tract level in York Region

**Table 5** Local parameter estimates of regression analysis of CVD mortality rates with significant variables in OLS and GWR analysis

| | Ordinary Least Squares (OLS) | | Geographically Weighted Regression (GWR) | |
|---|---|---|---|---|
| Variable | Parameter | Standard error | Average parameter | Average standard error |
| Intercept | 5.0917 | 0.9383 | −0.4901 | 0.2797 |
| Average age | 0.0114 | 0.0055 | 0.0118 | 0.0065 |
| Average length of time in Canada since immigration | −0.1141 | 0.0560 | −0.1357 | 0.0668 |
| Total recent immigrants | −0.1974 | 0.0500 | −0.2129 | 0.0619 |
| Distance-based accessibility index | −0.1341 (n.s.) | 0.1077 | −0.1602 | 0.1298 |
| Building density | 0.3462 | 0.0492 | 0.3521 | 0.0588 |
| Average number of opportunities | 0.3606 | 0.0719 | 0.3432 | 0.0878 |
| Number of street network connectivity | −0.1590 (n.s.) | 0.0844 | −0.1016 | 0.1020 |

All variables included in the OLS model were re-assessed in GWR analysis, and only variables significant at the $p < 0.05$ levels were included in this table

### *3.3 OLS and GWR Regression Analysis*

Table 5 lists the risk factors which were statistically significant ($p < 0.05$) for CVD mortality in classical ordinary least square (OLS) regression analysis and/or geographical weighted regression analysis (GWR). For CVD mortality rate, building density, average age and average number of fast-food restaurants, convenience stores, grocery store and recreational activities in each CT were positively associated with CVD mortality rate, and average length of time in Canada since immigration and total recent immigrants were negatively associated with mortality rate.

Compared with OLS, GWR analysis reports two additional parameter estimates (distance-based accessibility index and number of street network connectivity) which were statistically significant for CVD mortality rate, indicating a significant association between neighborhood environmental attributes and CVD local mortal ity rate. Overall, a greater variance in CVD mortality rate was observed in the GWR than OLS analysis (63% vs. 51%, respectively).

## 4 Discussion and Conclusion

This study shows that regional differences existed in risk factors and that several built environmental attributes – including high density of buildings, the long distance to the nearest fitness facilities, hospitals, recreation sites, long-term care facilities, bus stops, sidewalk, trails, bike paths, and green spaces – collectively increased CVD risk. These results suggest that neighborhood attributes such as building density, street connectivity, and the availablity and safety of recreational space and facilities that improve neighborhood walkability, biking, and other leisure activities, should be community-level targets for reducing the burden of CVDs. These findings are consistent with other studies on how safe pedestrain trails and recreational facilities would encourage walking and other physical activities to reduce the CVD risk (Kaczynski and Henderson 2008; Arango et al. 2013; Ferdinand et al. 2012; Malambo et al. 2016).

Aside from age, the most common individual-level CVD risk factors (e.g., obesity, hypertension, diabetes, heavy drinking, heavy smoking, sedentary lifestyle, low income, low consumption of fruit and vegetable, no physician access) involved in this study did not significantly contribute to the spatial variation of mortality rates at the CT level within our sample catchment area.

Preliminary analyses also found that high accessibility to fast-food restaurants, convenience stores, and grocery stores overall are associated with an increse in CVD risk. While the findings for grocery stores are not generally supported by other literature, it has been suggested that greater accessibility to fast food restaurants may incentivize people to choose unhealthy dietary or visit convenience stores or fast food restaurants, thus increasing the chance of consuming unhealthy foods may in turn increase CVD risk (Inagami et al. 2006; Burns and Inglis 2007).

This study also observed that neighborhoods with higher proportions of new immigrants tended to have higher rates of CVD and that time since immigration was inversely related to CVD risk in general. While differences in modifiable lifestyle factors are recent and longer-term immigrants have been shown (Langellier et al. 2012), the finding of regional hot spots for CVD outcomes has an important implication on the health policy focusing on the social determinant of health within newcomer groups.

Moreover, the results of GIS-based geospatial analyses suggest that health promotion strategies may need to be tailored to specific regions within a municipality, to account for variation in demographics and risk factor clusters. While OLS regression may be used to identify factors that are associated with mortality, accounting for shared features of the built environment can capture variations in health risk that would normally be left unaccounted for. When taken together, this method of analysis was able to identify variables associated with CVD mortality rate, while also using spatial analysis to identify regional clustering and hot/cold spots for intermediate risk factors.

These analyses demonstrate that incorporating multiple types and levels of data (i.e., variables from one survey provide the individual-level covariates, while GIS data is pooled to provide information for the CT) is feasible and will increase the variance in CVD mortality that can be accounted for. While these analyses were able to identify areas of hot/cold spots, clusters, and determine if spatial autocorrelation was present, results from this analysis suggest that any single method of geographic analysis may be insufficient to identify regions that are at greater risk of CVD outcomes. As the data for this study represent multiple waves of surveillance, the analyses and maps produced represent a period estimation of surveillance, as opposed to a specific point in time.

## 5 Limitations

As with any approach, this analysis must be interpreted in light of the limitations we identified while designing and evaluating this CVD geospatial surveillance system. For one, the analysis was based on a small sample of risk factor data (CCHS Cycle 3.1 2005). There were 1681 respondents, and not all data related to the built environment was obtainable for the York Region health unit area. While the CCHS did contain six-digit postal codes (the smallest geocoding available for CCHS), more precise address information (street and unit number) were not available. Using 2000–2005 morbidity data, 2003–2009 mortality data, 2006 census data, and 2005 CCHS data for indicators results in the data overlapping but not completely matching up. Moreover, there is no way to account for people moving in and out of the region, length of stay, or the lag time between people living in a particular region and the changes to their behavior or development of CVD-related outcomes. In this analysis, only one cycle of CCHS data was used. Due to the limitation of sampling size in one cycle, some CTs end up with few or no samples, which may lead to some biases on the robustness of analysis results. Multiple cycles of CCHS should be tested in the future to validate the results from this study, panning multiple urban, suburban, and rural regions.

It should also be noted that CVD and CVD mortality rate are highly age-dependent. Age-adjusted CVD mortality rate would be a better dependent variable. In addition, only GWR was tested to explore the impact of different factors on spatial variation of mortality rate due to the weak global spatial autocorrelation in the dataset. Other spatial regression models should be tested and used in the future for datasets showing strong spatial autocorrelation (Delmelle et al. 2016). In light of data quality and availability issues, the geospatial results described in this paper are exploratory. Public health units with more extensive GIS data sources could potentially see stronger effects between built environment indicators and CVD risk factors, morbidity, and mortality.

The findings from studies that explored neighborhood built environmental attributes and their association with CVD risks and major CVD outcomes will help guide policy-makers on the built environmental, transportation, and health planning to improve intervention programs at the local level. The spatial analyses framework outlined in this paper would be feasible to administer in other public health units. With analyses using data collected over multiple years, the surveillance system could detect trends with CVD risk factors through use of routinely collected data from provincial and federal health agencies. These databanks would be compiled largely based on aggregating local sources of health data from hospitals, thus representing the population of the local region.

**Acknowledgments** This study was funded by the Public Health Agency of Canada and involved the collaboration of partners from the Regional Municipality of York (Public Health and Geomatics Branches), Queen's University (Department of Geography), and York University (School of Kinesiology and Health Science) in the development of the current framework and conducting of the statistical and geospatial analysis. The authors would like to thank Dr. Eric Weir, Shelley

Stalker, Bill Kou, and Shanna Hoetmer at York Region Public Health for their help on this research. Three anonymous reviewers and the book editors have provided constructive suggestions for improving the quality of this chapter.

## References

Arango, C. M., Páez, D. C., Reis, R. S., Brownson, R. C., & Parra, D. C. (2013). Association between the perceived environment and physical activity among adults in Latin America: A systematic review. *International Journal of Behavioral Nutrition and Physical Activity, 10*(1), 122. https://doi.org/10.1186/1479-5868-10-122.

Bryan, S. N., Tremblay, M. S., Pérez, C. E., Ardern, C. I., & Katzmarzyk, P. T. (2006). Physical activity and ethnicity: Evidence from the Canadian Community Health Survey. *Canadian Journal of Public Health, 97*, 271–276.

Bennett GG, McNeill LH, Wolin KY, Duncan DT, Puleo E & Emmons KM. (2007). Safe to walk? Neighborhood safety and physical activity among public housing residents. PLoSMed. 4(10):1599–1607.

Berrigan D & Troiano RP. (2002). The association between urban form and physical activity in US adults. Am J Prev Med. 23(2S):74–79.

Ball K, Timperio A, & Crawford D. (2009). Neighbourhood socioeconomic inequalities in food access and affordability. Health & Place. 15:578–585.

Burns, DM & Inglis, AD. (2007). Measuring food access in Melbourne: access to healthy and fast foods by car, bus and foot in an urban municipality in Melbourne. Health Place. 2007 Dec; 13(4):877–85.

Caley, L. M. (2004). Using geographic information systems to design population-based interventions. *Public Health Nurse, 21*(6), 547–554.

Center for Disease Control and Prevention (CDC). (2017). Heart disease maps and data sources. Available at https://www.cdc.gov/heartdisease/maps_data.htm. Accessed 20 Mar 2018.

Cerin, E., Conway, T. L., Saelens, B. E., Frank, L. D., & Sallis, J. F. (2009). Cross-validation of the factorial structure of the Neighborhood Environment Walkability Scale (NEWS) and its abbreviated form (NEWS-A). *International Journal of Behavioral Nutrition and Physical Activity, 6*, 32. https://doi.org/10.1186/1479-5868-6-32.

Chow, C.-M., Donovan, L., Manuel, D., Johansen, H., & Tu, J. V. (2005). Regional variation in self-reported heart disease prevalence in Canada. *The Canadian Journal of Cardiology, 21*(14), 1265–1271.

Chum, A., & O'Campo, P. (2015). Cross-sectional associations between residential environmental exposures and cardiovascular diseases. *BMC Public Health, 15*, 438. https://doi.org/10.1186/s12889-015-1788-0.

Coombes E, Jones AP, & Hillsdon M. (2010) The relationship of physical activity and overweight to objectively measured green space accessibility and use. Social Science & Medicine. 70:816–822.

CCHS (2005), Canadian Community Health Survey Share File, 2005. Statistics Canada. Ontario Ministry of Health and Long-Term Care.

Delmelle, E., et al. (2016). A spatial model of socioeconomic and environmental determinants of dengue fever in Cali, Colombia. *Acta Tropica, 164*, 169–176.

Djietror, G. & Inungu, J. (2007). Spatial patterns and covariates of heart disease death rates in Michigan, 1998-2004. The Internet Journal of Health, Volume 8 Number 1.

Ezzati, M., Hoorn, S. V., Rodgers, A., Lopez, A. D., Mathers, C. D., & Murray, C. J. (2003). Estimates of global and regional potential health gains from reducing multiple major risk factors. *Lancet, 362*(9380), 271–280.

Evenson KR, Scott MM, Cohen DA, & Voorhees CC. (2007). Girls' Perception of Neighborhood Factors on Physical Activity, Sedentary Behavior, and BMI. Obesity. 15:430–445.

Ferdinand, A. O., Sen, B., Rahurkar, S., Engler, S., & Menachemi, N. (2012). The relationship between built environments and physical activity: A systematic review. *American Journal of Public Health, 102*(10), e7–e13. https://doi.org/10.2105/AJPH.2012.300740.

Filate, W. A., Johansen, H. L., Kennedy, C. C., & Tu, J. V. (2003). Regional variations in cardiovascular mortality in Canada. *The Canadian Journal of Cardiology, 19*(11), 1241–1248.

Frank LD, Schmid TL, Sallis JF, Chapman J, & Saelens BE. (2005). Linking objectively measured physical activity with objective measured urban form. Am J Prev Med. 28(2S2):117–125.

Gordon-Larsen P, Nelson MC, Page P, & Popkin BM. (2006). Inequality in the built environment underlies key health disparities in physical activity and obesity. Pediatrics. 117(2):417–424.

Hall, R. E., & Tu, J. V. (2003). Hospitalization rates and length of stay for cardiovascular conditions in Canada, 1994 to 1999. *The Canadian Journal of Cardiology, 19*(10), 1123–1131.

Handy, S., Boarnet, M. G., Ewing, R., & Killingsworth, R. E. (2002). How the built environment affects physical activity: Views from urban planning. *American Journal of Preventive Medicine, 23*, S64–S73.

Heart and Stroke Foundation of Canada. (2016). Report on the health of Canadians: The burden of heart failure. 12 pp. Available at https://www.heartandstroke.ca/-/media/pdf-files/canada/2017-heart-month/heartandstroke-reportonhealth-2016.ashx?la=en&hash=0478377DB7CF08A281E0D94B22BED6CD093C76DB. Accessed 20 Mar 2018.

Heath, G. W., Brownson, R. C., Kruger, J., Miles, R., Powell, K. E., Ramsey, L. T., & the Task Force on Community Preventive Services. (2006). The effectiveness of urban design and land use and transport policies and practices to increase physical activity: A systematic review. *Journal of Physical Activity and Health, 3*, S55–S76.

Holowaty, E. J., Norwood, T. A., Wanigaratne, S., Abellan, J. J., & Beale, L. (2010). Feasibility and utility of mapping disease risk at the neighbourhood level within a Canadian public health unit: An ecological study. *International Journal of Health Geographics., 9*, 21–35.

Hoehner CM, Brennan Ramirez LK, Elliott MB, Handy SL & Brownson RC. (2005) Perceived and objective environmental measures and physical activity among urban adults. Am J Prev Med. 28(2S2):105–116.

Humpel N, Owen N, Iverson D, Leslie E, & Bauman A. (2004). Perceived environment attributes, residential location, and walking for particular purposes. Am J Prev Med. 26(2):119–125.

Inagami, S., Cohen, D. A., Finch, B. K., & Asch, S. M. (2006). 2006. You are where you shop. Grocery store locations, weight, and neighborhoods. *American Journal of Preventive Medicine, 31*(1), 10–17. https://doi.org/10.1016/j.amepre.2006.03.019.

Jones J, Terashima M, & Rainham D. (2009). Fast Food and Deprivation in Nova Scotia. Can J Public Health. 100(1):32–35.

Kaczynski, A. T., & Henderson, K. A. (2008). Parks and recreation settings and active living: A review of associations with physical activity function and intensity. *Journal of Physical Activity and Health, 5*(4), 619–632.

Langellier, B. A., Garza, J. R., Glik, D., Prelip, M. L., Brookmeyer, R., Roberts, C. K., Peters, A., & Ortega, A. N. (2012). Immigration disparities in cardiovascular disease risk factor awareness. *Journal of Immigrant and Minority Health, 14*(6), 918–925. https://doi.org/10.1007/s10903-011-9566-2.

Leal, C., & Chaix, B. (2011). The influence of geographic life environments on cardiometabolic risk factors: A systematic review, a methodological assessment and a research agenda. *Obesity Reviews, 12*(3), 1–14.

Lee, D. S., Chiu, M., Manuel, D. G., Tu, K., et al. (2009). Trends in risk factors for cardiovascular disease in Canada: Temporal, socio-demographic and geographic factors. *CMAJ, 181*(3–4), LE55–LE66. https://doi.org/10.1503/cmaj.081629.

Li F, Harmer P, Cardinal BJ & Vongjaturapat. (2009). Built environment changes in blood pressure in middle aged and older adults. Prev Med. 48:237–241.

Malambo, P., Kengne, A. P., Villiers, A. D., Lambert, E. V., & Puoane, T. (2016). Built environment, selected risk factors and major cardiovascular disease outcomes: A systematic review. *PLoS One, 11*(11), e0166846.

McCormack, G., Giles-Corti, B., Lange, A., Smith, T., Martin, K., & Pikora, T. J. (2004). An update of recent evidence of the relationship between objective and self-report measures of the physical environment and physical activity behaviours. *Journal of Science and Medicine in Sport, 7*, S81–S92.

O'Donnell, C. J., & Elosua, R. (2008). Cardiovascular risk factors. Insights from Framingham heart study. *Revista Española de Cardiología, 61*(3), 299–310.

Odoi, A., Wray, R., Emo, M., Birch, S., Hutchison, B., Eyles, J., & Abernathy, T. (2005). Inequalities in neighborhood socioeconomic characteristics: Potential evidence-base for neighborhood health planning. *International Journal of Health Geography., 4*, 20.

Pickle, L. W. (2002). Spatial analysis of disease. In C. Beam (Ed.), *Biostatistical applications in cancer research* (pp. 113–150). Boston: Kluwer Academic Publishers.

Public Health Agency of Canada. (2016). Cardiovascular diseases. Available at http://cbpp-pcpe.phac-aspc.gc.ca/chronic-diseases/cardiovascular-diseases/. Accessed 12 Apr 2018.

Ross CE. (2000). Walking, exercising, and smoking: Does neighborhood matter? Soc Sci & Med. 51:265–274.

Sallis, J. F., Flyd, M. F., Rodriguez, D. A., & Saelens, B. E. (2012). The role of built environments in physical activity, obesity and CVD. *Circulation, 125*(5), 729–737.

Statistics Canada. (2016). Census profile – York region. Available at: http://www12.statcan.gc.ca/census-recensement/2011/dp-pd/prof/details/page.cfm?Lang=E&Geo1=CD&Code1=3519&Geo2=PR&Code2=35&Data=Count&SearchText=York&SearchType=Begins&SearchPR=01&B1=All&GeoLevel=PR&GeoCode=3519&TABID=2. Accessed 5 Feb 2018.

Statistics Canada. (2017). Prevalence of cardiovascular disease (CVD) risk, by sex, age and cardiovascular risk factors, household population aged 20 to 79, Canada excluding territories, 2007 to 2011. https://www.statcan.gc.ca/pub/82-003-x/2016001/article/14305/tbl/tbl01-eng.htm. Accessed 30 May 2018.

Shigematsu R, Sallis JF, Conway TL, Saelens BE, Frank LD, Cain KL, Capman JE., & King AC. (2009). Age differences in the relation of perceived neighborhood environment to walking. Med. Sci Sports Exerc. 41(2):314–321.

Saelens BE, Sallis JF, Black JB, & Chen D. (2003). Neighborhood-based differences in physical activity: An environment scale evaluation. Am J Public Health. 93(9):1552–1558.

Tanuseputro, P., Manuel, D. G., Leung, M., Nguyen, K., & Johansen, H. (2003). Risk factors for cardiovascular disease in Canada. *The Canadian Journal of Cardiology, 19*, 1249–1260.

Thornton, L. E., Pearce, J. R., & Kavanagh, A. M. (2011). Using geographic information systems (GIS) to assess the role of the built environment in influencing obesity: A glossary. *International Journal of Behavioral Nutrition and Physical Activity, 8*(1), 71. https://doi.org/10.1186/1479-5868-8-71.

Tremblay, M. S., Pérez, C. E., Ardern, C. I., Bryan, S. N., & Katzmarzyk, P. T. (2005). Obesity, overweight, and ethnicity. *Health Reports, 16*, 23–33.

Tremblay, M. S., Bryan, S. N., Pérez, C. E., Ardern, C. I., & Katzmarzyk, P. T. (2006). Physical activity and immigrant status: Evidence from the Canadian Community Health Survey. *Canadian Journal of Public Health, 97*, 277–282.

Tu, J. V., Ghali, W. A., Pilote, L., & Brien, S. (Eds.). (2006). *CCORT Canadian cardiovascular atlas*. Toronto: Pulsus Group Inc./Institute for Clinical Evaluative Sciences.

US Department of Health and Human Services. (1996). *Physical activity and health: a report of the Surgeon General*. Atlanta: US Department of Health and Human Services, Public Health Service, CDC, National Center for Chronic Disease Prevention and Health Promotion.

Yiannakoulias, N., Svenson, L. W., & Schopflocher, D. P. (2009). An integrated framework for the geographic surveillance of chronic disease. *International Journal of Health Geographics, 8*, 69.

**Lei Wang** received the Ph.D. degree in geography from York University. He is an associate professor at Institute of Remote Sensing and Digital Earth, Chinese Academy of Sciences, China. He was a postdoctoral fellow at the Department of Geography, Queen's University. His research interests include geospatial analysis, digital earth, digital ocean and internet GIS.

**Chris I. Ardern** is an Associate Professor in the School of Kinesiology and Health Science at York University, and Affiliated Investigator at Southlake Regional Health Centre. His primary research interests include the epidemiology of physical activity, obesity, and cardiometabolic risk. Most recently, his work has focused on the use of risk algorithms, behavioural profiling, and geospatial analysis for the identification of high-risk subgroups. Much of this work involves the analysis of routinely collected administrative and clinical data to examine patterns of movement behaviors and their interactions in relation to obesity phenotypes.

**DongMei Chen** received the B.A. in economic geography from Peking University, China; the master in GIS and remote sensing application from the Institute of Remote Sensing Application, Chinese Academic of Science; and the Ph.D. in geography from the joint doctoral program of San Diego State University and University of California at Santa Barbara. She is currently a professor at the Department of Geography and Planning, Queen's University, Canada. Her research interest focuses on spatial data analysis and modeling, GIS, remote sensing technology, and their applications in environmental management and public health. More details about Dr. Chen and her research laboratory can be found at gis.geog.queensu.ca.

# Evaluating the Effect of Domain Size of the Community Multiscale Air Quality (CMAQ) Model on Regional $PM_{2.5}$ Simulations

**Xiangyu Jiang and Eun-Hye Yoo**

**Abstract** A growing number of urban health impact studies use Community Multiscale Air Quality (CMAQ) models for air pollution exposure estimation, although the performance of CMAQ models is likely to be affected by multiple parameters, including the configuration setting of the study domain. We presented an approach for CMAQ model uncertainty assessment with respect to domain size and reported spatial and temporal variations of CMAQ model performance over two study domains, a relatively small domain ($D_S$) and a large domain ($D_L$). Specifically, we simulated daily $PM_{2.5}$ concentrations over two domains during 2011 and quantified the difference between the model predictions. The model performance was assessed by comparing modeled $PM_{2.5}$ against measured $PM_{2.5}$ values at monitoring sites located in the region of overlap for each domain. The results suggest that the CMAQ simulations over two domains were in good agreement across the study area except in southwestern areas. We also found that the overall model performance was better for CMAQ simulations with a large domain relative to the smaller domain. Based on our findings, we recommend applying a large domain for $PM_{2.5}$ simulations, particularly for urban health risk assessments conducted over summer months, which generally contain more emissions.

## Abbreviations

| | |
|---|---|
| AQS | Air Quality System |
| BCs | Boundary conditions |
| CMAQ | Community Multiscale Air Quality |
| EPA | Environmental Protection Agency |
| FB | Fractional bias |
| FE | Fractional error |

X. Jiang · E.-H. Yoo (✉)
Department of Geography, State University of New York at Buffalo, Buffalo, NY, USA
e-mail: eunhye@buffalo.edu



Y. Lu, E. Delmelle (eds.), *Geospatial Technologies for Urban Health*,
Global Perspectives on Health Geography, https://doi.org/10.1007/978-3-030-19573-1_4

| | |
|---|---|
| NEI | National Emission Inventory |
| NYC | New York City |
| $PM_{2.5}$ | Fine particulate matter with aerodynamic diameter less than or equal to 2.5 μm |
| SMOKE | Sparse Matrix Operator Kernel Emission |
| WRF | Weather Research and Forecasting |

# 1 Introduction

Fine particulate matter (hereafter referred to as $PM_{2.5}$) is defined as particles with aerodynamic diameters less than or equal to 2.5 μm, released primarily by electricity generation and vehicle emissions (Du et al. 2016). Because of their small size, these particles can easily pass through nose hairs and penetrate the gas-exchange areas of the lungs, thereby causing serious health problems (Dockery 2009; Xing et al. 2016). The association between $PM_{2.5}$ exposure and adverse health effects, including increased mortality (Bell et al. 2008; Kloog et al. 2013), higher risk of cardiovascular disease (Hoek et al. 2013; Wang et al. 2015), and aggravated respiratory symptoms (Weber et al. 2016; Xing et al. 2016), has been well established in numerous epidemiological and clinical studies. However, many of these epidemiological studies relied solely on data collected from sparse monitoring sites that operated only every 3 or 6 days (Murray et al. 2018). Consequently, it is challenging to adequately capture the spatial and temporal variability in $PM_{2.5}$ concentrations, particularly over metropolitan areas with no monitoring sites (Özkaynak et al. 2013; Krall et al. 2015). Furthermore, the lack of reliable monitoring data may lead to substantial biases in $PM_{2.5}$ exposure estimations for urban health studies (Ebisu and Bell 2012; Baxter et al. 2013).

The Model-3/Community Multiscale Air Quality modeling system (CMAQ, http://www.cmaq-model.org) has been considered as an alternative data source for regional air quality modeling. The CMAQ is a three-dimensional air quality model designed to describe chemical and physical processes in the atmosphere at multiple spatial scales over varying time periods (Samaali et al. 2009). This atmospheric model integrates information from meteorological fields, emission components, and a chemical transport model to simulate the gridded concentrations of various types of air pollutants across a study region (Byun and Schere 2006). Compared to the exposure models based on measurements from monitoring sites, the CMAQ model offers hourly air quality predictions with greater coverage in space and time (Bravo et al. 2012). Increasingly, CMAQ model has been used in urban health studies to estimate spatially varying air pollution exposure. For example, Bravo et al. (2016) utilized 24-hr average $PM_{2.5}$ concentrations from the CMAQ model to quantify the association between $PM_{2.5}$ and cardiovascular hospital admissions over 204 US urban counties. McGuinn et al. (2017) demonstrated that coronary artery disease was linked to $PM_{2.5}$ exposures, as obtained from CMAQ model simulations. More recently, Jiang and Yoo (2018) found that approximately 4,265 premature deaths in

New York City (NYC) were associated with annual $PM_{2.5}$ concentrations based on the CMAQ simulation model, and Karambelas et al. (2018) showed that a total of 117,200 premature deaths in urban areas of India were attributable to high $PM_{2.5}$ concentrations using the CMAQ model.

Meanwhile, the CMAQ model is subject to systematic bias and uncertainties arising from error-prone inputs and imperfect numerical representations of reality (Queen and Zhang 2008; Beddows et al. 2017), which are likely to affect the subsequent health effect estimates (Cefalu and Dominici 2014). One of well-known sources of uncertainty in air quality simulations includes the specification of boundary conditions (BCs) (Tang et al. 2007; Hogrefe et al. 2018). BCs prescribe the concentrations of air pollutant components along the boundaries of a modeling domain (Borge et al. 2010), which is one of the key parameters for CMAQ model simulations. In principle, BC values should be determined by direct observations or measurements, but it is not always feasible to obtain accurate and high-resolution measurements (Jiménez et al. 2007). Thus, BCs in CMAQ simulations are specified either by implementing a static BC concentration profile or a dynamic BC taken from larger scale CMAQ simulations/a global model, such as the Goddard Earth Observing System with Chemistry model (Borge et al. 2010). While some studies have argued that CMAQ with a dynamic BC improves air pollutant predictions relative to those using time-independent BC profiles (Samaali et al. 2009; Makar et al. 2010), others have shown that global models introduce additional uncertainties into CMAQ simulations and affect the accuracy of the model outputs (Tang et al. 2009). For example, Hogrefe et al. (2018) analyzed the impact of BCs on CMAQ simulations of ozone under seven scenarios, four of which were derived from different global models. They found substantial differences among the seven sets of ozone simulations, especially near the CMAQ domain boundaries. Moreover, the use of four dynamic BCs did not necessarily give consistent or optimal ozone predictions.

In order to minimize the influence of BCs on CMAQ simulations, the modeling domain can be extended; that is, the lateral boundaries can be pushed farther away from the region of interest (Seinfeld and Pandis 2016). Lee et al. (2008) simulated ozone concentrations using the CMAQ model over three different domains for 1 week. They concluded that the best prediction performance was obtained from the model with the largest domain based on 1-week-long evaluation. Barna and Knipping (2006) also showed that modeled sulfate concentrations were highly influenced by the BCs at monitoring sites close to the domain's boundaries, while the impact was small for sites at a distance from the boundaries. Similarly, Jiménez et al. (2007) found that the influence of BCs on ozone simulations was more significant for areas near the boundaries than in the middle of the modeling domain. Although these studies indicate the benefits of using a larger domain, there has been little work to systematically investigate the effect of domain size on air pollutant simulations in space and time, especially for $PM_{2.5}$. Moreover, CMAQ model simulations with a larger domain incur greater computational costs compared to modeling efforts for a smaller domain. According to Lee et al. (2008), 48-hour ozone simulations over the continental USA take 2.7 times more computation time than simulations over the northeastern USA. There is a greater need to improve our understanding on the extent to which the implementation of a larger domain improves $PM_{2.5}$ simulations and the sensitivity of $PM_{2.5}$ simulations to domain size.

The objective of this study is to present an approach to systematically evaluate the effect of domain size on CMAQ model performance. We ran CMAQ models over two study domains with different sizes, a relatively smaller domain ($D_S$) and a larger domain ($D_L$), for the year 2011. Each domain included the State of New York, but $D_L$ was about 2.4 times larger than the size of $D_S$. We assessed the effect of CMAQ domain size on both daily and annually averaged $PM_{2.5}$ simulations. More specifically, we compared the annual average of modeled $PM_{2.5}$ over $D_S$ and $D_L$ to determine whether domain size had any substantial effect on the modeled outputs. We also evaluated the effect of CMAQ domain size on model performance by comparing modeled with measured daily $PM_{2.5}$ concentrations at each monitoring site. Finally, we investigated the overall benefits gained by CMAQ simulations with the larger domain in different regions and over different time periods.

# 2 Method

## *2.1 CMAQ Model Setup*

The CMAQ model version 5.1 was executed to simulate hourly $PM_{2.5}$ concentrations from January 1 to December 31, 2011, at the horizontal resolution of 12 km, over two domain configurations. The larger domain ($D_L$) covered an area of $1116 \times 1260$ km$^2$, while the smaller domain ($D_S$) was situated within $D_L$, covering $708 \times 828$ km$^2$ areas (see Fig. 1). The distance between $D_S$ and $D_L$ was roughly 200 km in each direction. Both domains were centered on the State of New York on a Lambert projection. Their eastern boundaries adjoined the North Atlantic Ocean, while parts of the northern and western lateral boundaries were situated in southern Canada. The CMAQ model requires three inputs for $PM_{2.5}$ simulations: meteorological fields generated by a meteorological modeling system, emission data processed by an emission processor, and air pollutant components simulated by a chemical component transport model (Byun and Schere 2006).

We employed the Weather Research and Forecasting (WRF) model version 3.7 (http://www2.mmm.ucar.edu/wrf/users/downloads.html) to prepare meteorological parameters, such as air temperature, wind field, and humidity for each $12 \times 12$ km of the modeling domain. The inputs for the WRF model were obtained from the Global Forecast System model; these data were available at a $0.5 \times 0.5$ degree resolution every 6 hours. The physical options for both domains included the Noah land surface model, the Yonsei University planetary boundary layer scheme, and the rapid radiative transfer model scheme (Wang et al. 2016). The Meteorology-

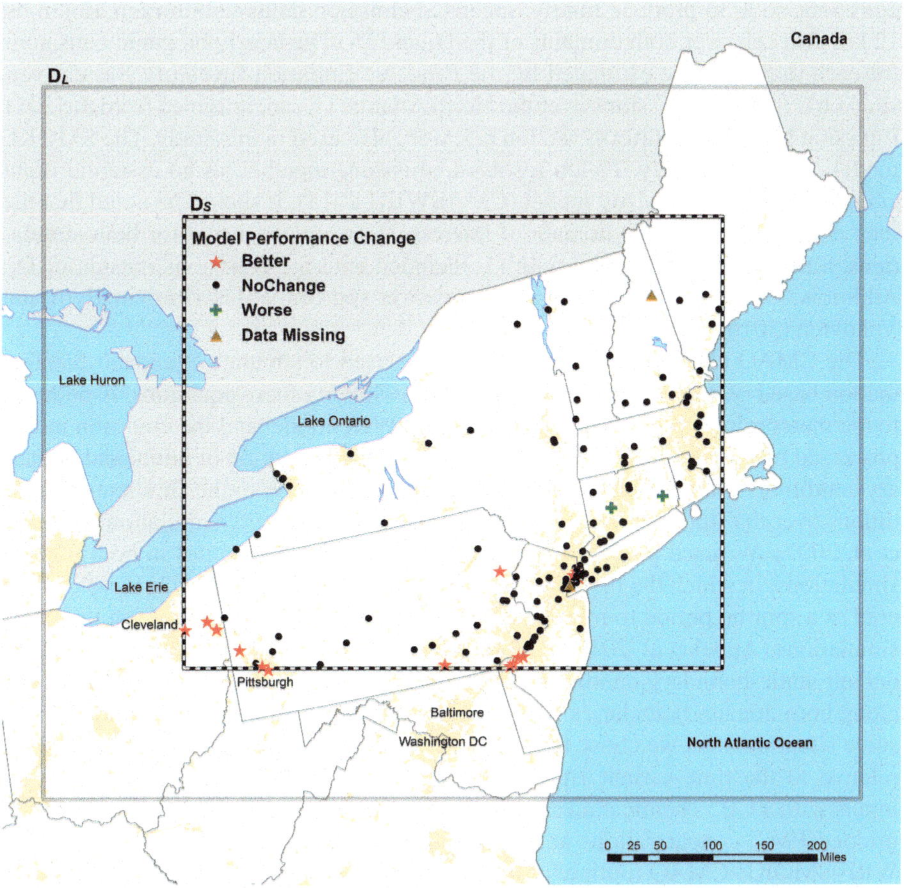


**Fig. 1** The CMAQ modeling of $D_S$ (dashed line), $D_L$ (solid line), and monitor stations (stars, circles, plus signs, and triangles), urban areas (shaded yellow polygons), lakes, and ocean (shaded blue polygons)

Chemistry Interface Processor version 4.3 was used for horizontal and vertical interpolation of the WRF outputs in order to generate CMAQ-required hourly meteorological fields over both domains (Appel et al. 2011). Further details regarding the WRF setup can be found in Jiang and Yoo (2018).

Emissions sources for both simulations were obtained from the 2011 US Environmental Protection Agency (EPA) National Emission Inventory (NEI2011). The NEI2011 consists of four major sources of emissions for the entire continental USA, including point, stationary area, non-road/onroad mobile, and biogenic emissions. It also includes parts of point and area emission sources over Canada (Eyth and Vukovich 2016). The Sparse Matrix Operator Kernel Emission (SMOKE) model version 3.7 (http://www.smoke-model.org/) was employed to process these emissions so as to produce hourly, speciated emission fields within each 12 km by 12 km grid cell over both domains of the $D_S$ and $D_L$. The hourly biogenic emissions for each domain were estimated by the Biogenic Emission Inventory System version 3.61. Sea-salt emissions over the North Atlantic Ocean, retrieved from the 2011 Emission Modeling Platform version 6.3, were also used in this study. The SMOKE model combined hourly, 12-km resolved emissions together, as an essential input compatible with the CMAQ model (CMASWIKI 2015). It should be noted that the total emissions within the domain of interest, $D_S$, were consistent for both simulations, although the simulation with $D_L$ included external emissions outside of $D_S$. Additional information regarding the emission sources is described in Eyth and Vukovich (2016).

The CMAQ chemical transport model was used to simulate chemical transformation based on a set of differential equations. Solving these equations, in addition to the meteorological fields obtained from the WRF model and the emission inputs processed by the SMOKE model, also requires the prescription of initial and boundary conditions (Samaali et al. 2009). Initial conditions refer to the air chemical conditions over the study region at the beginning of the air quality simulation (Jiménez et al. 2007). We used predefined constant initial condition profiles in both CMAQ simulations, in which the first 10 days (January 1 to January 10, 2011) were considered as a spin-up period to minimize the influence of initial conditions on the $PM_{2.5}$ simulations (Appel et al. 2017). The chemical BCs were specified using the CMAQ default static boundary condition profile, assuming relatively clean air conditions along both domain boundaries (Lee et al. 2012). This assumption was untenable in some situations, but we chose the static BC profile for its lower computational cost relative to the time-variant BC. Last, the CMAQ model collected all necessary inputs with a CB5 chemical mechanism and an Aero6 aerosol mechanism to predict gridded $PM_{2.5}$ concentrations at hourly intervals for $D_S$ and $D_L$, respectively. The WRF-SMOKE-CMAQ routine for both domains was conducted at the University at Buffalo Center for Computational Research.

## 2.2 Effect of CMAQ Domain Size on $PM_{2.5}$ Simulations

We assessed the effect of the CMAQ domain size on model performance using both the annual and daily average of $PM_{2.5}$ simulations. For the assessment of the CMAQ domain size on the annual average of $PM_{2.5}$ predictions, we aggregated the hourly $PM_{2.5}$ simulations over the entire year for each 12-km grid cell. Initially, we examined the differences between annual average of modeled $PM_{2.5}$ over two domains through the paired maps to identify regions with unusually higher and lower $PM_{2.5}$ concentrations. We also quantified differences between the two modeled outputs at each 12-km grid cell across the entire study domain by subtracting the modeled $PM_{2.5}$ concentrations over $D_L$ from those simulated by $D_S$. A positive sign indicated that a higher $PM_{2.5}$ value was estimated by the CMAQ model with $D_L$ relative to $D_S$, while a negative sign indicated that a lower $PM_{2.5}$ prediction was obtained from $D_L$. A scatter plot of the annual average $PM_{2.5}$ values obtained from the different domains enabled us to visually evaluate the relationship between the two domain simulations; it also enabled us to form a hypothesis for a statistical test, the paired t-test.

The investigation of the differences between annual average $PM_{2.5}$ concentrations in two domains presented the effect of domain size on the CMAQ simulations, but it did not provide information on the model performance. We evaluated the CMAQ model performance on simulating daily average $PM_{2.5}$ concentrations at each monitoring site using daily $PM_{2.5}$ concentrations obtained from the EPA Air Quality System (AQS) network (https://aqs.epa.gov/aqsweb/airdata/download_files.html) as reference measurements. The AQS network had a total of 146 monitoring sites, of which 125 were urban and 21 rural areas. Each site was operated more than 60 days in 2011 to measure daily $PM_{2.5}$ concentrations. The locations of these 146 monitoring sites are presented in Fig. 1. The average distance between the monitoring sites and their nearest lateral boundary was 111.23 km (about nine to ten 12-km grid cells), with a standard deviation of 70.15 km. The CMAQ modeled $PM_{2.5}$ values were paired with measurements from the AQS in space and time, resulting in a total of 36,782 model/measurement pairs.

For the assessment of the impact of domain size on model performance, we used two statistical parameters, fractional bias (FB) and fractional error (FE). These metrics were chosen to evaluate model performance rather than the absolute difference between modeled $PM_{2.5}$ and measured values for the practical reason that they are commonly used to evaluate CMAQ model simulations (Morris et al. 2005; Zhang et al. 2014). For FB, a plus or minus sign represented whether the modeled outputs over- or under-estimated the measured $PM_{2.5}$ concentrations, respectively. The smaller the absolute FB and FE values were, the better the model performance was, as suggested by Boylan and Russell (2006). If both absolute FB and FE values were smaller than the cutoff values of 60% and 75%, respectively, the level of accuracy in predicting $PM_{2.5}$ concentrations was considered to be acceptable. Hereafter, we used these values to define "Acceptable" model performance. FB and FE values exceeding this acceptable range are denoted as "Poor" model performance. A "Poor" performance indicates that the CMAQ simulations might be problematic due to insufficient emissions data and inappropriate parameter settings (EPA 2014). A detailed description of the evaluation metric, including their definitions and interpretations is summarized in Jiang and Yoo (2018). We examined differences in CMAQ model performance between $D_S$ and $D_L$ using Cohen's kappa statistics (Cohen 1960), which are commonly used in remote sensing for a change of detection (Foody 2002). In this study, we used this statistic to determine whether the difference between model performance over the two domains was substantial or not. A relatively high kappa coefficient (greater than 0.6) indicated a substantial agreement between model performance for $D_S$ and $D_L$ (Cohen 1960). Finally, we identified the spatial and temporal factors that were influential on the daily $PM_{2.5}$ simulation differences associated with the different domain sizes.

The exploratory analysis above indicates that the extent of the domain size may have an impact on $PM_{2.5}$ simulations, but intensity and magnitude may vary over regions and time periods. To demonstrate our points, we developed a multinomial logistic regression model and determined the possible factors associated with model performance change, which is written as:

$$\text{logit}\left(y\left(s_i, t_j\right)\right) = \beta_0 + \beta_1 x_1 + \beta_2 x_2 + \beta_3 x_3 + \beta_4 x_4 + \beta_5 x_5 + \varepsilon \quad (1)$$

where $y(s_i, t_j)$ denotes the model performance change from $D_S$ to $D_L$ at site $i$ on day $j$. It includes three categories: "Better," "No Change," and "Worse." We used the "Better" performance as the reference category, indicating that the model performance (in terms of FB and FE values) at site $i$ on day $j$ improved from "Poor" to "Acceptable" performance after $D_S$ was extended to $D_L$. The "Worse" performance indicates that CMAQ with $D_S$ yielded more accurate $PM_{2.5}$ predictions relative to $D_L$. The "No Change" class indicates that the CMAQ model performance remained either "Acceptable" or "Poor" in both the $D_S$ and $D_L$ simulations. The term $\beta_0$ is the intercept, and $\beta_1$ to $\beta_5$ are regression coefficients of the explanatory variables. The distance (km) from monitoring site $i$ to its nearest lateral boundary is denoted by $x_1$, and $x_2$ is the percentage of urban areas outside of $D_S$ but within a 48-km buffer zone around monitoring site $i$. The term $x_4$ represents the land use type (urban or rural) where site $i$ is located. The variable $x_4$ refers to whether day $j$ is within a summer month (May through September) or not, and $x_5$ represents whether day $j$ is on a weekday or a weekend. For categorical explanatory variables $x_3$, $x_4$, and $x_5$, we treated urban area, non-summer month, and weekday, respectively, as the reference classes. It should be noted that all the model output comparisons and model performance evaluations presented in the following sections were conducted for the region of overlap, that is, $D_S$.

# 3 Results

## 3.1 Effect of CMAQ Domain on Annual Average of $PM_{2.5}$ Simulations

The spatial distributions of annual average $PM_{2.5}$ concentrations modeled by CMAQ with $D_S$ and $D_L$ are presented in Fig. 2a, b, respectively. Both figures show that high $PM_{2.5}$ concentrations (>12 μg/m$^3$) occurred in densely populated areas, including NYC, Toronto, and Montreal, whereas lower concentrations of $PM_{2.5}$ (≤2.72 μg/m$^3$) were observed in national/state parks, wilderness areas, and the North Atlantic Ocean. The CMAQ model with a larger domain predicted higher $PM_{2.5}$ concentrations as compared with the CMAQ run in a smaller domain. For example, most modeled $PM_{2.5}$ values derived from $D_S$ in Allegheny National Forest were less than 2.72 μg/m$^3$, while higher $PM_{2.5}$ concentrations were predicted by the

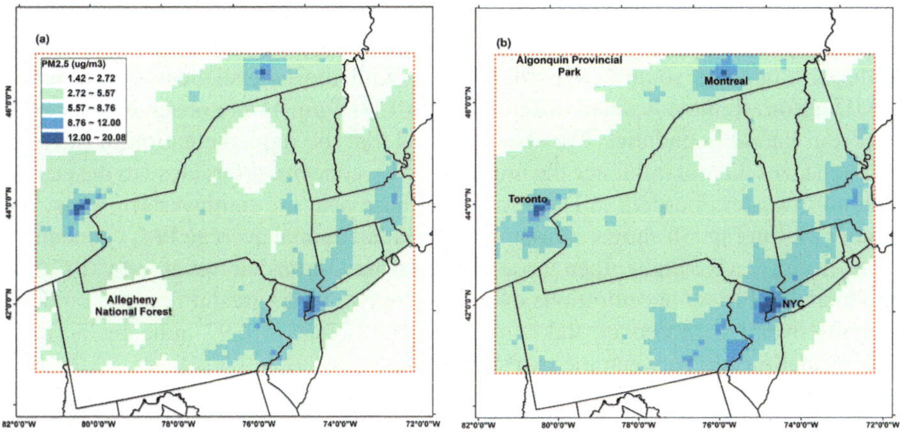


**Fig. 2** Annual average of modeled $PM_{2.5}$ concentrations derived from (**a**) $D_S$ and (**b**) $D_L$

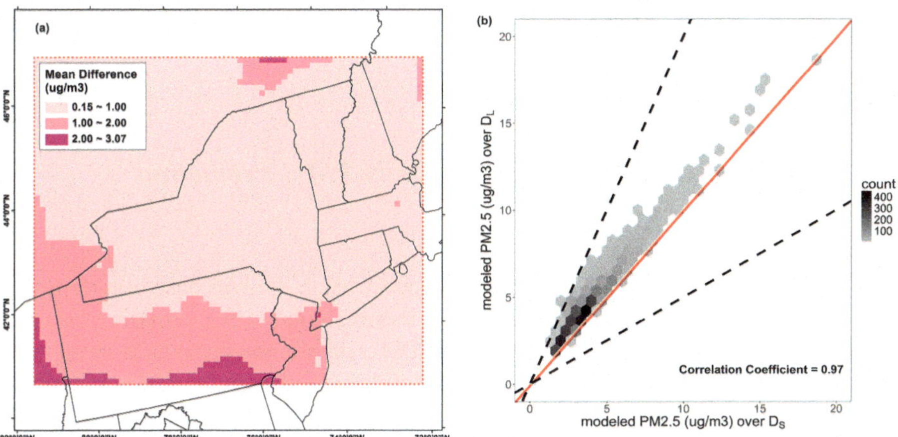


**Fig. 3** (**a**) The difference between the annual average of modeled $PM_{2.5}$ concentrations over $D_S$ and $D_L$. (**b**) Scatter plot of annual average of modeled $PM_{2.5}$ over $D_S$ and $D_L$. The red solid line represents the perfect linear relationship between two simulations, while the two dashed lines refer to the 1:2 and 2:1 relationship, respectively

CMAQ model with the larger domain. The lowest $PM_{2.5}$ values were 1.42 μg/m$^3$ for $D_S$ and 1.69 μg/m$^3$ for $D_L$; both were found in Algonquin Provincial Park in Canada. The peak $PM_{2.5}$ concentration (20.08 μg/m$^3$) simulated by the CMAQ model with $D_L$ was found in NYC, while the maximum concentration (18.52 μg/m$^3$) predicted by the CMAQ model with $D_S$ was observed in Toronto.

To facilitate comparison, we calculated the difference between annual average $PM_{2.5}$ concentrations from the larger domain and the smaller domain at each grid cell. The results are presented in Fig. 3a. All grid cells have positive values, showing that higher $PM_{2.5}$ predictions were obtained from the CMAQ model with $D_L$ relative to $D_S$ throughout the entire domain of interest. $PM_{2.5}$ values from both simulations were similar (≤+2 μg/m$^3$) over 97.2% of the overlapping areas, but considerably differed along the southwest border of $D_S$, with the largest mean difference of +3.07 μg/m$^3$. Relatively small differences (≤+1.00 μg/m$^3$) were observed over areas far from lateral boundaries. We also observed that the implementation of different domain sizes did not influence the modeled $PM_{2.5}$ concentrations close to the eastern and northwestern borders of $D_S$ as much as those near the southwestern areas. The scatter plot in Fig. 3b shows similar results that all annual average $PM_{2.5}$ concentrations from $D_L$ were higher than those derived from $D_S$ runs; however, only 0.27% of the modeled $PM_{2.5}$ concentrations over $D_L$ were two times greater than the modeled outputs over $D_S$. Moreover, the correlation coefficient of 0.97 indicates a strong linear relationship between the two simulations.

Lastly, we used a paired t-test to assess our hypothesis regarding the effect of domain size on CMAQ simulations. Our null hypothesis was that there would be no substantial difference between modeled outputs from $D_S$ and $D_L$, and 0.01 was chosen as the significance level. We calculated annual average $PM_{2.5}$ concentrations at each grid cell for both simulations and applied a paired t-test to the data. According to the test results ($t = -103.81$, $p < 2.2 \times 10^{-16}$), the differences between simulations with $D_S$ and $D_L$ were statistically significant.

### 3.2 Effect of CMAQ Domain on Daily $PM_{2.5}$ Prediction Performance

We evaluated the CMAQ model performance for each domain based on daily and site-specific FB and FE values. These values were calculated by comparing the modeled daily $PM_{2.5}$ against the collocated daily $PM_{2.5}$ measurements at each monitoring site. Based on the FB and FE values, we categorized the CMAQ model performance into "Acceptable" and "Poor" performance classes, as described in Sect. 2.2. Table 1 summarizes the statistical comparison results. Compared to the $D_S$ runs, approximately 10.54% of the modeled results improved the model performance from "Poor" to "Acceptable" when $D_L$ was used; however, roughly 5.82% of the modeled outputs were much closer to the $PM_{2.5}$ measurements over the smaller domain. We calculated the kappa index to estimate whether the domain sizes had a significant impact on model performance. The kappa coefficient was 0.67, indicating that CMAQ model

**Table 1** Percentage of "Acceptable/Poor" model performance obtained by $D_S$ and $D_L$ simulations

| | $D_S$ | | |
|---|---|---|---|
| $D_L$ | Acceptable | Poor | Sum |
| Acceptable | 42.21% | 10.54% | 52.75% |
| Poor | 5.82% | 41.43% | 47.25% |
| Sum | 48.03% | 51.97% | 100% |
| | Kappa coefficient = 0.67 | | |

performance between the two simulations was similar. Despite the difference between the two model performances not being substantial, we found that the CMAQ model with $D_L$ improved the overall model performance, as 4.72% more modeled results were classified into the "Acceptable" group for $D_L$.

## 3.3 Spatio-Temporal Variability of the CMAQ Domain Effect

We also explored the degree to which domain size affected CMAQ model performance in different regions of the study area, and for different time instances during the study period. For example, Fig. 1 illustrates the change in model performance from the $D_S$ to $D_L$ simulations at each monitoring site on April 3, 2011. The day of April 3, 2011 was chosen because the greatest number of monitoring sites were in operation. A total of 144 monitoring sites collected $PM_{2.5}$ data on that day, while only two sites (designated as "Data Missing" in Fig. 1) were closed. The CMAQ with $D_L$ showed clearly improved performance at 15 monitoring sites, as denoted by the "Better" symbols in Fig. 1. Out of the 15 monitoring sites with better performance, 11 were distributed along the southern border of $D_S$. Of the other four sites, one was located in a rural area and three were located close to, or in, NYC. The average distance between these monitoring sites and their nearest lateral boundaries was roughly 52.78 km. At two monitoring sites, the model performance with $D_L$ was worse than in the $D_S$ simulation. Both sites were located in urban areas with an average distance to their closest boundaries of 130.21 km; they are represented by the "Worse" symbols in Fig. 1. $D_L$ did not considerably affect simulation results at the remaining 127 sites, which had an average distance to their nearest lateral boundaries of 117.7 km. Among these 127 sites, the CMAQ model of both domains performed fairly well at 102 sites, but poorly at 25 sites. We used "No Change" to represent insubstantial changes in model performance obtained from both simulations. Compared to the "Worse" and "No Change" performances, a "Better" performance was more likely to be found at monitoring sites closer to the boundaries and near urban areas outside of $D_S$.

We also assessed the temporal variations of model performance for both simulations. As shown in Fig. 4a, b, simulations with $D_S$ exhibited a similar monthly pattern to $D_L$ runs that slightly overestimated $PM_{2.5}$ concentrations in January, February, and December, but largely underestimated $PM_{2.5}$ values during summer and early fall. Especially in June and July, CMAQ failed to reproduce the measured $PM_{2.5}$ concentrations, as their absolute FB and FE values were greater than their respective cutoff values of 60% and 75%. Compared to the CMAQ simulations with $D_S$, the implementation of $D_L$ greatly improved air quality predictions from May through September because their errors were much closer to 0. We also investigated model performance for each model domain over day of the week, as shown in Fig. 4c, d. Both simulations erred on the side of underestimation for daily $PM_{2.5}$ concentrations because most FB values were below 0. However, CMAQ with $D_L$ achieved slightly better model performance than the $D_S$ simulations, and the model improvement with $D_L$ was more apparent during the weekends.

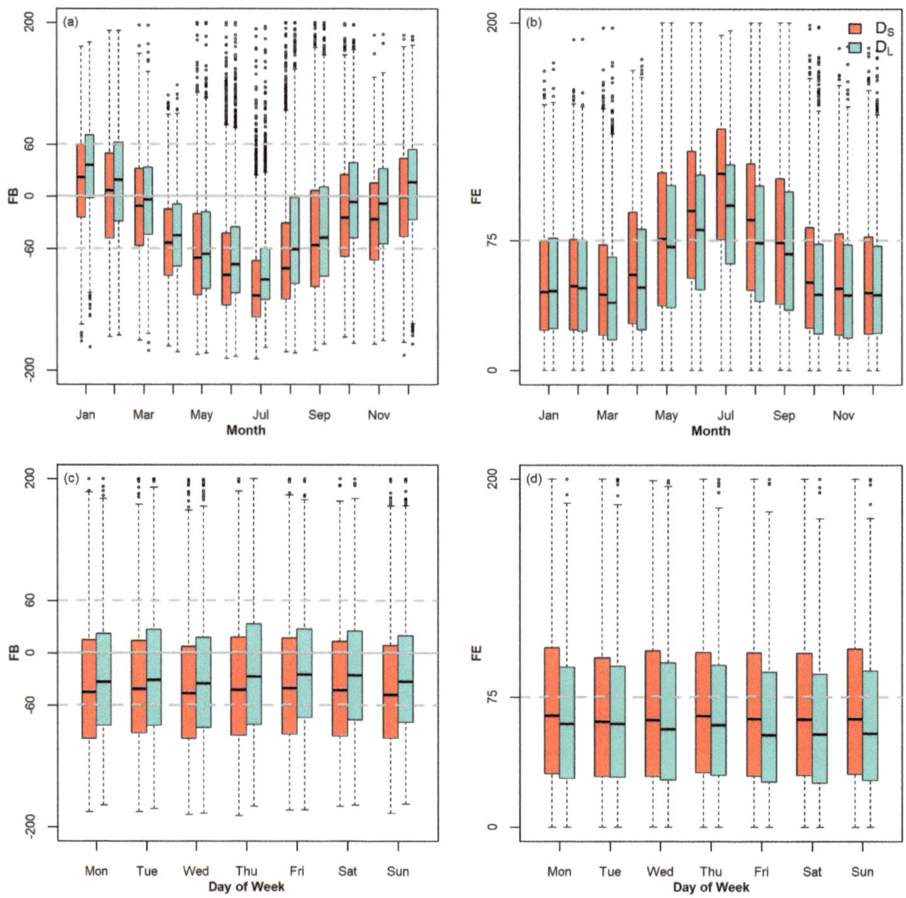


**Fig. 4** Daily (**a**) FB and (**b**) FE over months of the year. Daily (**c**) FB and (**d**) FE over days of the week

The descriptive analysis above indicates that the implementation of a larger domain improved overall CMAQ model performance on $PM_{2.5}$ simulations. In addition, the extent of model improvement might be associated with the spatial and temporal variables, such as distance to nearest boundary, proximity to metropolitan areas, land use type, month of the year, and day of the week. To demonstrate our points, we developed a multinomial logistic regression model (described in Sect. 2.2) to investigate the regions and time periods in which the CMAQ model with a larger domain was more likely to yield $PM_{2.5}$ concentrations with better accuracy. The regression model results are summarized in Table 2. The bold font indicates that the coefficients of the exploratory variables are significantly different from 0 using the 0.01 level, and the italic font represents statistical significance to the 0.05 level. According to the value of $\beta_1$, there was a greater likelihood of "No change" or "Worse" performance relative to "Better" performance for monitoring sites farther

**Table 2** Multinomial logistic regression model parameter estimations

| Model performance change | Worse | No change |
|---|---|---|
| ("'Better" is the reference category) | $\hat{\beta}_{\mathrm{Worse}}$ | $\hat{\beta}_{\mathrm{NoChange}}$ |
| Intercept | **−0.51** | **1.8** |
| Distance to boundary | **0.002** | **0.003** |
| % of urban areas outside $D_S$ | **−0.104** | **−0.057** |
| Rural | **−0.273** | *−0.098* |
| Summer | **−0.201** | **0.227** |
| Weekend | **−0.381** | **−0.135** |

from the domain boundary. In other words, the shorter the distance between a monitoring site and its nearest boundary, the greater the probability of getting "Better" performance. Moreover, the larger domain simulation had a greater chance of improving the predictions of $PM_{2.5}$ for sites surrounded by metropolitan regions outside of the domain of interest, based on the $\beta_2$ estimates and sites located in rural areas of $D_S$ according to the $\beta_3$ values. In addition, compared with the "Worse" model performance, model improvement was more likely to occur during the weekends and summer months. All explanatory variables included in the model were statistically significant to the 0.05 level. This indicates that the sensitivity of CMAQ to domain size is highly influenced by these spatial and temporal variables.

## 4 Discussion

We have demonstrated that the domain size of CMAQ models has a significant impact on the simulated annual average of $PM_{2.5}$ concentrations. We also found that CMAQ models with a larger domain are likely to predict higher $PM_{2.5}$ values in comparison to models over a smaller domain. The differences between the two simulations were more pronounced over the southwestern areas relative to other regions. One possible reason is that the southwestern regions of the study area were surrounded by highly populated cities such as Cleveland, Pittsburgh, Baltimore, and Washington, DC. These highly populated cities located outside $D_S$ emitted more anthropogenic emissions relative to rural areas and the polluted air were likely to flow into neighboring regions (Burr and Zhang 2011). The CMAQ simulations with $D_L$ captured some of the emissions from these urbanized areas transported to the regions inside of the smaller domain via southwesterly winds, while simulations with $D_S$ failed to capture these emissions sources. Meanwhile, external emissions decreased through downwind transport, which would lead to fewer emissions arriving in the central areas (Jiménez et al. 2007). Therefore, the difference between the CMAQ simulations with the two domains was more substantial over the southwestern areas than the central areas. In contrast, the different domain sizes did not considerably influence $PM_{2.5}$ simulations over the eastern and northwestern border of $D_S$. This can be explained by the fact that the eastern boundary bordered on the North

Atlantic Ocean and the northwestern area was located in or near the Algonquin Provincial Park in Canada, both of which had relatively low $PM_{2.5}$ levels. Although we pushed the lateral boundary eastward and northward by more than 200 km, there was not much air pollution from the rural and oceanic areas flowing into the $D_S$ (Burr and Zhang 2011). Our findings were consistent with previous studies (Warner et al. 1997; Barna and Knipping 2006; Pour-Biazar et al. 2011) in that areas near the domain boundaries and close to large emission sources outside of the model domain were more sensitive to the change in domain size. The high variability of $PM_{2.5}$ concentrations along a domain boundary can be considered as the "edge effects" or "boundary value problems" in spatial statistics denoting the situations where artificial boundaries delineated by researchers generate abrupt discontinuities in attributes values at borders (Griffith 1980; Ripley 1981; Griffith and Amrhein 1983; Yoo and Kyriakidis 2008; Zhu 2016). Griffith and Amrhein (1983), Yoo and Kyriakidis (2008), and other researchers proposed correction technique, but their applications to the CMAQ model need further investigation.

We also examined the CMAQ model performance associated with the domain specification at a finer temporal scale by comparing modeled daily $PM_{2.5}$ against daily $PM_{2.5}$ measurements at each AQS monitoring site. In the present study, we found that the larger domain attained better model performance, although the simulation with a larger domain may have lower prediction accuracy than the simulation with a smaller domain. Appel et al. (2017) explained the poor performance of CMAQ model with a large domain, specifically, the overestimation of $PM_{2.5}$ concentrations during the winter months, from uncertainties in gas and aerosol chemistry. The overestimation might not be significant in simulations of smaller domains; however, $PM_{2.5}$ simulations over a larger domain would predict higher $PM_{2.5}$ concentrations because the domain contains more emission sources. This result was also consistent with the findings from Lee et al. (2011) in which the largest domain produced the most frequent overestimation of ozone. In addition, the relatively high kappa coefficient of 0.67 indicated that the difference between the model performances of the $D_S$ and $D_L$ runs was not statistically significant. We suspect that $D_L$ might be not sufficiently large enough to make a significant improvement in the CMAQ model's performance (Lee et al. 2011; Borge et al. 2010). In summary, the evaluation schedule presented in this paper enabled researchers to identify an optimal domain size for a given application by performing CMAQ simulations with multiple domain configurations, repeating the analysis of the model performance.

A unique contribution of our study relative to previous studies is that we produced a continuous profile of modeled $PM_{2.5}$ concentrations over an entire year. This allowed us to evaluate not only spatial but also temporal variations of model performance over two domains. Our findings agreed with those of previous studies (Jiménez et al. 2007; Borge et al. 2010; Lee et al. 2011). That is, the use of a larger domain can greatly improve model performance in CMAQ simulations for areas close to the boundaries and/or near metropolitan cities outside of the smaller domain. Moreover, the model improvement of using a larger domain was more apparent during summer and at weekends. The monthly and day of the week variations might be explained by one of the major contributors to $PM_{2.5}$, secondary organic aerosol (SOA), which is greater during summer and weekends relative to other seasons and weekdays (Nolte et al. 2015; Gentner et al. 2017). The CMAQ with $D_L$ captured more SOA than with $D_S$, resulting in greater model performance differences between the two simulations in summer and at weekends. However, we also found that a clear underestimation existed in the month of June and July, even with the implementation of a larger domain. This large negative bias might be caused by insufficient emission sources and uncertainties in the meteorological fields (Fountoukis et al. 2013; Mancilla et al. 2015).

As an alternative source of air pollution modeling, the CMAQ model has drawn the attention of epidemiological studies in recent years (Xiao et al. 2016; Weber et al. 2016; Hu et al. 2017). Our findings would be useful for health studies, in particular for urban scale impact studies, for determining an appropriate domain size and the placement of lateral boundaries under different scenarios. As indicated by our results and those of previous studies (Lee et al. 2011; Seinfeld and Pandis 2016), a larger domain size is likely to yield accurate exposure estimates by the CMAQ model. However, considering the cost of computational burdens, we should place a minimum bounding rectangle over the health study areas and expand the domain toward the areas with larger emissions sources. Meanwhile, the lateral boundaries of the study domains should be placed in a region with relatively low air pollutant concentrations that is isolated from highly polluted areas outside of the model domain. In addition, a larger domain is recommended for use in health studies that focus on the summer and weekends.

Caution is warranted in the interpretation of our findings due to the limited experimental design (only two domains), but it represents an avenue for further investigation to improve our understanding of the effect of domain size on CMAQ performance. We have shown the process of "proof of a concept" in the present paper mainly due to the prohibitive computational cost. As noted by Samaali et al. (2009), the use of a larger domain is expensive with the possibility of better air quality predictions. Therefore, it is necessary to quantify the model improvement by using a larger domain, while accounting for computational costs. Second, $PM_{2.5}$ is a mixture of different components, such as organic carbon, black carbon, sulfate, and nitrate. Some components are more harmful to human health than others (Adams et al. 2015). Hence, tracking the influence of domain size on predictions of $PM_{2.5}$ components may be useful in identifying which component is more sensitive to the change in domain size, thereby assisting in higher accuracy of $PM_{2.5}$ component estimations. In addition, the current study failed to investigate CMAQ model performance over the regions or time periods where in situ measurements were not available. One solution to address this problem would be to utilize data with greater spatial and temporal coverage, such as satellite-based aerosol optical depth observations, so as to assess the performance of the CMAQ model (Roy et al. 2007). Alternatively, we could follow the Regionalized air Quality Model Performance method developed by Reyes et al. (2017) for a thorough evaluation of the CMAQ model's performance and the assessment of the systematic and random errors in predictions at any location.

# 5 Conclusion

We presented an approach for assessing the influence of domain size on CMAQ performance and reported the comparison results based on the two domain simulations. Our results suggest that domain size has a profound impact on $PM_{2.5}$ predictions. The inter-domain comparisons indicated that the CMAQ model results over a larger domain agreed with those from a smaller domain, except for the regions near the southwestern boundary. According to the model performance for each domain, the model performance of $PM_{2.5}$ simulations with the larger domain was superior to that of the smaller domain. However, the domain size did not have a substantial impact on the regions far from a boundary as much as it did for regions close to a boundary or near metropolitan areas outside of the study domain. We also found that the modeled $PM_{2.5}$ in rural areas was more sensitive to the change in domain size than in urban areas. The larger domain had more positive impacts on $PM_{2.5}$ simulations in summer and at weekends, which suggests that the specification of a large domain may yield more accurate $PM_{2.5}$ predictions. Under the consideration of computational burdens, however, we suggested that the extension of a model domain toward the areas containing emissions sources would guarantee improved model prediction rather than extending the domain toward areas that have relatively clean air conditions such as oceans, forests, and wilderness areas. Finally, the CMAQ model with a larger domain is highly recommended for summer months and weekends.

**Acknowledgments** The authors thank for the support provided by the Center for Computational Research (CCR) as well as the seed grant from University at Buffalo's Research and Education in Energy, Environment & Water (RENEW) Institute.

# References

Adams, K., Greenbaum, D. S., Shaikh, R., van Erp, A. M., & Russell, A. G. (2015). Particulate matter components, sources, and health: Systematic approaches to testing effects. *Journal of the Air & Waste Management Association, 65*(5), 544–558.

Appel, K. W., Foley, K., Bash, J., Pinder, R., Dennis, R., Allen, D., & Pickering, K. (2011). A multi-resolution assessment of the Community Multiscale Air Quality (CMAQ) model v4. 7 wet deposition estimates for 2002–2006. *Geoscientific Model Development, 4*(2), 357.

Appel, K. W., Napelenok, S. L., Foley, K. M., Pye, H. O., Hogrefe, C., Luecken, D. J., ... & Hutzell, W. T. (2017). Description and evaluation of the Community Multiscale Air Quality (CMAQ) modeling system version 5.1. *Geoscientific Model Development, 10*(4), 1703–1732.

Barna, M. G., & Knipping, E. M. (2006). Insights from the BRAVO study on nesting global models to specify boundary conditions in regional air quality modeling simulations. *Atmospheric Environment, 40*, 574–582.

Baxter, L. K., Dionisio, K. L., Burke, J., Sarnat, S. E., Sarnat, J. A., Hodas, N., ... & Kumar, N. (2013). Exposure prediction approaches used in air pollution epidemiology studies: Key findings and future recommendations. *Journal of Exposure Science and Environmental Epidemiology, 23*(6), 654.

Beddows, A. V., Kitwiroon, N., Williams, M. L., & Beevers, S. D. (2017). Emulation and sensitivity analysis of the Community Multiscale Air Quality Model for a UK ozone pollution episode. *Environmental Science & Technology, 51*(11), 6229–6236.

Bell, M. L., Ebisu, K., Peng, R. D., Walker, J., Samet, J. M., Zeger, S. L., & Dominici, F. (2008). Seasonal and regional short-term effects of fine particles on hospital admissions in 202 US counties, 1999–2005. *American Journal of Epidemiology, 168*(11), 1301–1310.

Borge, R., López, J., Lumbreras, J., Narros, A., & Rodríguez, E. (2010). Influence of boundary conditions on CMAQ simulations over the Iberian Peninsula. *Atmospheric Environment, 44*(23), 2681–2695.

Boylan, J. W., & Russell, A. G. (2006). PM and light extinction model performance metrics, goals, and criteria for three-dimensional air quality models. *Atmospheric Environment, 40*(26), 4946–4959.

Bravo, M. A., Fuentes, M., Zhang, Y., Burr, M. J., & Bell, M. L. (2012). Comparison of exposure estimation methods for air pollutants: Ambient monitoring data and regional air quality simulation. *Environmental Research, 116*, 1–10.

Bravo, M. A., Ebisu, K., Dominici, F., Wang, Y., Peng, R. D., & Bell, M. L. (2016). Airborne fine particles and risk of hospital admissions for understudied populations: Effects by urbanicity and short-term cumulative exposures in 708 U.S. counties. *Environmental Health Perspectives, 125*(4), 594–601.

Burr, M. J., & Zhang, Y. (2011). Source apportionment of fine particulate matter over the Eastern US Part I: Source sensitivity simulations using CMAQ with the Brute Force method. *Atmospheric Pollution Research, 2*(3), 300–317.

Byun, D., & Schere, K. L. (2006). Review of the governing equations, computational algorithms, and other components of the Models-3 Community Multiscale Air Quality (CMAQ) modeling system. *Applied Mechanics Reviews, 59*(2), 51–77.

Cefalu, M., & Dominici, F. (2014). Does exposure prediction bias health effect estimation? The relationship between confounding adjustment and exposure prediction. *Epidemiology (Cambridge, Mass.), 25*(4), 583.

CMAQ version 5.0 (February 2010 release) OGD. (2015, December 4). CMASWIKI, Retrieved 14:35, May 5, 2019 from https://www.airqualitymodeling.org/index.php?title=CMAQ_version_5.0_(February_2010_release)_OGD&oldid=682.

Cohen, J. (1960). A coefficient of agreement for nominal scales. *Educational and Psychological Measurement, 20*(1), 37–46.

Dockery, D. W. (2009). Health effects of particulate air pollution. *Annuals of Epidemiology, 19*(4), 257–263.

Du, Y., Xu, X., Chu, M., Guo, Y., & Wang, J. (2016). Air particulate matter and cardiovascular disease: The epidemiological, biomedical and clinical evidence. *Journal of Thoracic Disease, 8*(1), E8.

Ebisu, K., & Bell, M. L. (2012). Airborne $PM_{2.5}$ chemical components and low birth weight in the northeastern and mid-Atlantic regions of the United States. *Environmental Health Perspectives, 120*(12), 1746.

EPA. (2014). Modeling guidance for demonstrating attainment of air quality goals for ozone, $PM_{2.5}$, and regional haze-December 2014 DRAFT. US Environmental Protection Agency, Office of Air Quality Planning and Standards. https://www3.epa.gov/scram001/guidance/guide/Draft_O3-PM-RH_Modeling_Guidance-2014.pdf.

Eyth, A., & Vukovich, J. (2016). *Technical Support Document (TSD) preparation of emissions inventories for the version 6.3, 2011 emissions modeling platform.* US Environmental Protection Agency, Office of Air Quality Planning and Standards.

Foody, G. M. (2002). Status of land cover classification accuracy assessment. *Remote Sensing of Environment, 80*(1), 185–201.

Fountoukis, C., Koraj, D., van der Gon, H. D., Charalampidis, P., Pilinis, C., & Pandis, S. (2013). Impact of grid resolution on the predicted fine PM by a regional 3-D chemical transport model. *Atmospheric Environment, 68*, 24–32.

Gentner, D. R., Jathar, S. H., Gordon, T. D., Bahreini, R., Day, D. A., El Haddad, I., ... & Goldstein, A. H. (2017). Review of urban secondary organic aerosol formation from gasoline and diesel motor vehicle emissions. *Environmental Science & Technology, 51*(3), 1074–1093.

Griffith, D. A. (1980). Towards a theory of spatial statistics. *Geographical Analysis, 12*(4), 325–339.

Griffith, D. A., & Amrhein, C. G. (1983). An evaluation of correction techniques for boundary effects in spatial statistical analysis: Traditional methods. *Geographical Analysis, 15*(4), 352–360.

Hoek, G., Krishnan, R. M., Beelen, R., Peters, A., Ostro, B., Brunekreef, B., & Kaufman, J. D. (2013). Long-term air pollution exposure and cardio-respiratory mortality: A review. *Environmental Health, 12*(1), 43.

Hogrefe, C., Liu, P., Pouliot, G., Mathur, R., Roselle, S., Flemming, J., Lin, M., & Park, R. J. (2018). Impacts of different characterizations of large-scale background on simulated regional-scale ozone over the continental United States. *Atmospheric Chemistry and Physics, 18*(5), 3839.

Hu, J., Li, X., Huang, L., Qi, Y., Zhang, Q., Zhao, B., Wang, S., & Zhang, H. (2017). Ensemble prediction of air quality using the WRF/CMAQ model system for health effect studies in China. *Atmospheric Chemistry and Physics, 17*(21), 13103.

Jiang, X., & Yoo, E.-h. (2018). The importance of spatial resolutions of Community Multiscale Air Quality (CMAQ) models on health impact assessment. *Science of the Total Environment, 627*, 1528–1543.

Jiménez, P., Parra, R., & Baldasano, J. M. (2007). Influence of initial and boundary conditions for ozone modeling in very complex terrains: A case study in the northeastern Iberian Peninsula. *Environmental Modelling & Software, 22*(9), 1294–1306.

Karambelas, A., Holloway, T., Kinney, P. L., Fiore, A. M., DeFries, R., Kiesewetter, G., & Heyes, C. (2018). Urban versus rural health impacts attributable to $PM_{2.5}$ and $O_3$ in northern India. *Environmental Research Letters, 13*(6), 064010.

Kloog, I., Ridgway, B., Koutrakis, P., Coull, B. A., & Schwartz, J. D. (2013). Long- and short-term exposure to $PM_{2.5}$ and mortality: Using novel exposure models. *Epidemiology (Cambridge, Mass.), 24*(4), 555.

Krall, J. R., Chang, H. H., Sarnat, S. E., Peng, R. D., & Waller, L. A. (2015). Current methods and challenges for epidemiological studies of the associations between chemical constituents of particulate matter and health. *Current Environmental Health Reports, 2*(4), 388–398.

Lee, P., Kang, D., McQueen, J., Tsidulko, M., Hart, M., DiMego, G., Seaman, N., & Davidson, P. (2008). Impact of domain size on modeled ozone forecast for the northeastern United States. *Journal of Applied Meteorology and Climatology, 47*(2), 443–461.

Lee, H., Liu, Y., Coull, B., Schwartz, J., & Koutrakis, P. (2011). A novel calibration approach of MODIS AOD data to predict $PM_{2.5}$ concentrations. *Atmospheric Chemistry and Physics, 11*(15), 7991–8002.

Lee, D., Wang, J., Jiang, X., Lee, Y., & Jang, K. (2012). Comparison between atmospheric chemistry model and observations utilizing the RAQMS–CMAQ linkage. *Atmospheric Environment, 61*, 85–93.

Makar, P. A., Gong, W., Mooney, C., Zhang, J., Davignon, D., Samaali, M., ... & Chen, J. (2010). Dynamic adjustment of climatological ozone boundary conditions for air-quality forecasts. *Atmospheric Chemistry and Physics, 10*(18), 8997–9015.

Mancilla, Y., Herckes, P., Fraser, M. P., & Mendoza, A. (2015). Secondary organic aerosol contributions to $PM_{2.5}$ in Monterrey, Mexico: Temporal and seasonal variation. *Atmospheric Research, 153*, 348–359.

McGuinn, L. A., Ward-Caviness, C., Neas, L. M., Schneider, A., Di, Q., Chudnovsky, A., ... & Kraus, W. E. (2017). Fine particulate matter and cardiovascular disease: Comparison of assessment methods for long-term exposure. *Environmental Research, 159*, 16–23.

Morris, R. E., McNally, D. E., Tesche, T. W., Tonnesen, G., Boylan, J. W., & Brewer, P. (2005). Preliminary evaluation of the Community Multiscale Air Quality model for 2002 over the Southeastern United States. *Journal of the Air & Waste Management Association, 55*(11), 1694–1708.

Murray, N., Chang, H. H., Holmes, H., & Liu, Y. (2018). Combining satellite imagery and numerical model simulation to estimate ambient air pollution: An ensemble averaging approach. *arXiv preprint arXiv: 1802.03077.*

Nolte, C., Appel, K., Kelly, J., Bhave, P., Fahey, K., Collett, J., Jr., Zhang, L., & Young, J. (2015). Evaluation of the Community Multiscale Air Quality (CMAQ) model v5. 0 against size-resolved measurements of inorganic particle composition across sites in North America. *Geoscientific Model Development, 8*(9), 2877–2892.

Özkaynak, H., Baxter, L. K., Dionisio, K. L., & Burke, J. (2013). Air pollution exposure prediction approaches used in air pollution epidemiology studies. *Journal of Exposure Science and Environmental Epidemiology, 23*(6), 566–572.

Pour-Biazar, A., Khan, M., Wang, L., Park, Y.-H., Newchurch, M., McNider, R. T., Liu, X., Byun, D. W., & Cameron, R. (2011). Utilization of satellite observation of ozone and aerosols in providing initial and boundary condition for regional air quality studies. *Journal of Geophysical Research: Atmospheres, 116*(D18).

Queen, A., & Zhang, Y. (2008). Examining the sensitivity of MM5–CMAQ predictions to explicit microphysics schemes and horizontal grid resolutions, Part III–The impact of horizontal grid resolution. *Atmospheric Environment, 42*(16), 3869–3881.

Reyes, J. M., Xu, Y., Vizuete, W., & Serre, M. L. (2017). Regionalized $PM_{2.5}$ Community Multiscale Air Quality model performance evaluation across a continuous spatiotemporal domain. *Atmospheric Environment, 148*, 258–265.

Ripley, B. D. (1981). *Spatial statistics*. New York: Wiley.

Roy, B., Mathur, R., Gilliland, A. B., & Howard, S. C. (2007). A comparison of CMAQ-based aerosol properties with IMPROVE, MODIS, and AERONET data. *Journal of Geophysical Research: Atmospheres*, (D14), 112.

Samaali, M., Moran, M. D., Bouchet, V. S., Pavlovic, R., Cousineau, S., & Sassi, M. (2009). On the influence of chemical initial and boundary conditions on annual regional air quality model simulations for North America. *Atmospheric Environment, 43*(32), 4873–4885.

Seinfeld, J. H., & Pandis, S. N. (2016). *Atmospheric chemistry and physics: From air pollution to climate change*. Wiley.

Tang, Y., Carmichael, G. R., Thongboonchoo, N., Chai, T., Horowitz, L. W., Pierce, R. B., ... & Sachse, G. W. (2007). Influence of lateral and top boundary conditions on regional air quality prediction: A multiscale study coupling regional and global chemical transport models. *Journal of Geophysical Research: Atmospheres, 112*(D10).

Tang, Y., Lee, P., Tsidulko, M., Huang, H. C., McQueen, J. T., DiMego, G. J., ... & Kang, D. (2009). The impact of chemical lateral boundary conditions on CMAQ predictions of tropospheric ozone over the continental United States. *Environmental Fluid Mechanics, 9*(1), 43–58.

Wang, C., Tu, Y., Yu, Z., & Lu, R. (2015). $PM_{2.5}$ and cardiovascular disease in the elderly: An overview. *International Journal of Environmental Research and Public Health, 12*(7), 8187–8197.

Wang, W., Barker, D., Bray, J., Bruyere, C., Duda, M., Dudhia, J., Gill, D., & Michalakes, J. (2016). User's guide for the Advanced Research WRF (ARW) modeling system version 3.7. http://www2.mmm.ucar.edu/wrf/users/docs/user_guide_V3.7/ARWUsersGuideV3.7.pdf.

Warner, T. T., Peterson, R. A., & Treadon, R. E. (1997). A tutorial on lateral boundary conditions as a basic and potentially serious limitation to regional numerical weather prediction. *Bulletin of the American Meteorological Society, 78*(11), 2599–2617.

Weber, S. A., Insaf, T. Z., Hall, E. S., Talbot, T. O., & Huff, A. K. (2016). Assessing the impact of fine particulate matter ($PM_{2.5}$) on respiratory-cardiovascular chronic diseases in the New York City Metropolitan area using Hierarchical Bayesian Model estimates. *Environmental Research, 151*, 399–409.

Xiao, Q., Liu, Y., Mulholland, J. A., Russell, A. G., Darrow, L. A., Tolbert, P. E., & Strickland, M. J. (2016). Pediatric emergency department visits and ambient air pollution in the US State of Georgia: A case-crossover study. *Environmental Health, 15*(1), 115.

Xing, Y.-F., Xu, Y.-H., Shi, M.-H., & Lian, Y.-X. (2016). The impact of PM2.5. On the human respiratory system. *Journal of Thoracic Disease, 8*(1), E69.

Yoo, E.-H., & Kyriakidis, P. (2008). Area-to-point prediction under boundary conditions. *Geographical Analysis, 40*(4), 355–379.

Zhang, H., Chen, G., Hu, J., Chen, S.-H., Wiedinmyer, C., Kleeman, M., & Ying, Q. (2014). Evaluation of a seven-year air quality simulation using the Weather Research and Forecasting (WRF)/Community Multiscale Air Quality (CMAQ) models in the eastern United States. *Science of the Total Environment, 473*, 275–285.

Zhu, X. (2016). *GIS for environmental applications: A practical approach*. Routledge.

**Xiangyu Jiang** is a PhD candidate at the Department of Geography, College of Arts and Sciences, University at Buffalo. Her research interests are in the fields of GIScience, public health, and environmental modeling. Her current research focuses on the wildland fire-related air pollution exposure modeling and health impact assessments.

**Eun-Hye Yoo** is a PhD graduate. She is an Associate Professor at the Department of Geography, College of Arts and Sciences, University at Buffalo. She is a geographer with a special training in GIScience. Her research interests include spatial scale issues and error/uncertainty in geographic data, as well as their effects on statistical analyses. Her past research has examined these issues in relation to diverse topics, such as hedonic price models, population density, mosquito abundance, presettlement vegetation, air pollution, and respiratory disease. Her current research projects focus on fine-scale air pollution exposure modeling, geospatial health effect assessments, and human time-activity analysis.

# Part II
# Urban Health Service Access

# Serving a Segregated Metropolitan Area: Disparities in Spatial Access to Primary Care Physicians in Baton Rouge, Louisiana

**Fahui Wang, Michael Vingiello, and Imam M. Xierali**

**Abstract** This study examines spatial accessibility of primary care in the Baton Rouge Metropolitan Statistical Area, Louisiana. Two popular accessibility measures are used: the proximity method focuses on the travel time from the nearest facility and the two-step floating catchment area (2SFCA) method considers the match ratio between providers and population as well as the complex spatial interaction between them. The two methods capture different elements of spatial accessibility: one being physically close to a facility and another adding availability of service. Both properties can be valuable for residents. In the study area, residents in urban areas generally enjoy shorter travel time from their nearest service providers as well as higher accessibility scores measured by the 2SFCA method (i.e., physicians per 1000 residents) than rural residents. Overall, disproportionally higher percentages of African Americans are in areas with shorter travel time to the nearest primary care providers and higher accessibility scores; so are residents in areas of higher poverty rates. This "reversed racial advantage" in spatial accessibility does not capture nonspatial obstacles related to financial and other socioeconomic factors for African Americans (and population in poverty) and nevertheless represents one fewer battle to fight in reducing healthcare disparities for various disadvantaged population groups. Such an advantage disappears or is even reversed in remote rural areas with high concentration of African Americans, who suffer from double disadvantages in both spatial and nonspatial access to primary care.

F. Wang (✉)
Department of Geography and Anthropology, Louisiana State University,
Baton Rouge, LA, USA
e-mail: fwang@lsu.edu

M. Vingiello
The Water Institute of the Gulf, Baton Rouge, LA, USA

I. M. Xierali
Department of Family and Community Medicine, University of Texas Southwestern Medical Center, Dallas, TX, USA



Y. Lu, E. Delmelle (eds.), *Geospatial Technologies for Urban Health*,
Global Perspectives on Health Geography, https://doi.org/10.1007/978-3-030-19573-1_5

## 1 Introduction

Accessibility refers to the relative ease by which activities or services – in this case, healthcare – can be reached by someone at a given location (Penchansky and Thomas 1981). Accessibility can be related to spatial and nonspatial factors (Khan 1992). Spatial accessibility emphasizes geographic barriers between service centers (supply) and residents (demand) and how they are connected in space (Joseph and Phillips 1984). Nonspatial factors include various demographic and socioeconomic variables that affect one's ability to obtain the services. In short, spatial accessibility is because of "where you are," and nonspatial accessibility is because of "who you are." This chapter focuses on spatial accessibility, i.e., place-based barriers that impede residents from reaching their service providers. Such barriers include being in a remote area with absence or paucity of the service, poor road conditions, overwhelming traffic, and poorly designed and disconnected road networks. Additionally, the chapter also examines how the spatial and nonspatial factors interact, such as racial disparity in spatial accessibility.

The American Academy of Family Physicians defines primary care as "that care provided by physicians specifically trained for and skilled in comprehensive first contact and continuing care for persons with any undiagnosed sign, symptom, or health concern (the 'undifferentiated' patient) not limited by problem origin (biological, behavioral, or social), organ system, or diagnosis" (AAFP 2018). Primary care includes health promotion, disease prevention, health maintenance, counseling, patient education, and diagnosis and treatment of acute and chronic illnesses in a variety of healthcare settings such as office, inpatient, critical care, long-term care, home care, and day care and is usually provided by a primary care physician who is a specialist in family medicine, general internal medicine, or pediatrics. Although non-primary care physicians (e.g., cardiologists, ophthalmologists) as well as non-physician healthcare providers (e.g., nurse practitioners, physician assistants) can also provide certain primary care, an effective system of primary care may utilize them as members of the healthcare team with a primary care physician maintaining responsibility for the function of the healthcare team and the comprehensive, ongoing healthcare of the patient (ibid).

Primary care is thus an integral component of a rational and efficient health delivery system and is critical for the success of preventive care (Lee 1995). Access to primary care varies spatially because it is affected by where health professionals are located and where people reside; neither health professionals nor population is uniformly distributed. Maldistribution of the primary care workforce leads to the "shortages amid surplus paradox" (Hart et al. 2002: 212). The U.S. Department of Health and Human Services (DHHS) has implemented various programs including the designations of Health Professional Shortage Areas (HPSAs) for improving access to primary care for the underserved (DHHS 2018). The effectiveness of such programs relies on an appropriate and accurate measure of accessibility so that resources can be allocated to areas of the greatest need. Among others, adequate spatial access to primary care is a major factor for ensuring delivery of quality services.

Our survey of the literature indicates that the majority of early studies employ the simple proximity method. In other words, distance or travel time to the nearest service provider reflects one's convenience or accessibility of obtaining the service. Proximity is considered the most influential component in a community for healthcare services (Law et al. 2011). Most recently, Yin et al. (2018) used the proximity method to measure the spatial accessibility to medical facilities at the county level in China and by using the Theil Index to quantify the inequality in access. However, the proximity method assumes that the service is available in abundance and does not account for possible crowdedness patients may experience in seeking the care. The two-step floating catchment area (2SFCA) developed by Luo and Wang (2003) considers the match ratio between supply and demand as well as the complex spatial interaction between them and has become the most popular method in measuring spatial accessibility. With recent advancements in integrating diverse travel distance decay behaviors of patients (Wang 2012) and the method's automation as an ArcGIS toolkit (Wang 2015: 112–113), the 2SFCA is an optimal choice for capturing availability of a service when scarcity of its provision is a concern. A recent study by Luo et al. (2018) used a modified version of 2SFCA, termed E2SFCA developed by Luo and Qi (2009), to measure the spatial accessibility of medical services for the elderly in Wuhan, China. Various versions of 2SFCA all yield an accessibility score that can be interpreted as a supply-demand ratio, e.g., number of physicians per capita. When the ratio is small, it is typically inflated 1000 times to represent number of physicians per 1000 population. The proximity and availability of a service are two distinctive and related properties in accessibility, and both are valued by people (e.g., Ikram et al. 2015; Luo et al. 2017). This study uses both measures for a more comprehensive assessment of primary care accessibility based on primary care physician distribution.

Three aspects differentiate this study from existing work on primary care accessibility:

1. Both proximity and 2SFCA-based accessibility methods are used to evaluate whether and how the two measures differ in geographic patterns.
2. Methods are employed to examine whether the disparities across geographic areas (e.g., areas of various urbanicity) and between demographic groups are statistically significant.
3. A regionalization method is used to divide the study area into several regions that are relatively homogenous in racial structure, and disparities in accessibility are assessed across these regions.

## 2 Study Area and Data Sources

The Baton Rouge Metropolitan Statistical Area (BRMSA) is selected as the study area due to its socioeconomic diversity and full rural–urban continuum, and hereafter simply referred to as "Baton Rouge." As shown in Fig. 1, this region consists of

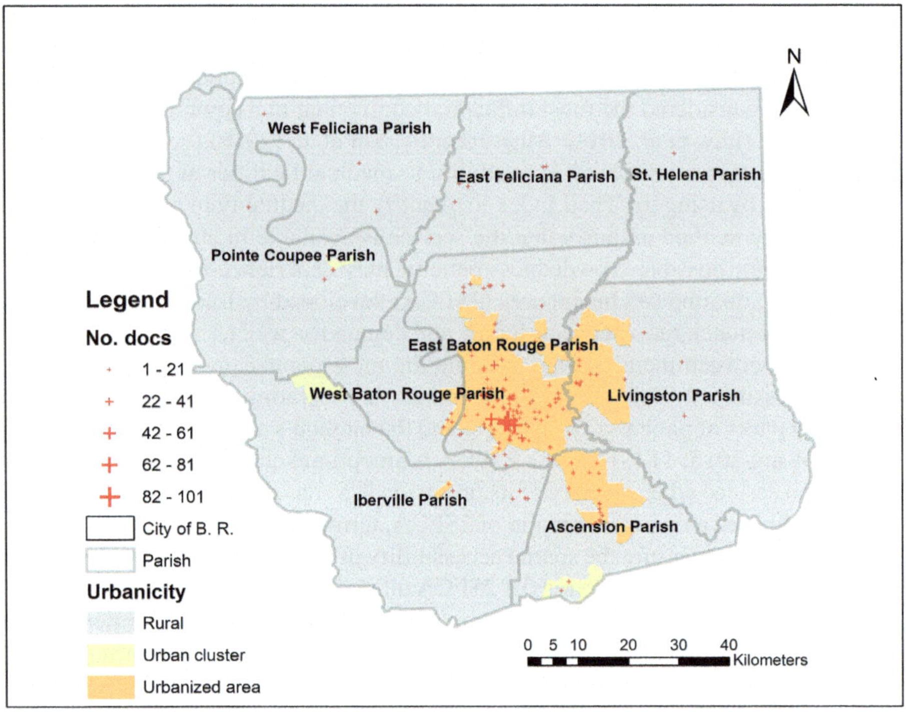


**Fig. 1** Primary care physicians and urban areas in Baton Rouge MSA

nine parishes. "Parish" is the county equivalent unit in Louisiana. The City of Baton Rouge, the state's capital city, resides in East Baton Rouge Parish. According to a recent report by East Baton Rouge Parish (2016), it has double the national benchmarks in both low birth-weight rate and uninsured population rate, ranks second in the nation for new HIV/AIDS cases, and has six times the national average rate of sexually transmitted diseases. All highlight the importance of research and understanding of health-related issues, including primary care accessibility.

Data for the analysis are composed of three parts: supply (facilities), demand (population), and the road network linking them. For the supply side, data of individual physicians (including specialty and geographic location) in Louisiana in 2016 (October 9 snapshot) were obtained from the National Plan and Provider Enumeration System (NPPES) of the Centers for Medicare and Medicaid Services (CMS). In a previous study, physician practice location data in the NPPES was shown to be comparable in enumeration of providers but to have less spatial uncertainty than other data sources such as the American Medical Association Physician Masterfile or state medical licensure data. This may be possibly due to the fact that providers are required to include their NPPES IDs on claims in order to receive payment for Medicare services from the CMS (Xierali et al. 2016). Spatial uncertainty in physician workforce data generally refers to the uncertainty in locating the exact practice location of physicians due to errors in address data collection, errors in the address reference database, uncertainty of whether an address is the practice address or home address, and/or whether physician engaged in multi-site practicing (Shi et al. 2016; Xierali 2018). The role of CMS in issuing NPPES IDs is independent of its role as a payer for Medicare services, and health-care providers are required to have NPPES IDs in order to transfer claims and other healthcare information electronically (Bindman 2013). The provider practice location data from the NPPES were processed and geocoded for health workforce analysis. Only 955 doctors classified as primary care physicians (PCP) were extracted for the BRMSA. For privacy concerns, the PCP data are aggregated to the block level identified by whether a PCP falls inside a block, and the BRMSA has 204 blocks with at least 1 PCP. The location of each block is represented by the average coordinates of all PCPs within its boundary, and thus is more accurate than its geographic centroid. One common concern in accessibility studies is the edge effect, referring to less reliable results near the edge of a study area where interactions with neighboring areas are not considered. For example, residents in the study area may visit doctors beyond the study area and vice versa. Such an edge effect is limited since most of physicians are located in the urban areas away from the boundary of the study area (Fig. 1), and interactions beyond the boundary are considered minor.

On the demand side, the 2011–2015 Five-Year American Community Survey (ACS) data at the census block group level are used to define population and related socio-demographic variables (U.S. Bureau of Census 2018a). The 2011–2015 ACS data was the most recent available when the research was conducted and is a reasonable match for the 2016 PCP data in time. Block group is the smallest area unit that comes with socioeconomic variables such as poverty status from the ACS data. Similarly, the location of each census block group is represented by the population-weighted centroid based on the block-level data for better spatial accuracy. Readers may consult Wang (2015, 78) for technical detail of calibrating weighted centroids. There are 483 block groups in the study area with total population of 787,961. Excluding two block groups with zero population or household, 481 block groups are used for the study.

The BRMSA is composed of 60.5% white residents and 35.7% African American residents. Other racial-ethnic groups are not considered because their percentages are below 5%. For socioeconomic factors, this research only considers households under poverty, and the BRMSA has an average poverty rate of 15.60%. Figure 2 shows the geographic distribution of African Americans across the BRMSA. The highest concentration of African Americans (with rates higher than 80%) is in the northwest part of the City of Baton Rouge. Most of the three northern parishes, the northwest part of East Baton Rouge Parish and a relatively small area in the middle-south of the MSA have African Americans rates of 40–80%. Figure 3 shows the statistical distribution of block groups in terms of African American percentage (each bar indicates the frequency of observations in a 5% range such as 0–5%, 5–10%, and so on). The U shape highlights a segregated pattern of the BRMSA with the highest numbers of block groups being either 0–5% or 95–100%. More in-depth analysis is explored in a later section.

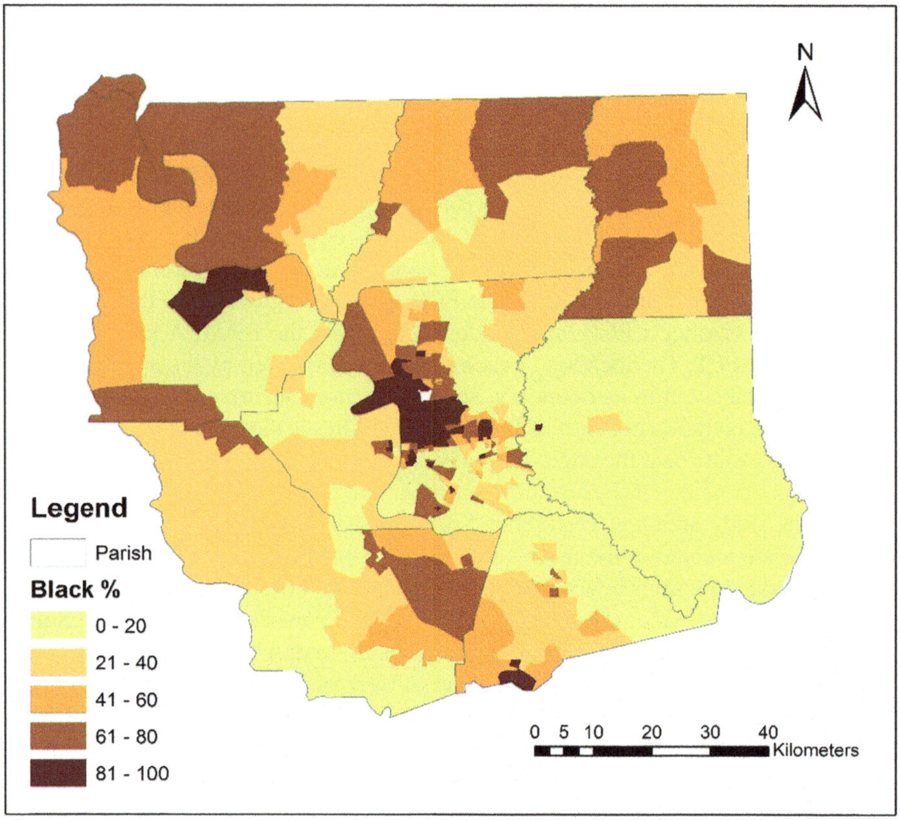


**Fig. 2** African American percent across block groups in Baton Rouge MSA

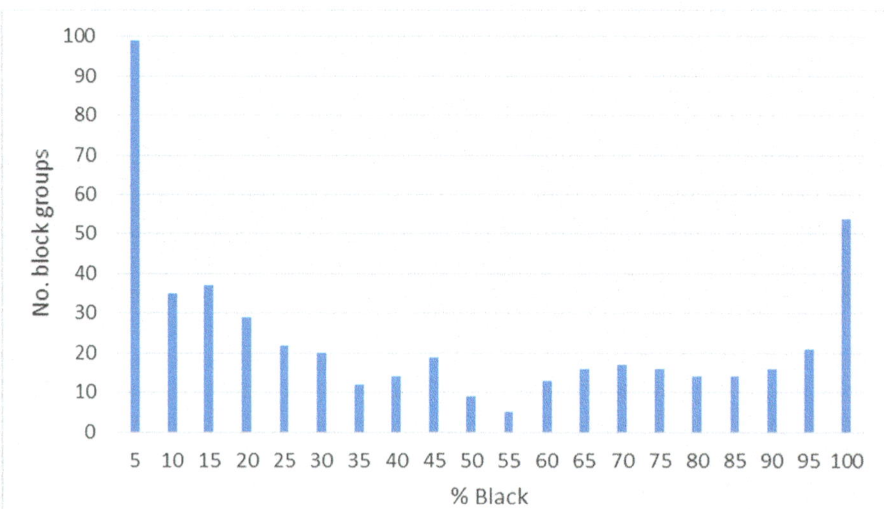


**Fig. 3** Numbers of block groups in various African American percent ranges

**Table 1** Demography and primary care access across areas of urbanicity

| Area (No. block groups) | Population | White % | African American % | Household under poverty % | Average time to nearest PCP (minutes) | Average accessibility score (physicians per 1000) | |
|---|---|---|---|---|---|---|---|
| | | | | | | $D_0 = 25$ | $D_0 = 30$ |
| *Whole study area (n = 481)* | *787,961* | *60.5* | *35.7* | *15.6* | *5.31* | *1.2120* | *1.2120* |
| Rural (*n* = 323) | 187,462 (33.8%) | 70.4 | 27.1 | 13.4 | 10.82 | 0.5928 | 0.7142 |
| Urban cluster (*n* = 9) | 13,680 (1.7%) | 34.6 | 64.6 | 25.2 | 6.04 | 0.4248 | 0.3461 |
| Urbanized area (*n* = 149) | 508,114 (64.5%) | 56.1 | 39.4 | 16.4 | 2.53 | 1.5994 | 1.5004 |

The road-network dataset is also downloaded from the U.S. Census Bureau (2018b) web site. Based on the data, the ArcGIS Network Analysts is used to estimate the shortest-path travel time from each demand location (i.e., population-weighted centroid of non-zero-population block group) to each supply location (i.e., average location of non-zero PCPs in a block) and the Network Analyst then produces a travel time matrix of 481 × 204 or 98,124 O-D pairs.

As we are interested in examining the variability of several factors across the rural–urban continuum, data of urban areas are also downloaded from the U.S. Census Bureau (2018c) web site. Based on the 2010 Census Urban and Rural Classification (U.S. Census Bureau 2018c), a census block group is assigned to an urbanized area (UA) (50,000 or more people) or an urban cluster (UC) (at least 2500 and less than 50,000 people) if its centroid falls within an UA or UC, respectively. The remaining areas are classified as rural. As shown in Fig. 1 and Table 1, most of the block groups (323) in the BRMSA are rural, followed by UA (149), and the fewest (9) in UC. However, UA has the most population (508,114 or 64.5%), followed by rural (187,462 or 33.8%) and UC (13,680 or 1.7%). Rural areas have the highest percentage in of white residents (70.4%), followed by UA (56.1%) and then UC (34.6%).The order is reversed for percentages of African American and household under poverty.

## 3 Methods of Spatial Accessibility Measures

As stated previously, most early measures of accessibility emphasize proximity to supply locations and do not consider the number of individuals competing for the service. Minimum distance or travel time to the closest facility is often used to measure accessibility of healthcare, such as general practitioners (Brabyn and Gower 2003), cancer screening (Wang et al. 2008), cancer care (Onega et al. 2017), and others.

Termed the "proximity method", it assumes that residents only use the nearest service provider. As was previously mentioned, this study uses the ArcGIS Network Analyst to estimate the travel time between a census block group and its nearest primary care location through the road network as a way of measuring proximity.

In our study area, the average travel time for residents to the nearest primary care physician (PCP) is 5.31 minutes – a considerably short time. Note that the travel time is estimated by ArcGIS network analysis module assuming that people follow the speed limits with free flow traffic. See Wang (2015:38–40) for its step-by-step implementation. However, estimation of travel impedance is far more complex than the above assumption (Delmelle et al. 2013, 2018). For the same study area, Wang and Xu (2011) found that ArcGIS tended to underestimate about 5 minutes below what travelers usually experience (e.g., derived by the Google Maps API). Therefore, average residents in the BRMSA are expected to spend slightly over 10 minutes reaching their closest PCP. Figure 4 shows the estimated travel time to the closest PCP across block groups. For the aforementioned tendency of underestimated travel time by ArcGIS, one may add 5 minutes to time displayed in Fig. 4 to better reflect

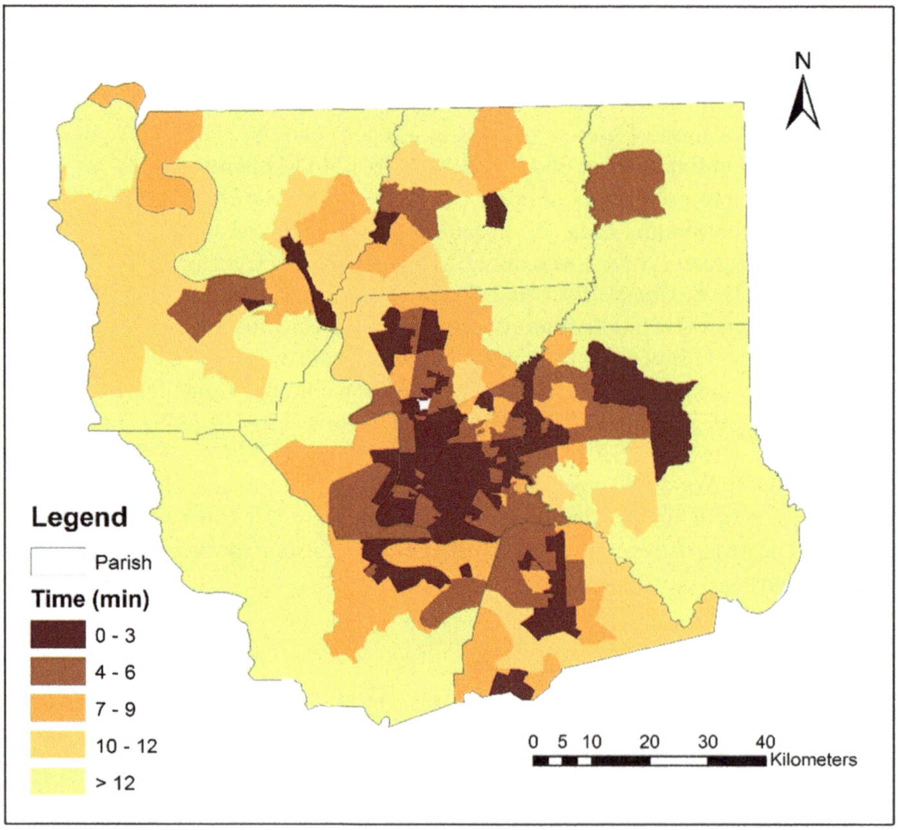


**Fig. 4** Estimated travel time to the nearest primary care physicians in Baton Rouge MSA

actual travel time. When referencing to Fig. 1, the pattern shows a clear urban advantage with better proximity enjoyed by those in the central city (City of Baton Rouge) and its urbanized extensions toward east and southeast.

The second method of accessibility is the popular 2SFCA method (Luo and Wang 2003). The first step computes the supply–demand ratio $R_j$ within the catchment area around each facility $j$ to capture its availability. Only demands $D_k$ across locations $k$ within the catchment contribute to the facility's crowdedness. The second step sums up those ratios at supply locations $j$ that are within the same catchment range from a demand location $i$. The availability of supply locations $j$ within the catchment contributes to the demand location's accessibility. The formula is written as:

$$A_i = \sum_{j \in \{d_{ij} \le d_0\}}^{n} R_j = \sum_{j \in \{d_{ij} \le d_0\}}^{n} \left( \frac{S_j}{\sum_{k \in \{d_{kj} \le d_0\}}^{m} D_k} \right) \tag{1}$$

where $d_{ij}$ (or $d_{kj}$) is the distance between $i$ and $j$ (or $k$), $D_k$ is the demand at location $k$ that falls within the catchment from supply location $j$ (i.e., $d_{kj} \le d_0$) with a capacity $S_j$, and $R_j$ is the supply to demand ratio at supply location $j$ that falls within the catchment centered at $i$ (i.e., $d_{ij} \le d_0$), $n$ and $m$ are the total numbers of supply locations and demand locations, respectively. Equation (1) is essentially the ratio of supply to demand that interacts within a threshold distance or filtering window. A larger value of $A_i$ indicates a better accessibility at a location.

The catchment area or threshold travel time $d_0$ is a critical parameter in the 2SFCA method. The literature suggests using a threshold travel time of 30 minutes (Lee 1991; Luo and Wang 2003). Considering that the estimated time in ArcGIS is about 5 minutes short, this study adopts an estimated time of 25 minutes as threshold $d_0$, and also uses the 30-minute threshold to test sensitivity. The 2SFCA is implemented in a customized ArcGIS toolkit (Wang 2015, 112–113).

One property of the accessibility scores is that their weighted average score is about the ratio of total supply and total demand (Wang 2015: 110–111). In our case, it is the ratio of total number of PCPs and total population, i.e., 955/787,961 = 0.001212. The raw scores are then multiplied by 1000 to avoid small numbers. That is to say, the average accessibility score for the BRMSA is 1.2120 PCPs per 1000 residents. Figure 5 shows how the 2SFCA-derived accessibility score varies across block groups. The urban advantage in Fig. 5 is similar to the geographic pattern of proximity measure in Fig. 4 and perhaps even stronger. The concentric decline in accessibility score away from the city center of Baton Rouge is distinguishable. The 2SFCA uses catchment twice (first on facility and second on residents) to emphasize neighboring effect, and thus the resulting accessibility scores are smoothed to some extent.

The results are further analyzed and discussed in the next three sections. First, we examine the variability in accessibility across geographic areas of various urbanicities. We then analyze how space and race (and another nonspatial factor such as

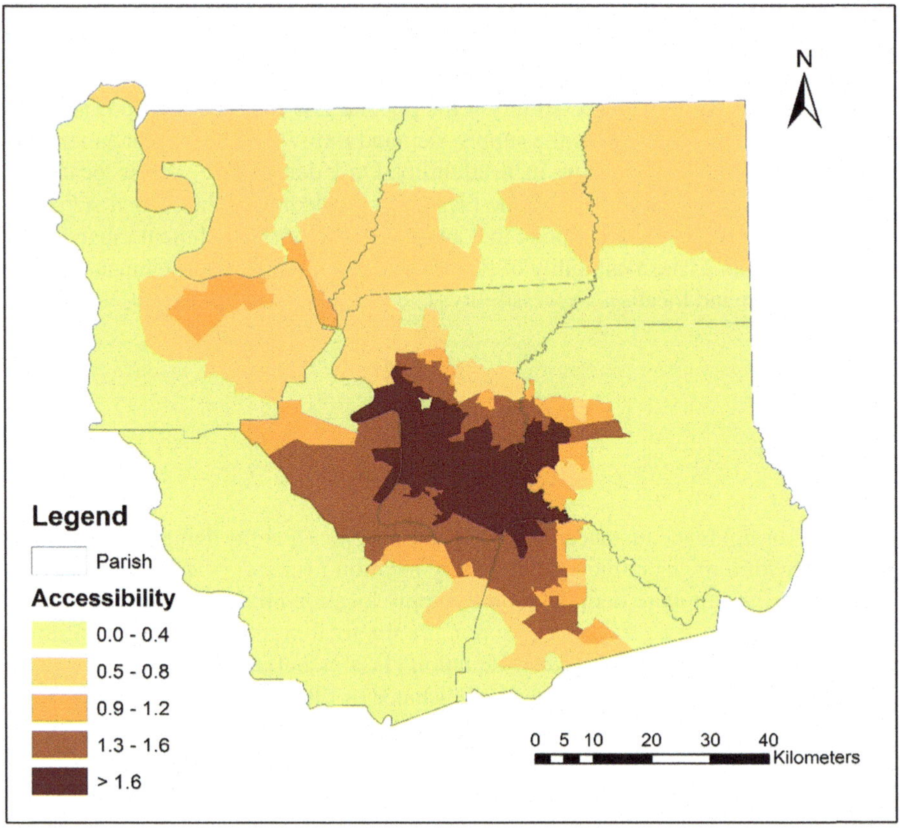


**Fig. 5** 2SFCA-derived accessibility score for primary care in Baton Rouge MSA

poverty status) interact and affect disparities in accessibility. The third part moves the analysis of place-based racial disparity a step further by delineating the study area into a small number of regions based on racial structure (e.g., percentage of African Americans) and examining the variability of accessibility across these regions.

## 4 Variation of Spatial Accessibility by Urbanicity

Examining the variability of healthcare environment across geographic areas of different urbanities is a common theme in public health studies (Larson et al. 2003). This section examines how spatial accessibility of primary care differs across areas of various urbanicity levels.

The last three columns in Table 1 present the average travel time to the nearest PCP and average 2SFCA-based accessibility scores (using catchments of estimated time of 25 and 30 minutes, respectively) across areas of three urbanicity types. By "average," each is calculated as the population-weighted average. On average, travel time increases from 2.53 minutes in Urbanized Areas (UA) to 6.04 minutes in Urban Clusters (UC), and again to 10.82 minutes in rural areas. Based on the 25-minute-catchment 2SFCA, UA also enjoys the highest accessibility score (1.5994), but rural areas have a slightly better accessibility score (0.5928) than UC (0.4248). The same trend is confirmed in the result by the 30-minute-catchment 2SFCA. In general, the results confirm the urban advantage observed from the maps. The slight edge in 2SFCA scores in rural over UC needs to be verified statistically.

A simple regression model with dummy variables is formulated to examine whether above observed disparities in accessibility are statistically significant across urban–rural categories. The variable of interest, accessibility value, defines the dependent variable in the regression, and the independent variables are the dummy variables that code the urban–rural categories. Here, two dummy variables are used to code three urbanicity categories: the reference category "rural" is coded as $x_1 = x_2 = 0$; the category "UC" is coded as $x_1 = 1$, $x_2 = 0$; and the category "UA" is coded as $x_1 = 0$, $x_2 = 1$. The model is written as:

$$A = b_0 + b_1 x_1 + b_2 x_2 \tag{2}$$

where $A$ is an accessibility measure (time or 2SFCA score) in a block group, the constant term $b_0$ is the average score in areas of *reference category* "rural," the coefficient $b_1$ (or $b_2$) is the difference of average $A$ values between reference category and UC (or UA), and the $t$-values for the corresponding coefficients indicate whether the average of accessibility score of a specific category differs from that of the reference category significantly.

The results are reported in Table 2. Taking the model for travel time as an example, the intercept from the regression model ($b_0 = 10.82$) is the average time for the reference category (i.e., rural, also reported in Table 1). The coefficient $b_1 = -4.78$ for UC indicates that the average time of UC is 4.78 below that of rural; moreover, its associated $t$-value (−2.96) indicates that the difference is statistically significant at 0.05. The coefficient $b_2 = -8.29$ for UA indicates that the average time of UA is 8.29 below that of rural, and its associated $t$-value (−17.87) indicates that the difference is statistically significant at 0.001. Similar interpretation applies to the results on accessibility scores (third and fourth columns in Table 2). Note that the $t$-values indicate

**Table 2** Statistical test on disparity in average travel time and accessibility across areas of urbanicity

| | Travel time from the nearest primary care provider (minutes) | Accessibility score (physicians per 1000) | |
|---|---|---|---|
| | | $D_0 = 25$ | $D_0 = 30$ |
| *Rural (reference category)* | *10.82* | *0.5928* | *0.7142* |
| Urban cluster (UC) | −4.78*<br>(−2.96) | −0.1680<br>(−1.32) | −0.3681**<br>(−3.26) |
| Urbanized area (UA) | −8.29***<br>(−17.87) | 1.0116***<br>(27.55) | 0.7909***<br>(24.40) |

Note: $t$-value in parenthesis; *Significant at 0.05; **Significant at 0.01; ***Significant at 0.001

that the difference between rural and UC is not significant in the 25-minute-catchment accessibility scores while the difference is significant in the 30-minute-catchment accessibility scores. The relatively small sample size for the UC (9) contributes to the less reliable differences between rural and UC detected. For this reason, we refrain from reading too much into the results on UC and focus on the big picture of urban–rural disparity. Nevertheless, it is interesting to point out that a recent study on the spatial accessibility of National Cancer Institute Cancer Centers reports the order of average accessibility is UA > rural > UC (Xu et al. 2017), consistent with our result of 30-minute-catchment accessibility scores. Another study cited previously (Yin et al. 2018) reports that more urbanized areas enjoy better access to medical facilities than less urbanized areas in China.

## 5 Disparities of Spatial Accessibility Between Demographic Groups

This section examines disparities in accessibility by race (white vs. African American) and by socioeconomic status (here poverty status is chosen as an example). Note that the accessibility index is based on census block groups, not individuals, and has an ecological nature. In other words, various racial-ethnical groups and both people above and below the poverty line may be present in a census block group. It is the variability of their concentrations (i.e., percentages) across block groups that leads to disparity. Our approach here is to assess whether one group is disproportionally represented in areas of different levels of accessibility.

We begin with comparing the average accessibility values for different demographic groups to gain some preliminary understanding of the issue. Specifically, the weighted average for each group is calibrated by using the number of individuals in that group in an area as the weight across all block groups. As reported in Table 3, African Americans in BRMSA on average have a shorter travel time from the nearest PCP and also a higher accessibility score (using either a 25-minute or 30-minute catchment) than whites. Similarly, households under poverty enjoy a shorter travel time on average from the nearest PCP and a higher average accessibility score than the population in general. This observation indicates that in terms of spatial accessibility, minorities, such as African Americans, and disadvantaged

**Table 3** Weighted average travel time and accessibility by demographic groups

| | Travel time from the nearest primary care provider (minutes) | Accessibility score (physicians per 1000) | |
|---|---|---|---|
| | | $D_0 = 25$ | $D_0 = 30$ |
| *All population* | *5.31* | *1.2120* | *1.2120* |
| White | 6.00 | 1.1234 | 1.1586 |
| African-American | 4.29 | 1.3402 | 1.2852 |
| Household under poverty | 4.75 | 1.2956 | 1.2579 |

groups, such as those in poverty, actually come ahead of whites and those above the poverty line. It may be termed "reversed racial advantage" (Xu et al. 2017: 203). This may be attributable to that a disproportionally high number of African Americans and people under the poverty line tend to concentrate more in central city areas, and thus have better spatial accessibility in either accessibility measure.

Are these differences in average accessibility measures across demographic groups statistically significant? We formulate it as a weighted ordinary-least-squares (OLS) regression model such as

$$Y = a + b * Flag \tag{3}$$

where the dependent variable $Y$ stands for ratios of demographic groups across block groups, the independent variable $Flag$ is a binary dummy variable (= 0 or 1, corresponding to whether a block has an accessibility value above or below the average), and $a$ and $b$ are parameters to be estimated. By employing a weight term (i.e., population in each area) in the regression, the error term is weighted heavier in an area with more population than one with less population.

For example, the average travel time across all 481 block groups is 5.3. These block groups are divided into two groups: block groups in Group 1 are coded as "Flag = 0" when their travel times are higher than or equal to 5.3, the rest in Group 2 with values lower than 5.3 are coded as "Flag = 1". As shown in Table 4, the model result for whites indicates that the sample mean of white percentages in above-average-time areas is 73.58 (when Flag = 0), whereas the sample mean of white ratios in below-average-time tracts is 73.58–18.53 = 55.05 (when Flag = 1). The corresponding $t$-value (−5.77) indicates that the difference is statistically significant ($p < 0.001$). That is to say, disproportionally higher percentages of whites are concentrated in above-average-travel-time areas.

As the results for accessibility scores are consistent between using 25-minute and 30-minute catchment areas, only the former is presented in Table 4. Percentages

**Table 4** Statistical test on disparity in average time and accessibility across demographic groups

| | Travel time from the nearest primary care provider (minutes) | | | Accessibility score (physicians per 1000) ($D_0 = 25$) | | |
|---|---|---|---|---|---|---|
| | ≤5.3 (Flag = 1) | >5.3 (Flag = 0) | Difference ($t$-value) | >1.2120 (Flag = 1) | ≤1.2120 (Flag = 0) | Difference ($t$-value) |
| *No. block groups (n)* | *347* | *134* | | 322 | 159 | |
| White % | 55.05 | 73.58 | −18.53 (−5.77***) | 53.63 | 72.60 | −19.07 (−6.27***) |
| African-American % | 40.46 | 24.22 | 16.24 (4.97***) | 41.54 | 25.27 | 16.27 (5.24***) |
| Household under poverty % | 17.49 | 11.53 | 5.96 (4.22***) | 17.22 | 13.05 | 4.17 (3.08*) |

Note: $t$-value in parenthesis; **Significant at 0.01; ***Significant at 0.001

of white residents are higher in the block groups with above average travel time and in areas of below average accessibility scores, and the opposite is observed for African Americans. By poverty status, percentages of households below poverty line are higher in below-average-travel-time areas or above-average-accessibility-score areas. These findings are consistent with the results from Table 3, and all disparities are statistically significant. That is to say, the reverse advantages for African American and households under poverty are validated by the statistical test. Such an advantage in spatial accessibility may not be realized in true advantage in access since these demographic groups often have lower vehicle ownership and thus lack transportation means of overcoming spatial barriers. Readers are encouraged to look into the literature on nonspatial factors in healthcare accessibility (Wang 2012).

## 6 Segregation and Spatial Accessibility Disparity Through the Lens of Regionalization

As shown in Figs. 2 and 3, the residential segregation of African Americans is prevalent in the BRMSA. This section analyzes this issue in more depth by a GIS-automated regionalization method and further discusses the intersection of racial concentration and spatial accessibility of PCP.

Regionalization is to group a large number of areas into a smaller number of spatially contiguous regions while maximizing a homogeneity (or minimizing a heterogeneity) measure of the derived regions. This study uses a popular automated regionalization method termed "regionalization with dynamically constrained agglomerative clustering and partitioning (REDCAP)" (Guo 2008; Guo and Wang 2011). REDCAP groups contiguous areas of similar attribute values to produce a set of regions with a minimum total homogeneity. The method has several advantages over existing regionalization methods such as spatial compactness, attribute homogeneity, and scale flexibility (i.e., generating a user-defined number of derived regions), and thus is chosen for this study. Readers interested in exploring other GIS-automated regionalization methods are referred to Wang (2015:193–215) for a review of recent development in this area, e.g., the mixed-level regionalization (MLR) method by Mu et al. (2015). The purpose of regionalization is to generate a series of areal units, through which the spatial patterns of residential segregation by race and associated disparity in access can be examined at multiple geographic scales.

The REDCAP method has two steps. Step 1 is a bottom-up contiguity-constrained hierarchical clustering process, i.e., grouping two adjacent and most similar (least dissimilar) areas[1] to form the first cluster and continuing until the whole study area is one cluster. A clustering tree is established to record the cluster hierarchy. Step 2 is a top–down tree partitioning process, i.e., removing the best edge in the clustering

[1]Dissimilarity between two neighboring areas $i$ and $j$ is measured by their attribute distance $D_{ij}$ such as $D_{ij} = (x_i - x_j)^2$, where $x_i$ and $x_j$ are standardized attribute values (here African American percentage) for $i$ and $j$, respectively.

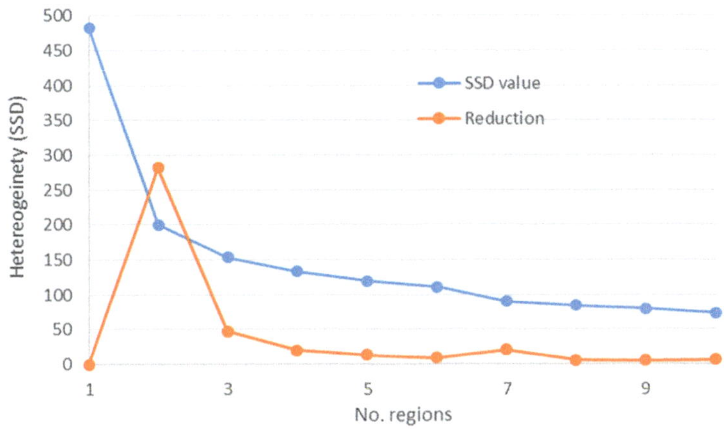


**Fig. 6** Heterogeneity values in REDCAP-derived regions

tree to create two regions with maximal homogeneity (i.e., minimal heterogeneity)[2], and continuing the partitioning until the desired number of regions is reached.

Our study is interested in generating regions with various levels of concentration of African Americans. Therefore the percentage of African Americans in each block group is used as the attribute variable in regionalization. As stated previously, the REDCAP can generate any number of regions defined by a user (up to the total number of block groups). A common measure for the quality of regionalization result is total sum of squared deviations (SSD) for the overall heterogeneity in derived regions (see Footnote 2 for its formula). Figure 6 shows how SSD declines as the number of regions increases (truncated at 10 regions) and suggests that the reduction in SSD value peaks at two and then seven regions. Therefore, two or seven regions may be considered as good regionalization scenarios.

Figure 7 shows the corresponding regionalization results, and the numbers (1–7) label the order of derived regions. When only two regions are produced, region 1 (southeast quad) comes out first with very low African American rate (9.2%, see Table 5) and covers areas in the highly urbanized south and east parts of the City of Baton Rouge, the entirety of its eastern neighbor Livingston Parish, the northeastern part of Ascension Parish, and the remaining areas form another region. The following focuses on the result of seven regions, which provide a finer resolution of geographic variability.

Table 5 summarizes the demographic information and spatial accessibility measures across the seven REDCAP-derived regions. Among the four regions with above-average African American percentages, Region 5 is highly urbanized (mostly

---

[2] The total sum of squared deviations (SSD) measures overall heterogeneity in derived regions such as $\mathrm{SSD} = \sum_{r=1}^{k} \sum_{i=1}^{nr} (x_i - \overline{x})^2$, where $k$ is the number of regions, $n_r$ is the number of small areas in region $r$; $x_i$ is the standardized attribute value, and $\overline{x_t}$ is the regional mean.

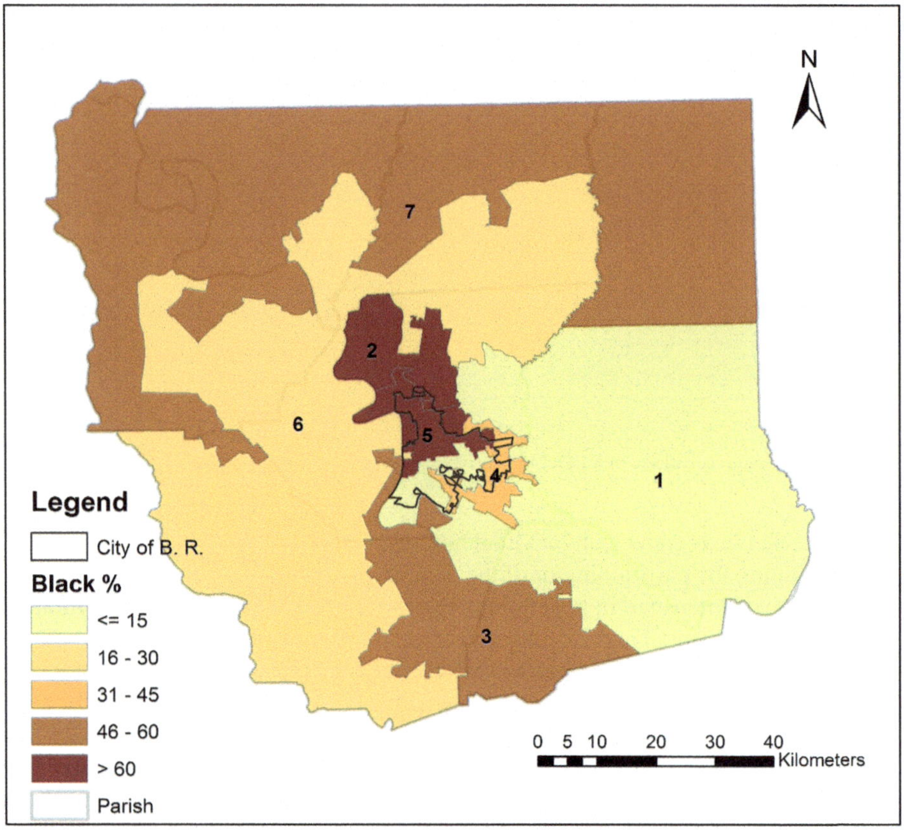


**Fig. 7** REDCAP-derived regions based on African American Percent in Baton Rouge MSA

in the City of Baton Rouge), Region 2 is considered suburban to its northwest, and Regions 7 and 3 are rural in the north and south of the BRMSA, respectively. All these four regions have higher than 55% African Americans but display very different spatial accessibility of PCP. Only Region 5 (central urban) enjoys above-average accessibility in both proximity to PCP and 2SCFA accessibility score. The other three regions all suffer from below-average 2SCFA accessibility score. The worst is experienced by Region 7 (north rural) which suffers from the longest travel time from their nearest PCP (13.7 minutes) and the lowest accessibility score (0.4255 per 1000 residents) in the whole BRMSA. This is an important finding and reveals that the so-called overall "reversed racial advantage" for African American in spatial access of PCP is not necessarily transferrable to those in suburban areas – and certainly not rural areas. Also note the high poverty rates in the two rural regions (Regions 7 and 3).

Among the three regions with below-average African American percentages, Region 1 (southeast quad) stands out as the largest region with a population of

**Table 5** Characteristics of seven REDCAP-derived regions based on racial composition

| Region No. | Population | African American (%)[a] | Household under poverty (%) | Time from primary care (minutes) | Accessibility score (physicians per 1000) | Comments |
|---|---|---|---|---|---|---|
| *Study area* | *787,961* | *35.7* | *15.6* | *5.3* | *1.2120* | |
| 5 | 129,551 | 89.2 | 29.1 | 2.1 | 1.6956 | Central urban w/ very high African American concentration |
| 2 | 34,600 | 65.0 | 11.6 | 4.3 | 1.1871 | B.R. northwest suburban w/ high African American concentration |
| 7 | 46,937 | 60.4 | 23.5 | 13.7 | 0.4255 | North rural w/ high African American concentration |
| 3 | 70,308 | 55.5 | 19.9 | 3.8 | 0.9648 | South rural w/ high African American concentration |
| 4 | 87,451 | 33.7 | 10.7 | 2.5 | 1.7193 | B.R. southeast suburban w/ moderate African American rate |
| 6 | 66,713 | 20.9 | 12.0 | 12.2 | 0.5433 | Southwest rural w/ low African American rate |
| 1 | 352,401 | 9.2 | 11.5 | 5.1 | 1.2688 | Southeast quad w/ very low African American rate |

Note: [a]Rows in a descending order of African American %

352,401 or 45% of the total population and the lowest African American percentage (9.2%). As discussed previously, this region covers areas across the whole rural–urban spectrum and has both accessibility measures slightly better than the averages. Region 6 (southwest rural) also has a relatively low African American percentage (20.9%), but poor accessibility in both measures (only second to the north rural Region 7). Region 4 (southeast suburban Baton Rouge), with an African American percentage at about the average level of the BRMSA (33.7%), enjoys the highest accessibility score (1.7193 PCPs per 1000 residents) as well as very good proximity to PCP (2.5 minutes or second best).

To recap the above discussion, the best spatial accessibility areas include a region of the highest concentrations of African Americans and another about the average level of African Americans and both are urbanized areas in or around the central city. The worst are represented by a region with a relatively high African American percentage and another with a low African American percentage, and both are rural. That is to say, racial composition is not the only story in variability of spatial access of PCP, rather the intersection of race and location presents a complex picture of access disparity. For example, even within the City of Baton Rouge (Fig. 7), its north side with high concentration of African Americans (forming most of Region 5) has 20% of the MSA's population but only 8% of PCPs, and its south side, mostly whites (part of Region 1), has 30% of the MSA's population and 66% of PCPs. It is a city of two tales.

## 7 Concluding Comments

Spatial accessibility reflects the relative ease by which activities or services can be accessed from a given location. It is an important location amenity for residents. This study examines accessibility to primary medical care in Baton Rouge MSA, Louisiana, in 2016. Two measures of spatial accessibility are used for residents at the census block group level. The proximity method assumes that residents use the nearest primary care physicians (PCP), measured in travel time. The two-step floating catchment area (2SFCA) method accounts for the ratio of supply (physicians) and demand (population) that interact within a threshold travel time and yields an accessibility score interpreted as physicians per 1000 residents. Based on results from both methods, we examine the disparities in spatial accessibility across geographic areas of various urbanicity levels and across major racial-ethnic groups and validate whether the disparities are statistically significant. Furthermore, the study area is divided into multiple regions by a GIS-automated regionalization method and an in-depth analysis of intersection of racial makeup and accessibility across these regions.

There are several interesting findings from the study. First, the urban advantage is evident in both measures of PCP accessibility and is validated in a statistical test. Secondly, overall, African Americans (or population under poverty) are disproportionally concentrated in areas closer to their nearest PCP in terms of travel time and also in areas with above-average accessibility scores (i.e., higher ratios of physicians per 1000 residents), termed "reversed racial advantage". Such differences are also validated by a statistical test. Such an advantage in accessibility may not pan out when multimodal transportation is considered (Dony et al. 2015; Mao and Nekorchuk 2013) since a disproportionally higher ratio of African Americans rely on much slower public transits. Thirdly, the analysis of accessibility measures across regionalization-derived regions reveals significant variability. Most importantly, the aforementioned racial advantage for African Americans is not applicable to those in suburban – and certainly not rural areas.

**Acknowledgement** We are grateful for the supports by the National Institutes of Health (Grant No. R21CA212687, Wang) and the ASPIRE undergraduate research program in the College of Humanities and Social Sciences at Louisiana State University (Vingiello).

## References


American Academy of Family Physicians (AAFP). Primary care. Available at https://www.aafp.org/about/policies/all/primary-care.html. Accessed 19 Nov 2018.

Bindman, A. B. (2013). Using the national provider identifier for health care workforce evaluation. *Medicare & Medicaid Research Review, 3*. pii: mmrr.003.03.b03, E1.

Brabyn, L., & Gower, P. (2003). Mapping accessibility to general practitioners. In O. Khan & R. Skinner (Eds.), *Geographic information system and health applications* (pp. 289–307). Hershey: Idea Group Publishing.

Delmelle, E. M., et al. (2013). Modeling travel impedance to medical care for children with birth defects using Geographic Information Systems. *Birth Defects Research Part A: Clinical and Molecular Teratology, 97*, 673–684.

Delmelle, E. M., et al. (2018). Travel impedance agreement among online road network data providers. *International Journal of Geographical Information Science, 32*, 1–19.

Dony, C. C., Delmelle, E. M., & Delmelle, E. C. (2015). Re-conceptualizing accessibility to parks in multi-modal cities: A variable-width floating catchment area (VFCA) method. *Landscape and Urban Planning, 143*, 90–99.

East Baton Rouge Parish. (2016). 2015 Community health needs assessment. Available at http://www.healthybr.com/assets/uploads/docs/CHNA.FINAL.6.24.16.pdf. Accessed 6-1-2018.

Guo, D. (2008). Regionalization with dynamically constrained agglomerative clustering and partitioning (REDCAP). *International Journal of Geographical Information Science, 22*, 801–823.

Guo, D., & Wang, H. (2011). Automatic region building for spatial analysis. *Transactions in GIS, 15*(s1), 29–45.

Hart, L. G., Salsberg, E., Phillips, D. M., & Lishner, D. M. (2002). Rural health care providers in the United States. *Journal of Rural Health, 18*, 211–232.

Ikram, S. Z., Hu, Y., & Wang, F. (2015). Disparities in spatial accessibility of pharmacies in Baton Rouge, Louisiana. *Geographical Review, 105*, 492–510.

Joseph, A. E., & Phillips, D. (1984). *Accessibility and utilization—Geographical perspectives on healthcare delivery*. New York: Harper & Row.

Khan, A. A. (1992). An integrated approach to measuring potential spatial access to health care services. *Socio-Economic Planning Sciences, 26*(4), 275–287.

Larson, E., Lin, S. X., & Gomez-Durate, C. (2003). Antibiotic use in Hispanic households, New York City. *Emerging Infectious Diseases, 9*, 1096–1102.

Law, M., Dijkstra, A., Douillard, J., & Morgan, S. (2011). Geographic accessibility of community pharmacies in Ontario. *Healthcare Policy, 6*(3), 36–46.

Lee, R. C. (1991). Current approaches to shortage area designation. *Journal of Rural Health, 7*(4), 437–450.

Lee, P. R. (1995). Health system reform and generalist physician. *Academic Medicine, 70*, S10–S13.

Luo, W., & Qi, Y. (2009). An enhanced two-step floating catchment area (E2SFCA) method for measuring spatial accessibility to primary care physicians. *Health and Place, 15*, 1100–1107.

Luo, W., & Wang, F. (2003). Measures of spatial accessibility to health care in a GIS environment: Synthesis and a case study in the Chicago region. *Environment and Planning B: Planning and Design, 30*, 865–844.

Luo, J., Tian, L., Luo, L., Yi, H., & Wang, F. (2017). Two-step optimization for spatial accessibility improvement: A case study of health care planning in rural China. *BioMed Research International, 2017*, Article ID 2094654.

Luo, J., Chen, G., Li, C., et al. (2018). Use of an E2SFCA method to measure and analyse spatial accessibility to medical services for elderly people in Wuhan, China. *International Journal of Environmental Research in Public Health, 15*, 1503.

Mao, L., & Nekorchuk, D. (2013). Measuring spatial accessibility to healthcare for populations with multiple transportation modes. *Health & Place, 24*, 115–122.

Mu, L., Wang, F., Chen, V. W., & Wu, X. (2015). A place-oriented, mixed-level regionalization method for constructing geographic areas in health data dissemination and analysis. *Annals of the Association of American Geographers, 105*, 48–66.

Onega, T., Alford-Teaster, J., & Wang, F. (2017). Population-based geographic access to National Cancer Institute (NCI) Cancer Center parent and satellite facilities. *Cancer, 123*, 3305–3311.
Penchansky, R., & Thomas, J. W. (1981). The concept of access: Definition and relationship to consumer satisfaction. *Medical Care, 19*(2), 127–140.
Shi, X., Xue, B., & Xierali, I. M. (2016). Identifying the uncertainty in physician practice location through spatial analytics and text mining. *International Journal of Environmental Research and Public Health, 13*(9), 930.
U.S. Bureau of Census. (2018a). TIGER/Line® with selected demographic and economic data. Available at https://www.census.gov/geo/maps-data/data/tiger-data.html. Accessed 6-1-2018.
U.S. Bureau of Census. (2018b). TIGER products. Available at https://www.census.gov/geo/maps-data/data/tiger.html. Accessed 6-1-2018.
U.S. Bureau of Census. (2018c). 2010 Census urban and rural classification and urban area criteria. Available at https://www.census.gov/geo/reference/ua/urban-rural-2010.html. Accessed 6-1-2018.
U.S. Department of Health and Human Services (DHHS). (2018). Health professional shortage areas (HPSAs). Available at https://bhw.hrsa.gov/shortage-designation/hpsas. Accessed 6-1-2018.
Wang, F. (2012). Measurement, optimization, and impact of health care accessibility: A methodological review. *Annals of the Association of American Geographers, 102*(5), 1104–1112.
Wang, F. (2015). *Quantitative methods and socioeconomic applications in GIS*. Boca Raton: CRC Press.
Wang, F., & Xu, Y. (2011). Estimating O-D travel time matrix by Google Maps API: Implementation, advantages and implications. *Annals of GIS, 17*, 199–209.
Wang, F., McLafferty, S., Escamilla, V., & Luo, L. (2008). Late-stage breast cancer diagnosis and health care access in Illinois. *The Professional Geographer, 60*, 54–69.
Xierali, I. M. (2018). Physician multisite practicing: Impact on access to care. *Journal of the American Board of Family Medicine, 31*(2), 260–269.
Xierali, I. M., Nivet, M. A., & Bazemore, A. B. (2016). Modeling physician distribution uncertainty in three common health workforce data. Paper presented at the 12th association of American Medical Colleges (AAMC) annual health workforce research conference, Hyatt Regency, Chicago, IL, May 4–May 6.
Xu, Y., Fu, C., Onega, T., Shi, X., & Wang, F. (2017). Disparities in geographic accessibility of National Cancer Institute Cancer centers in the United States. *Journal of Medical Systems, 41*, 203.
Yin, C., He, Q., Liu, Y., et al. (2018). Inequality of public health and its role in spatial accessibility to medical facilities in China. *Applied Geography, 92*, 50–62.

**Fahui Wang** is James J. Parsons Professor and Chair of the Department of Geography and Anthropology, Louisiana State University (LSU). Dr. Wang's research focuses on applications of GIS and computational methods in human geography (including urban, economic, transportation, and historical geography) and public policy (including urban planning, public health, and public safety).

**Michael Vingiello** is a Louisiana State University (LSU) Geography and Interdisciplinary Studies graduate with minors in economics, history, and Spanish and now a Research Associate on the Human Dimensions team at the Water Institute of the Gulf in Baton Rouge, Louisiana. His areas of interest are community well-being, equality of access, socioeconomic disparities, environmental justice, fluvial trends and processes, design, and disaster planning and response.

**Imam M. Xierali** is an Associate Professor at the Department of Family and Community Medicine, University of Texas Southwestern Medical Center (UT Southwestern). Dr. Xierali's research focuses on relationship among population health, health workforce, and primary care using GIS and computational methods in health geography.

# Considerations When Using Individual GPS Data in Food Environment Research: A Scoping Review of 'Selective (Daily) Mobility Bias' in GPS Exposure Studies and Its Relevance to the Retail Food Environment

**Reilley Plue, Lauren Jewett, and Michael J. Widener**

**Abstract** Advancements in geospatial technologies including geographic information systems and global positioning system (GPS) devices have provided insights on how the retail food environment might be contributing to the ongoing obesity epidemic. Caution has been raised, however, around the potential for research that uses GPS-captured activity spaces to overestimate the impact that exposure to food retailers has on food choices and behaviour. This phenomenon, where it is difficult to discern whether an individual is passively exposed to a space or actively seeks it out, is referred to as a 'selective (daily) mobility bias'. Researchers' understanding of this bias is relatively new and understudied, particularly in the food environment literature, where the bias could have serious implications. This chapter reviews 14 peer-reviewed papers and two doctoral theses to identify and critique the methods proposed for handling this bias and offer recommendations to consider as the use of GPS-activity space studies continues to grow.

R. Plue · L. Jewett
Department of Geography & Planning, University of Toronto, Toronto, ON, Canada

M. J. Widener (✉)
Department of Geography & Planning, University of Toronto, Toronto, ON, Canada

Dalla Lana School of Public Health, University of Toronto, Toronto, ON, Canada
e-mail: michael.widener@utoronto.ca

 
Y. Lu, E. Delmelle (eds.), *Geospatial Technologies for Urban Health*,
Global Perspectives on Health Geography, https://doi.org/10.1007/978-3-030-19573-1_6

## Abbreviations

| | |
|---|---|
| FFR | Fast food retailer |
| GIS | Geographical information systems |
| GPS | Global positioning system |
| HPF | Highly processed food |
| SDMB | Selective daily mobility bias |
| SMB | Selective mobility bias |

# 1 Introduction

Food is an essential aspect of everyday life, but understanding what motivates food choice is extremely complicated. Despite the tireless efforts of the public health community to educate consumers, diet quality remains suboptimal, and the rates of metabolic illness continue to climb (Hall 2017; 'WHO | Obesity and overweight' n.d.). Highly processed foods (HPF) are those which have been significantly changed from their original state with the addition of salt, sugar, additives, and/or preservatives, and include items such as sweetened breakfast cereals, packaged soups, and processed meats (Moubarac et al. 2017). With high caloric content but low nutrient value, regular consumption of HPF is a well-known risk factor for weight gain and other chronic illnesses such as diabetes, high cholesterol, and cancer (Moubarac et al. 2017; Steele et al. 2016).Typical obesity interventions aim to change lifestyle habits through education and behaviour modification but have been shown to have minimal long-term success, indicating that improving knowledge alone is not enough (Camacho and Ruppel 2017; Cawley and Wen 2018; Teixeira et al. 2015). Researchers are increasingly understanding that the social and environmental factors leading to obesity are more complex, and no single variable is responsible (Teixeira et al. 2015; White 2016).

Focus has more recently shifted to the analysis of and interventions in the built environment, specifically referred to as the food environment, to better understand and intervene in drivers of food behaviour (Caspi et al. 2012). For example, across many urban areas there is an over-abundance of cheap, fast, and processed food options, and research has indicated that this makes it more difficult for consumers to prioritize healthy options that may not be as economical, easy, or enticing (Clary et al. 2017; Minaker et al. 2016). This is true both when individuals are trying to eat healthy, but struggle to refrain from the abundance of unhealthy options, and also when we consider how 'junk food' is designed to encourage addictive behaviour, making it harder for people to *want* to prioritize healthy food in the first place (Boswell and Kober 2016; Drewnowski and Kawachi 2015). This is supported by decades-worth of research in nutritional anthropology, biochemistry, biopsychology, and behavioural sciences that provide evidence of an evolutionary and psychological motivation for humans to consume foods high in sugar, fat, and salt (as is the case with HPF) (Cornelsen et al. 2015; Crézé et al. 2018; Hebebrand et al. 2014; Ma et al. 2017; Ridder et al. 2016; Ventura and Mennella 2011).

Interventions including taxation of processed foods (Bes-Rastrollo et al. 2016; Cawley and Wen 2018), adding restrictions on where food advertisements and fast food retailers (FFR) can locate (Boone-Heinonen et al. 2011; Cawley and Wen 2018; Sturm and Cohen 2009; White 2016), and limiting the density of FFRs have all been suggested as ways to limit unwanted exposure to HPF (Eckert and Shetty 2011; Hager et al. 2017; Sturm and Cohen 2009). These examples point to the fact that researchers are keen to utilize geospatial technologies in ways that help them better understand the complex problem of food behaviour and obesity. Advancements in geographic information systems (GIS) and portable global positioning systems (GPS) devices are providing new insights on how the retail food environment might be contributing to this epidemic (Ahalya et al. 2017; Christian 2012; Clary et al. 2017; Drewnowski and Kawachi 2015).

Over the past decade, food environment research has become more sophisticated and is increasingly drawing upon individuals' daily geographies with the use of GPS-enabled devices and increased access to activity and travel surveys. These data collection tools are allowing researchers to study how people navigate their food environments with finer detail, capturing nuanced differences in behaviour and associated health outcomes within various contexts. As this field of research grows, there is increased collaboration occurring between disciplines such as geography, transportation, public health, and nutrition science; all of which provide unique contributions to the study of individual-level dietary, health, and movement behaviours (Ahalya et al. 2017).

With the ubiquity of GPS-enabled mobile phones, collecting individual geographies is becoming increasingly streamlined and less burdensome for participants. However, standardized data collection, analysis, and reporting guidelines are needed to compare studies, increase external validity of research, and most importantly, design and implement successful interventions that encourage choices that improve health and overall quality of life. A survey of the literature has found that food environment researchers have used over 500 different methodologies (including GPS, time-use surveys, etc.) (Health Canada 2013), and there remains no standard approach for work using GPS devices (Cetateanu and Jones 2016).

Inevitably, inconsistent methodologies have led to inconsistent results. Of particular relevance to this chapter is the potential for research that uses GPS-captured activity spaces to overestimate the impact that exposure to food retailers actually has on food choices and behaviour (Chaix et al. 2013). Are people choosing to consume fast food because they are exposed to it? Or are they seeking out those environments specifically to engage in that activity? This idea is being referred to as 'selective daily mobility bias' (SDMB) in the literature (sometimes referred to simply as 'selective mobility bias' (SMB)) and was first proposed by Chaix et al. in 2012. Both terms (SDMB and SMB) are used interchangeably in the literature, though a small note of caution should be made to distinguish selective daily mobility bias from 'selective residential mobility bias' when using 'selective mobility bias' (Kwan 2018). For this reason, and to promote consistency, we suggest 'selective daily mobility bias' be used. We will refer to the bias as such for the remainder of the paper, defined as ' (a bias in) exposure studies, where a person is found to be (more) exposed to some place because they make an active choice to go to that place' (Widener et al. 2018, pg. 9).

While we recognize that this term is relatively new amongst exposure research, it appears to be understudied and poorly addressed in literature published to date. Despite the limited research on its' significance or the methods used to account for it, selective daily mobility bias has appeared listed as a study limitation in a number recent GPS-based exposure studies, including food environment and greenspace exposure research (Kwan 2018; Fong et al. 2018; Zenk et al. 2018; Widener et al. 2018). This chapter will therefore review literature in *all* fields of exposure research in order to answer the question of who and how researchers are addressing selective daily mobility bias. The goal is to enhance understanding how the bias is currently being addressed, and to provide a preliminary framework for researchers using GPS data, so that research in this field may move closer towards a consistent approach for measuring and accounting for SDMB.

While the topic is important across many fields interested in understanding the impacts of various types of exposure, it is a critical issue for researchers studying obesity and food environments, as more and more public health interventions are considering the built environment, so consistency in analyses and interpretation are key. This will allow researchers and policymakers to actually develop evaluations of food systems and monitor the effectiveness of interventions for obesity using these advanced and novel geospatial technologies.

## 2 Background

Much has been written on the concept of the 'food environment' in the past (Caspi et al. 2012; Cetateanu and Jones 2016; Giskes et al. 2011). The retail food environment specifically refers to the geographic distribution of food retail but implicitly considers the relationship between the locations of these retailers and the individuals who utilize them. Put simply, the goals of this body of research are to document inequities in access to healthy food and to inform policy by providing evidence that supports the development of healthier food environments (Minaker 2016).

Retail food environment research is complicated by the fact that it is embedded within complex social and geographic contexts, so similar spatial configurations of food retail in two distinct regions may result in different effect on diets. Despite this, researchers hypothesize that there may be general trends in the ways that the mix and distribution of food retailers affect how and what food is purchased and consumed. Because of this, replicability of work and consistency in approach are of key importance, but as stated in the previous section, such consistency has yet to be achieved.

If food environments are intuitively understood to play a role in food purchasing and consumption choices, it is important to ask why study results are not more consistent. One reason is likely due to limitations in data availability, as early research tended to focus on access to food retail from residential locations, and often relied on generalizations of locations by using aggregated population counts in census zones. Researchers recognize that people spend significant time outside of their home, however, and the recent advancement of GPS technology is now allowing for more complex studies that account for the complete activity space of an individual (Christian 2012; Kestens et al. 2012; Perchoux et al. 2016).

One approach for incorporating other relevant locations (e.g. work, shopping centres, and school) is through standard mobility surveys (Kestens et al. 2012), commuting data (Widener et al. 2015; Widener and Shannon 2014), or electronic mapping tools with an embedded survey of regular destinations (Chaix et al. 2012). These can be used to generate daily activity paths (Zenk et al. 2011). Beyond these spatial data collection tools, studies have recently been incorporating individual-level GPS data to identify these activity spaces to gain an understanding of access and exposure (Clary et al. 2017). Using activity space data generated by GPS devices or other surveys has proven to be effective in showing how health outcomes may vary based on differences between individuals' access and exposure throughout their daily travel patterns (Burgoine and Monsivais 2013; Cebrecos et al. 2016; Cetateanu and Jones 2016; Christian 2012; Kestens et al. 2012). Additionally, the use of GPS devices in particular allows for the objective identification of precise locations where individuals spend time and is typically not subject to limitations, like faulty memory, of self-reported activity spaces obtained through surveys.

For the purposes of this chapter, the recent turn towards using GPS devices and activity surveys in food environment research is of interest. However, as is the case with the selective daily mobility bias, any advancement in methodology brings a chance that new sources of error may be inadvertently introduced. The following section will review the literature published on this new concept, and specifically, the methods that have been suggested and used to mitigate it.

## 3 Review of the Literature

### *3.1 Goals of This Literature Review*

As the field of retail food environment research continues to advance, this scoping review is intended to provide a better understanding of how the term 'selective daily mobility bias' is currently being used and handled in the literature and offer guidance towards a standardized method for conducting multi-place exposure research that will more seriously attempt to identify and account for this potential bias. This is particularly important to do now as food environment and other exposure-based research shifts towards using more GPS data over strictly GIS-based approaches.

### *3.2 Search Methods and Identified Studies*

Google Scholar, Medline, and PubMed databases were searched using combinations of the terms 'selective mobility bias', 'daily mobility bias', 'mobility bias', 'daily mobility', 'bias', 'geography', 'geographic information systems', 'exposure', and 'health' on May 30, 2018. Steps were taken after reviewing articles' titles, abstracts, and text for any mention of 'bias' or 'error', and 14 peer-reviewed publications and two doctoral theses were found that mention, discuss, or evaluate the phenomenon being referred to as selective (daily) mobility bias. The main focus of these papers, elaborated on in Table 1, was to look broadly at the methods being used to study the exposure ($n = 10$) and examine the effects of the built environment on food consumption ($n = 9$), greenspace use ($n = 5$), and physical activity ($n = 4$). While not all of this research is concerned with the food environment, it is included in our analysis of the literature to better understand how selective daily mobility bias is identified and handled.

Of the three papers that played a key role in identifying this potential source of bias and developing the term, 'selective daily mobility bias' (Zenk et al. 2011; Chaix et al. 2012, 2013), Chaix et al. (2012, 2013) are most frequently cited. This work does give credit to Zenk et al. (2011) for first identifying this potential bias within their discussion of study limitations (p. 1158). Papers that include quantitative approaches to understanding or addressing selective daily mobility bias were published very recently (2015–2018).

In their commentary of selected literature, Chaix et al. (2013) suggest three methods for addressing the confounding from selective daily mobility bias (p. 49–50):

1. Researchers can exclude activity sites that are related to the behaviour of interest from the data collected by regular destination surveys (GIS-based) or GPS tracking. In the context of the food environment, this could be achieved by considering the exposure to food retailers only after the removing activity sites that result in food purchases, as demonstrated in Fig. 1. With GPS data, this would involve first identifying all activity sites where participants spent a minimum amount of time (for example, ≥10 min), and then generating a count of retailers located within a given buffer around each site. The exposure metric for each activity site would be a combination of time at each activity site ($t$) multiplied by the count ($n$) of fast food retailers within each buffer; where activity sites that are visited specifically to engage with fast food would be removed from the overall sum of exposure for an individual. This method is referred to as a 'truncated activity space' and is said to be the most robust approach. However, in order for this method to work, information about the purpose of trips and places visited need to be reliably reported.
2. A less technological approach is to calculate exposure around a few key spatial anchor points, including major and minor daily life centres (e.g. home, work, daycare, or school). These locations can be geocoded, and therefore, there is no need to collect GPS data with this method. There is, however, a risk of missing important information, including variation in activity paths and exposure that occurs outside of these few locations.
3. Including additional survey questions that capture reasons why individuals choose their particular daily activity sites is suggested as a complimentary approach to better understand how an individuals' personal preferences can introduce another, but related, form of bias (Chaix et al. 2013).

**Table 1** Results of literature review

| Study | Where (p. #)/how selective daily mobility bias (SDMB) is addressed in article | Term used | Study focus | | | |
|---|---|---|---|---|---|---|
| | | | Food environment | Greenspace | Physical activity | Methods to measure exposure |
| Kestens et al. (2010) | Discussion/limitations (p. 1101) | NA | X | | | |
| Zenk et al. (2011) | Discussion/limitations (p. 1158) | NA | X | | X | X |
| Chaix et al. (2012) | Conceptual overview of MB (p. 444) | Selective daily mobility bias | | | | X |
| Kestens et al. (2012) | Discussion/limitations (p. 11) | Selective daily mobility bias | X | | | |
| Chaix et al. (2013) | Review article of SDMB (p. 46–50) | Selective daily mobility bias | | | | X |
| McCrorie et al. (2014) | Discussion/limitations (pg. 11) | Selective daily mobility bias | | X | X | |
| Burgoine et al. (2015) | Test significance of bias (p. 1–11) | Selective daily mobility bias | X | X | X | X |
| Byrnes et al. (2016) | Discussion/limitations (p. 68) | Selective mobility bias | X[a] | | | |
| Cetateanu and Jones (2016) | Discussion/limitations (p. 203) | Selective mobility bias | X | | | |
| Mitchell (2016)[b] | Accounted for in study design (p. 36–37, 101) | Selective mobility bias | | X | X | |
| Scully (2016)[b] | Accounted for in study design[c](p. 42, 61, 118) | Selective/spatial mobility bias | X | | | X |
| Perchoux et al. (2016) | Test significance of bias[c] (p. 116–121) | Selective daily mobility bias | | X | | X |
| Kwan (2018) | Discussion/limitations (p. 5–6) | Selective daily mobility bias | | | | X |
| Fong et al. (2018) | Discussion/limitations (p. 84) | Daily selective mobility bias | | X | | |
| Zenk et al. (2018) | Discussion/limitations (p. 53) | Selective mobility bias | | | | X |
| Widener et al. (2018) | Discussion/limitations (p. 11) | Selective mobility bias | X | | | |

[a]Alcohol outlets
[b]Thesis dissertation
[c]Used a method proposed by Chaix et al. (2013)

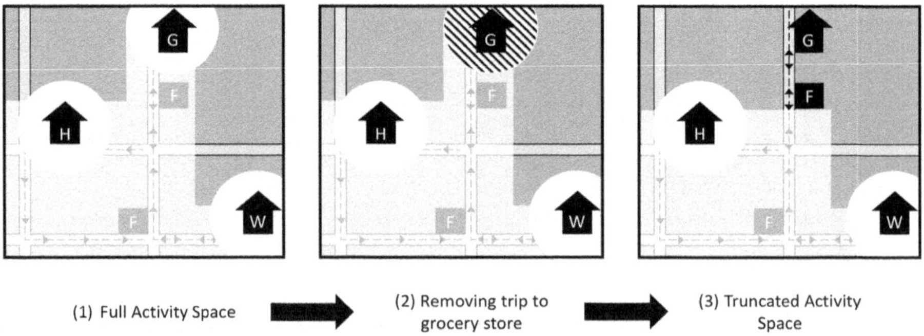


**Fig. 1** Diagram showing (1) a full activity space (lighter) including daily activity path between home (H), work (W), and a grocery store (G), with exposure to two fast food retailers (F); (2) the removal (truncation) of the part of the activity space that included the trip to the grocery store (G); (3) the truncated activity space with exposure to one fast food retailers after removing the trip to the grocery store

As previously noted in the introduction, the majority of the literature simply describes selective daily mobility bias as a potential limitation to their study ($n = 10$), and we found only two papers which have actually tested for its significance (Burgoine et al. 2015; Perchoux et al. 2016), only one of which used GPS (Burgoine et al. 2015).The earliest attempt to do this compared actual GPS routes of school children with the shortest street network route (GIS-based) (Burgoine et al. 2015). This was based on previous research that had found children who walk, tend to take longer, more obesogenic routes to and from home and school (Harrison et al. 2014), possibly seeking out these environments with more opportunities to purchase HPF. Burgoine et al. (2015) assumed that children whose actual GPS routes were more 'obesogenic' than their GIS-modelled shortest route were seeking out those environments to partake in fast food retail opportunities and would therefore have a higher consumption of these foods and, thus, a higher BMI as a result. However, there was no significant difference in the mean BMI of students whose GPS paths differed from the GIS-modelled shortest street network route.

Thus, this study found no evidence of a selective daily mobility bias where BMI (as a proxy for the consumption of junk food) was higher amongst students who selected the route with higher exposure/opportunity. The authors did suggest that further research using autonomous adults is necessary.

The second study to examine the effects of selective daily mobility bias did find evidence of the bias using the truncated activity space approach, proposed by Chaix et al. (Perchoux et al. 2016). This study did not use GPS trajectories, however. Instead, adult participants identified the locations of places they often visit (e.g. home, work, school, parks) in a web-based GIS application, with the help of a technician. These locations were then geocoded, and exposure to greenspace was compared between GIS-generated activity spaces of (1) all reported destinations (full activity space), and (2) destinations unrelated to the exposure of interest (truncated activity space). Amongst other conclusions, they found that exposure to greenspace was higher using full activity spaces (GIS-modelled), leading to the conclusion that a truncated activity space approach could be useful in mitigating a selective daily mobility bias. It is important to emphasize that this study used self-reported locations, whereas GPS trajectories would show all locations actually visited. The study could therefore be limited by participants who selectively omitted or forgot about locations.

Beyond the two papers just described, two recent geography doctoral theses accounted for selective daily mobility bias within their study design (Mitchell 2016; Scully 2016). Both studies used GPS-generated activity spaces to analyse behaviour outcomes associated with exposure to different features of the built environment but took different approaches to account for selective daily mobility bias, based on the activity of interest. In alignment with the truncated activity space method, Scully (2016) removed 'GPS/GIS data that [was] associated with travel-log-reported visits to [fast food restaurants]' (Scully 2016, pg. 42).

Mitchell (2016), on the other hand, included more information from their collected GPS dataset. The objective of Mitchell's research was to understand how neighbourhood-built environment features offer children opportunities to engage in moderate to vigorous physical activity (MVPA). Data was collected with an accelerometer and personal GPS device, and instead of removing trips that ended in the activity of interest, GPS data from all levels of activity (sedentary, light, moderate, and vigorous) were included. This was intentionally done to avoid including only the children who were most physically active. Unlike determining how exposure to a predetermined feature of the built environment influences the population, such as FFR and obesity, the objective of this particular research was to determine *which* features within the built environment are actually associated with MVPA. Thus, removing trips that ended in this activity would have discarded the key data being collected. However, removing these trips and comparing the truncated activity space to the full activity space would have allowed researchers to determine (a) if being exposed to these features at times unrelated to MVPA had an influence on the overall level of MVPA and (b) if there was a significant difference in the level of MVPA related to exposure comparing the truncated and full activity spaces.

Because neither of these two dissertations compared results to those obtained without accounting for a selective daily mobility bias, no conclusions can be drawn on whether or not their approaches made any difference to their final outcomes.

### 3.3 *Limitations*

As a relatively new concept in food environment research (est. 2011/2012), the terminology used to describe the selective daily mobility bias is not consistent, making it a challenge to identify key search terms. Therefore, despite using a standardized approach to search the literature, the review presented in Sect. 3 may be missing literature that tackles the issue using different language or conceptual frameworks. A next step in developing a more standardized approach to test and account for a selective daily mobility bias (and related concepts) is to conduct a full systematic review of the literature. As such, the search conducted for this chapter should only be interpreted as preliminary scan of the literature, akin to a scoping review, and is not a full systematic review. The approach used here returned studies that included some variation of the term selective daily mobility bias in the field of exposure research, including food environment, green space, and physical activity. This chapter can serve as a starting point for more robust and formal reviews of the literature in the future, as more researchers begin to grapple with the issue.

Beyond this, the use of Google Scholar as a database increased the likelihood of capturing papers that referred to the bias anywhere within an article, chapter, or book's body of text (Bramer et al. 2017; Haddaway et al. 2015; Zientek et al. n.d.). Starting this particular literature search with Google Scholar further increased the likelihood of capturing new or unpublished research by including access to grey literature. While more formal literature reviews of well-established concepts may exclude these sources, at least acknowledging the existence of the two theses (Mitchell 2016; Scully 2016) was important, given the limited number of peer-reviewed papers ($n = 14$) that were returned in the search.

# 4 Discussion and Recommendations

## *4.1 Overview of the Literature Review*

This review of selective daily mobility bias included 14 papers and two theses. Of the two papers that sought to evaluate the bias, only one found evidence supporting its significance (Perchoux et al. 2016). Unfortunately, at this point, generalizations on the usefulness of the methods being used to address SDMB cannot really be made with only two papers having compared the outcomes of both accounting for and ignoring a SDMB. The following section therefore acts as a starting point for future research in this field to work towards understanding when and how to study the effects of SDMB in exposure research.

### 4.1.1 Providing a Common Definition

In-text definitions of this concept consistently cite Chaix et al. (2012, 2013), but are highly modifiable to the specific behaviour being studied. For that reason, we reiterate the definition given in the introduction, with the intent of giving future researchers a common understanding of selective daily mobility bias within any context. Generally, selective daily mobility bias can be understood as: 'a bias … in GPS-based exposure studies, where a person is found to be (more) exposed to some place because they make an active choice to go to that place' (Widener et al. 2018, pg. 9).

## 4.2 Is Important Information Being Discarded When Truncating Activity Spaces?

The potential for a selective daily mobility bias to overestimate the influence that exposure has on behaviour is a valid concern, particularly when using GPS data in isolation. Incorporating a travel log diary to confirm the purpose of trips can help to identify locations related to the activity of interest. Removing these sites from the final analysis would provide information on the level of exposure that occurs throughout the day, outside of those times when an individual is specifically seeking out that environment. However, similar to the concern of using anchor points outlined in Sect. 3.2, it is important to ask: Is important information being discarded when these trips are being removed when using this method for a truncated activity space approach?

This concern can be illustrated using Fig. 1. In this simplified schematic, the count of fast food retailers in the full activity space is 2. On the way to work, Individual X passes by one FFR, but on the way home they divert their route in order to stop at the grocery store, which has them pass by an additional FFR. In this example, Individual X is not necessarily partaking in the FFR opportunities, but the question remains; is this exposure consequential? It is well known that food cues (sight, smell, taste) trigger an automatic desire to eat (Nederkoorn & Jansen, 2002; Tang, Fellows, Small, & Dagher, 2012), so the real question is how significantly this additional exposure changes their behaviour at the grocery store, or later that day? There is also the concern about long-term exposure and how that shapes perceptions and ideas of normal behaviour. The 'visual normalization theory' has been used to describe the phenomenon of normalizing obesity (Robinson 2017), but could also be used to describe the normalization of fast food consumption, where the repetitive visual exposure to FFRs may in turn reinforce the feeling of social appropriateness and lead to changes in attitudes towards FF. So, whether or not individuals succumb to the appeal of fast food at these intervening opportunities, researchers in the fields of geography and biopsychology should increase collaboration efforts in order to more significantly investigate how this visual and olfactory exposure actually influences subsequent food purchases, along with daily and long-term eating patterns.

One important consideration for researchers to understand is that the biggest concern for confounding from SDMB comes from studying non-residential environments, because there is flexibility for individuals to choose to visit locations that support certain behaviours, such as visiting fast food retailers to purchase highly processed food, or visiting greenspace to engage in physical activity. In this sense, the ability to choose to engage in these activities is potentially both an outcome of exposure, as well as a driver of exposure. There is less concern around other types of exposures that are less associated with behaviours of choice, such as being exposed to air pollution while commuting to work (Chaix et al. 2012).

This distinction of 'choice' is precisely why truncating trips in the method proposed by Chaix et al. (2013) and outlined in Sect. 4.2 may not be the best option when it comes to understanding food behaviours. Where a person shops and what they eat can be thought of as a constrained choice; Unlike engaging in regular physical activity, the act of eating itself is a basic requirement of life. The ways in which people are interacting with the food environment is changing over recent years; People are making more frequent, smaller trips to the grocery store (Yoo et al. 2006) and regularly consuming food prepared out of the home (Laska et al. 2015; Monsivais et al. 2014; Pelletier et al. 2014). By removing trips that end in food purchases, important information like the compounding exposure experienced on food-specific trips is being lost.

A more flexible approach to truncation that includes the exposure along the daily activity paths between activity sites should also be considered and determined by the hypothesis being tested. For example, in the example scenario outlined earlier in this section, knowing the accumulated number of FFRs a person is exposed to along paths and at activities could be a key variable in understanding if general exposure to these places is more predictive of consumption of fast food than exposure only at activity sites. This can also be understood using the trip-chaining framework proposed by Chen and Kwan (2012), where people will frequently combine multiple stops into one trip in order to improve time efficiency, as is the case of the individual in Fig. 1, where the trip to the grocery store is part of the journey home and not its own specific trip. By keeping the paths between activity sites on multi-purpose trips, researchers can better account for the true exposure that occurs. The downside of this approach is the increased complexity that accompanies the identification and evaluation of activity paths, rather than just activity sites.

## *4.3 Recommendations for Future Work*

It is important to remember that the built environment is just one component in a much more complex equation (Teixeira et al. 2015; White 2016). Removing trips that end in the activity of interest ignores the biggest question of all: Why are people seeking out those environments? We recommend that more emphasis be placed on determining this. Future work that aims to evaluate the effects of exposure to the food environment, or any built environment for that matter, should therefore include:

1. A combination of GPS technologies and GIS to track and map participants daily activity paths. GPS trajectories are known to be more predictive of health outcomes than GIS models that make assumptions about routes and destinations, and most research in this field is already moving in this direction (Cetateanu and Jones 2016). We also suggest that pathways between activity sites be considered when possible to enhance the understanding of exposure that occurs

during these trips, and reduce the likelihood of discarding important information using a truncated activity space approach to accounting for SDMB.

2. Some form of activity diary should complement GPS data and travel surveys to identify the precise activity and reason for each trip. This could be in the form of a travel log diary (Sadler and Gilliland 2015). Better yet, using GPS-enabled phone applications that include momentary/ecological dietary assessment in their design would allow attribute data to be collected instantaneously through embedded survey prompts (Spook et al. 2013). An example of this would be a participant entering meal data in the exact location and time that it was consumed.
3. In line with the third recommendation put forth by Chaix et al. (2013), information on 'why' and 'how' people choose the locations they frequently visit should accompany the collection of GPS and activity log diaries. Including complementary surveys or prompts that ask questions such as:

   - Why did you choose to shop/eat here?
   - Would you have preferred to shop/eat somewhere else?
   - How many days a week do you purchase breakfast/lunch/dinner?
   - Why do you purchase and consume food away from home?
   - Would you prefer to eat food prepared at home?

We also recommend that researchers stop only citing SDMB as a study limitation, and actually start testing for it within the study design. At present, there are no studies comparing different methods of accounting for selective daily mobility bias using GPS. Though the truncated activity space method appears to be a promising approach in collaboration with travel diaries and preference surveys, it is important that future studies compare the truncated activity space approach(es) to a control obtained without removing these behaviour-specific trips. This is important to (a) learn more about the bias and refine the methods used to account for it and (b) advance our understanding of how and why certain environments are being chosen. In an effort to continue the momentum towards more interdisciplinary collaboration, not seeking to understand the motivations for selecting certain activity sites would be an oversight to advancing the understanding of food environment psychology. It is crucial that methods for understanding and handling selective daily mobility bias be developed over the next decade as researchers continue to grapple with the impact of food environments so that we can make informed policy decisions and assess intervention impacts.

## 5 Conclusions

Geospatial technologies are reshaping the ways researchers explore a wide range of topics, including dietary behaviours. As more food retailers populate streets (Thompson 2017) and mounting time pressures (Beshara et al. 2010; Widener et al. 2015) continue to drive people towards quick food options, individuals may be more exposed to food retailers and subsequently may more frequently seek out these environments for the purpose of making out-of-home food purchases. With very little division of healthy and unhealthy food environments in urban areas, it can be argued that exposure to FFRs while seeking out food is a particularly potent source of exposure that has a strong capacity to influence both short- and long-term food behaviour (Clary et al. 2017; Drewnowski and Kawachi 2015). So, while methods like truncating activity spaces may be appropriate in some studies, in the realm of retail food environment research, some questions may only be answered through a close examination of all types of trips.

## 5.1 Where to Go Next with Selective Daily Mobility Bias?

Of the two papers that sought to evaluate the bias, only one found evidence supporting its significance (Perchoux et al. 2016). So while the above text provides a starting point for researchers using spatiotemporal data in food environment research going forward, selective daily mobility bias is clearly in its infancy as a methodological term and will require significantly more work to determine a standardized approach that allows for comparison across studies and populations. It is crucial that the data that comes from the advancements in GPS technologies are produced and used in ways that can lead to reproducible and robust findings. At this point, only one study using the truncated approach has actually compared results with exposures calculated using the full activity space (Perchoux et al. 2016), and this approach alone may not be appropriate in all types of exposure research (Chaix et al. 2012). More work is needed to confirm when, and in what contexts, truncation is a suitable method.

It is also important to continue to test for differences in exposure measures and associated behaviour outcomes using the different methods currently being used in research on exposure to food environments, including routes and activity spaces. By examining the effects of different methods using the same datasets and variables of interest, such as food consumption, a more robust understanding of how the food environment affects diet and health can be derived. From the work presented here, it is clear that while geospatial technologies have provided advances in and new opportunities for analysis, the food environment research community must continue to develop and critique methods, so they may confidently interpret and appropriately generalize their findings.

## References


Ahalya, M., Jane, Y. P., Éric, R., Marc, L., Tina, M., & Leia, M. M. (2017). Geographic retail food environment measures for use in public health. *Health Promotion and Chronic Disease Prevention in Canada: Research, Policy and Practice, 37*(10), 357–362.

Beshara, M., Hutchinson, A., & Wilson, C. (2010). Preparing meals under time stress. The experience of working mothers. *Appetite, 55*(3), 695–700. https://doi.org/10.1016/j.appet.2010.10.003.

Bes-Rastrollo, M., Sayon-Orea, C., Ruiz-Canela, M., & Martinez-Gonzalez, M. A. (2016). Impact of sugars and sugar taxation on body weight control: A comprehensive literature review. *Obesity, 24*(7), 1410–1426. https://doi.org/10.1002/oby.21535.

Boone-Heinonen, J., Gordon-Larsen, P., Kiefe, C. I., Shikany, J. M., Lewis, C. E., & Popkin, B. M. (2011). Fast food restaurants and food stores: Longitudinal associations with diet in young adults: The CARDIA Study. *Archives of Internal Medicine, 171*(13), 1162–1170. https://doi.org/10.1001/archinternmed.2011.283.

Boswell, R. G., & Kober, H. (2016). Food cue reactivity and craving predict eating and weight gain: A meta-analytic review. *Obesity Reviews: An Official Journal of the International Association for the Study of Obesity, 17*(2), 159–177. https://doi.org/10.1111/obr.12354.

Bramer, W. M., Rethlefsen, M. L., Kleijnen, J., & Franco, O. H. (2017). Optimal database combinations for literature searches in systematic reviews: A prospective exploratory study. *Systematic Reviews, 6*. https://doi.org/10.1186/s13643-017-0644-y.

Burgoine, T., & Monsivais, P. (2013). Characterising food environment exposure at home, at work, and along commuting journeys using data on adults in the UK. *International Journal of Behavioral Nutrition and Physical Activity, 10*, 85. https://doi.org/10.1186/1479-5868-10-85.

Burgoine, T., Jones, A. P., Namenek Brouwer, R. J., & Benjamin Neelon, S. E. (2015). Associations between BMI and home, school and route environmental exposures estimated using GPS and GIS: Do we see evidence of selective daily mobility bias in children? *International Journal of Health Geographics, 14*, 8. https://doi.org/10.1186/1476-072X-14-8.

Byrnes, H. F., Miller, B. A., Morrison, C. N., Wiebe, D. J., Remer, L. G., & Wiehe, S. E. (2016). Brief report: Using global positioning system (GPS) enabled cell phones to examine adolescent travel patterns and time in proximity to alcohol outlets. *Journal of Adolescence, 50*, 65–68. https://doi.org/10.1016/j.adolescence.2016.05.001.

Camacho, S., & Ruppel, A. (2017). Is the calorie concept a real solution to the obesity epidemic? *Global Health Action, 10*(1), 1289650. https://doi.org/10.1080/16549716.2017.1289650.

Caspi, C. E., Sorensen, G., Subramanian, S. V., & Kawachi, I. (2012). The local food environment and diet: A systematic review. *Health & Place, 18*(5), 1172–1187. https://doi.org/10.1016/j.healthplace.2012.05.006.

Cawley, J., & Wen, K. (2018). Policies to prevent obesity and promote healthier diets: A critical selective review. *Clinical Chemistry, 64*(1), 163–172. https://doi.org/10.1373/clinchem.2017.278325.

Cebrecos, A., Díez, J., Gullón, P., Bilal, U., Franco, M., & Escobar, F. (2016). Characterizing physical activity and food urban environments: A GIS-based multicomponent proposal. *International Journal of Health Geographics, 15*. https://doi.org/10.1186/s12942-016-0065-5.

Cetateanu, A., & Jones, A. (2016). How can GPS technology help us better understand exposure to the food environment? A systematic review. *SSM - Population Health, 2*, 196–205. https://doi.org/10.1016/j.ssmph.2016.04.001.

Chaix, B., Kestens, Y., Perchoux, C., Karusisi, N., Merlo, J., & Labadi, K. (2012). An interactive mapping tool to assess individual mobility patterns in neighborhood studies. *American Journal of Preventive Medicine, 43*(4), 440–450. https://doi.org/10.1016/j.amepre.2012.06.026.

Chaix, B., Méline, J., Duncan, S., Merrien, C., Karusisi, N., Perchoux, C., et al. (2013). GPS tracking in neighborhood and health studies: A step forward for environmental exposure assessment, a step backward for causal inference? *Health & Place, 21*, 46–51. https://doi.org/10.1016/j.healthplace.2013.01.003.

Chen, X., & Kwan, M.-P. (2012). Choice set formation with multiple flexible activities under space–time constraints. *International Journal of Geographical Information Science, 26*(5), 941–961. https://doi.org/10.1080/13658816.2011.624520.

Christian, W. J. (2012). Using geospatial technologies to explore activity-based retail food environments. *Spatial and Spatio-Temporal Epidemiology, 3*(4), 287–295.

Clary, C., Matthews, S. A., & Kestens, Y. (2017). Between exposure, access and use: Reconsidering foodscape influences on dietary behaviours. *Health & Place, 44*, 1–7. https://doi.org/10.1016/j.healthplace.2016.12.005.

Cornelsen, L., Green, R., Dangour, A., & Smith, R. (2015). Why fat taxes won't make us thin. *Journal of Public Health, 37*(1), 18–23. https://doi.org/10.1093/pubmed/fdu032.

Crézé, C., Notter-Bielser, M.-L., Knebel, J.-F., Campos, V., Tappy, L., Murray, M., & Toepel, U. (2018). The impact of replacing sugar- by artificially-sweetened beverages on brain and behavioral responses to food viewing – An exploratory study. *Appetite, 123*, 160–168. https://doi.org/10.1016/j.appet.2017.12.019.

Drewnowski, A., & Kawachi, I. (2015). Diets and health: How food decisions are shaped by biology, economics, geography, and social interactions. *Big Data, 3*(3), 193–197. https://doi.org/10.1089/big.2015.0014.

Eckert, J., & Shetty, S. (2011). Food systems, planning and quantifying access: Using GIS to plan for food retail. *Applied Geography, 31*(4), 1216–1223. https://doi.org/10.1016/j.apgeog.2011.01.011.

Fong, K. C., Hart, J. E., & James, P. (2018). A review of epidemiologic studies on greenness and health: Updated literature through 2017. *Current Environmental Health Reports, 5*(1), 77–87. https://doi.org/10.1007/s40572-018-0179-y.

Giskes, K., van Lenthe, F., Avendano-Pabon, M., & Brug, J. (2011). A systematic review of environmental factors and obesogenic dietary intakes among adults: Are we getting closer to understanding obesogenic environments? *Obesity Reviews, 12*(5), e95–e106. https://doi.org/10.1111/j.1467-789X.2010.00769.x.

Haddaway, N. R., Collins, A. M., Coughlin, D., & Kirk, S. (2015). The role of Google Scholar in evidence reviews and its applicability to grey literature searching. *PLoS One, 10*(9), e0138237. https://doi.org/10.1371/journal.pone.0138237.

Hager, E. R., Cockerham, A., O'Reilly, N., Harrington, D., Harding, J., Hurley, K. M., & Black, M. M. (2017). Food swamps and food deserts in Baltimore City, MD, USA: Associations with dietary behaviours among urban adolescent girls. *Public Health Nutrition, 20*(14), 2598–2607. https://doi.org/10.1017/S1368980016002123.

Hall, K. D. (2017). Did the food environment cause the obesity epidemic? *Obesity, 26*(1), 11–13. https://doi.org/10.1002/oby.22073.

Harrison, F., Burgoine, T., Corder, K., van Sluijs, E. M., & Jones, A. (2014). How well do modelled routes to school record the environments children are exposed to?: A cross-sectional comparison of GIS-modelled and GPS-measured routes to school. *International Journal of Health Geographics, 13*(1), 5. https://doi.org/10.1186/1476-072X-13-5.

Health Canada. (2013, October 9). Measuring the Food Environment in Canada [research]. Retrieved May 3, 2018, from https://www.canada.ca/en/health-canada/services/food-nutrition/healthy-eating/nutrition-policy-reports/measuring-food-environment-canada.html.

Hebebrand, J., Albayrak, Ö., Adan, R., Antel, J., Dieguez, C., de Jong, J., et al. (2014). "Eating addiction", rather than "food addiction", better captures addictive-like eating behavior. *Neuroscience & Biobehavioral Reviews, 47*, 295–306. https://doi.org/10.1016/j.neubiorev.2014.08.016.

Kestens, Y., Lebel, A., Daniel, M., Thériault, M., & Pampalon, R. (2010). Using experienced activity spaces to measure foodscape exposure. *Health & Place, 16*(6), 1094–1103. https://doi.org/10.1016/j.healthplace.2010.06.016.

Kestens, Y., Lebel, A., Chaix, B., Clary, C., Daniel, M., Pampalon, R., et al. (2012). Association between activity space exposure to food establishments and individual risk of overweight. *PLoS One, 7*(8), e41418. https://doi.org/10.1371/journal.pone.0041418.

Kwan, M.-P. (2018). The limits of the neighborhood effect: Contextual uncertainties in geographic, environmental health, and social science research. *Annals of the American Association of Geographers, 0*(0), 1–9. https://doi.org/10.1080/24694452.2018.1453777.

Laska, M. N., Hearst, M. O., Lust, K., Lytle, L. A., & Story, M. (2015). How we eat what we eat: Identifying meal routines and practices most strongly associated with healthy and unhealthy dietary factors among young adults. *Public Health Nutrition, 18*(12), 2135–2145. https://doi.org/10.1017/S1368980014002717.

Ma, Y., Ratnasabapathy, R., & Gardiner, J. (2017). Carbohydrate craving: Not everything is sweet. *Current Opinion in Clinical Nutrition & Metabolic Care, 20*(4), 261. https://doi.org/10.1097/MCO.0000000000000374.

McCrorie, P. R., Fenton, C., & Ellaway, A. (2014). Combining GPS, GIS, and accelerometry to explore the physical activity and environment relationship in children and young people - a

review. *International Journal of Behavioral Nutrition and Physical Activity, 11*(1), 93. https://doi.org/10.1186/s12966-014-0093-0.
Minaker, L. M. (2016). Retail food environments in Canada: Maximizing the impact of research, policy and practice. *Canadian Journal of Public Health = Revue Canadienne De Sante Publique, 107*.(Suppl 1, 5632.
Minaker, L. M., Shuh, A., Olstad, D. L., Engler-Stringer, R., Black, J. L., & Mah, C. L. (2016). Retail food environments research in Canada: A scoping review. *Canadian Journal of Public Health, 107*(0), 4–13.
Mitchell, C. (2016). Children's physical activity and the built environment: The impact of neighbourhood opportunities and contextual environmental exposure. *Electronic Thesis and Dissertation Repository*. Retrieved from https://ir.lib.uwo.ca/etd/3524
Monsivais, P., Aggarwal, A., & Drewnowski, A. (2014). Time spent on home food preparation and indicators of healthy eating. *American Journal of Preventive Medicine, 47*(6), 796–802. https://doi.org/10.1016/j.amepre.2014.07.033.
Moubarac, J.-C., Batal, M., Louzada, M. L., Martinez Steele, E., & Monteiro, C. A. (2017). Consumption of ultra-processed foods predicts diet quality in Canada. *Appetite, 108*(Suppl C), 512–520. https://doi.org/10.1016/j.appet.2016.11.006.
Nederkoorn, C., & Jansen, A. (2002). Cue reactivity and regulation of food intake. *Eating Behaviors, 3*(1), 61–72. https://doi.org/10.1016/S1471-0153(01)00045-9.
Pelletier, J. E., Graham, D. J., & Laska, M. N. (2014). Social norms and dietary behaviors among young adults. *American Journal of Health Behavior, 38*(1), 144. https://doi.org/10.5993/AJHB.38.1.15.
Perchoux, C., Chaix, B., Brondeel, R., & Kestens, Y. (2016). Residential buffer, perceived neighborhood, and individual activity space: New refinements in the definition of exposure areas – The RECORD Cohort Study. *Health & Place, 40*(Suppl C), 116–122. https://doi.org/10.1016/j.healthplace.2016.05.004.
Ridder, D. D., Manning, P., Leong, S. L., Ross, S., Sutherland, W., Horwath, C., & Vanneste, S. (2016). The brain, obesity and addiction: An EEG neuroimaging study. *Scientific Reports, 6*(34122). https://doi.org/10.1038/srep34122.
Robinson, E. (2017). Overweight but unseen: A review of the underestimation of weight status and a visual normalization theory. *Obesity Reviews, 18*(10), 1200–1209. https://doi.org/10.1111/obr.12570.
Sadler, R. C., & Gilliland, J. A. (2015). Comparing children's GPS tracks with geospatial proxies for exposure to junk food. *Spatial and Spatio-Temporal Epidemiology, 14–15*, 55–61. https://doi.org/10.1016/j.sste.2015.09.001.
Scully, J. Y. (2016). *Human Mobility, Exposure to the Built Environment, and Health* (Thesis). Retrieved from https://digital.lib.washington.edu:443/researchworks/handle/1773/36862.
Spook, J. E., Paulussen, T., Kok, G., & Empelen, P. V. (2013). Monitoring dietary intake and physical activity electronically: Feasibility, usability, and ecological validity of a mobile-based ecological momentary assessment tool. *Journal of Medical Internet Research, 15*(9), e214. https://doi.org/10.2196/jmir.2617.
Steele, E. M., Baraldi, L. G., Louzada, M. L. d. C., Moubarac, J.-C., Mozaffarian, D., & Monteiro, C. A. (2016). Ultra-processed foods and added sugars in the US diet: Evidence from a nationally representative cross-sectional study. *BMJ Open, 6*(3), e009892. https://doi.org/10.1136/bmjopen-2015-009892.
Sturm, R., & Cohen, D. A. (2009). Zoning for health? The year-old ban on new fast-food restaurants in South LA. *Health Affairs (Project Hope), 28*(6), w1088–w1097. https://doi.org/10.1377/hlthaff.28.6.w1088.
Tang, D. W., Fellows, L. K., Small, D. M., & Dagher, A. (2012). Food and drug cues activate similar brain regions: A meta-analysis of functional MRI studies. *Physiology & Behavior, 106*(3), 317–324. https://doi.org/10.1016/j.physbeh.2012.03.009.
Teixeira, P. J., Carraça, E. V., Marques, M. M., Rutter, H., Oppert, J.-M., De Bourdeaudhuij, I., et al. (2015). Successful behavior change in obesity interventions in adults: A systematic review of self-regulation mediators. *BMC Medicine, 13*, 84. https://doi.org/10.1186/s12916-015-0323-6.

Thompson, D. (2017, June 20). *The Golden Age of Restaurants Is Stranger Than It Seems*. Retrieved July 30, 2018, from https://www.theatlantic.com/business/archive/2017/06/its-the-golden-age-of-restaurants-in-america/530955/.

Ventura, A. K., & Mennella, J. A. (2011). Innate and learned preferences for sweet taste during childhood. *Current Opinion in Clinical Nutrition & Metabolic Care, 14*(4), 379. https://doi.org/10.1097/MCO.0b013e328346df65.

White, M. (2016). Population approaches to prevention of type 2 diabetes. *PLoS Medicine, 13*(7), e1002080. https://doi.org/10.1371/journal.pmed.1002080.

WHO | Obesity and overweight. (n.d.). Retrieved November 29, 2017, from http://www.who.int/mediacentre/factsheets/fs311/en/.

Widener, M. J., & Shannon, J. (2014). When are food deserts? Integrating time into research on food accessibility. *Health & Place, 30*, 1–3. https://doi.org/10.1016/j.healthplace.2014.07.011.

Widener, M. J., Farber, S., Neutens, T., & Horner, M. (2015). Spatiotemporal accessibility to supermarkets using public transit: An interaction potential approach in Cincinnati, Ohio. *Journal of Transport Geography, 42*, 72–83. https://doi.org/10.1016/j.jtrangeo.2014.11.004.

Widener, M. J., Minaker, L. M., Reid, J. L., Patterson, Z., Ahmadi, T. K., & Hammond, D. (2018). Activity space-based measures of the food environment and their relationships to food purchasing behaviours for young urban adults in Canada. *Public Health Nutrition, 21*, 1–14. https://doi.org/10.1017/S1368980018000435.

Yoo, S., Baranowski, T., Missaghian, M., Baranowski, J., Cullen, K., Fisher, J. O., et al. (2006). Food-purchasing patterns for home: A grocery store-intercept survey. *Public Health Nutrition, 9*(3), 384–393. https://doi.org/10.1079/PHN2005864.

Zenk, S. N., Schulz, A. J., Matthews, S. A., Odoms-Young, A., Wilbur, J., Wegrzyn, L., et al. (2011). Activity space environment and dietary and physical activity behaviors: A pilot study. *Health & Place, 17*(5), 1150–1161. https://doi.org/10.1016/j.healthplace.2011.05.001.

Zenk, S. N., Matthews, S. A., Kraft, A. N., & Jones, K. K. (2018). How many days of global positioning system (GPS) monitoring do you need to measure activity space environments in health research? *Health & Place, 51*, 52–60. https://doi.org/10.1016/j.healthplace.2018.02.004.

Zientek, L. R., Werner, J. M., Campuzano, M. V., & Nimon, K. (n.d.). The use of Google Scholar for research and research dissemination. *New Horizons in Adult Education and Human Resource Development, 30*(1), 39–46. https://doi.org/10.1002/nha3.20209.

**Reilley Plue** is a master's student at the University of Toronto in the Department of Geography and Planning, where she is completing an MA in Human Geography with a collaboration in Environment and Health. Her interests lie at the complex intersection of food and nutrition, agro-ecology, human behaviour, and public health and preventative medicine.

**Lauren Jewett** is a PhD Student at the University of Toronto in the Department of Geography and Planning. Lauren completed a BSc from McMaster University in Life Sciences and Geospatial Science (2012) and a Master of Geographic Information Systems from the University of Calgary (2017). Lauren specializes in spatial statistics and modelling for understanding disparities in health services across large geographies and vulnerable populations.

**Michael J. Widener** is a Canada Research Chair (Tier 2) in Transportation and Health at the University of Toronto – St. George. He is an Assistant Professor in Geography and Planning, with a cross-appointment in Epidemiology at the Dalla Lana School of Public Health. Dr. Widener's research focuses on how public health affects, and is affected by, transportation systems.

# Dynamic Emergency Medical Service Dispatch: Role of Spatiotemporal Machine Learning

**Sunghwan Cho and Dohyeong Kim**

**Abstract** Previous research has suggested that providing prompt access to emergency medical services (EMS) may greatly improve the health outcomes of patients with urgent conditions. However, there has not been enough research on ways in which planning resources for ambulance dispatch may enhance the response time of EMS. GIS has been used to manage and visualize the spatial distribution of EMS demand, but there is still a need for more empirical evidence from spatiotemporal demand-based prediction techniques, such as machine learning. We applied the long short-term memory (LSTM) method to forecast EMS demands based on past records and reallocated service locations using a dynamic maximal covering location model. The training of the prediction models and validation were conducted with 323,993 emergency calls in the Gyeongnam Province in Korea in 2014. We found that conventional hotspot-based emergency dispatch systems, ignoring temporal variations of service demands, could fail to fulfill a desired coverage standard. This study shows an evidence that demand-based spatiotemporal demand prediction and dynamic dispatch protocol based on machine learning algorithm have the potential to support more efficient allocation of resources, especially when resources are limited.

## Abbreviations

| | |
|---|---|
| EMS | Emergency medical services |
| GIS | Geographic information systems |
| LSTM | Long short-term memory |

S. Cho
Korea Land and Geospatial Informatrix Corporation,
Deokjin-gu, Jeonju-si, Jeollabuk-do, South Korea

D. Kim (✉)
University of Texas at Dallas, Richardson, TX, USA
e-mail: dohyeong.kim@utdallas.edu

 
Y. Lu, E. Delmelle (eds.), *Geospatial Technologies for Urban Health*,
Global Perspectives on Health Geography, https://doi.org/10.1007/978-3-030-19573-1_7

| | |
|---|---|
| MLPs | Multilayer perceptrons |
| MSE | Mean squared error |
| RT | Response time |
| SC | Safety center |
| MCLM | Maximal covering location model |
| OLS | Ordinary least squares |

# 1 Introduction

The "golden hour" refers to the importance of transporting a critically injured person to a hospital within the first hour after injury. Similarly, the "platinum 5 (or 10) minutes" of response time (RT) for the arrival of emergency medical service (EMS) has been accepted as a critical prehospital norm highly associated with survival of trauma patients (Rogers et al. 2015). Many studies report that emergency calls with RTs less than "on-scene time" were associated with improved survival, when compared to calls with longer RTs (Pell et al. 2001; Pons et al. 2005; Blackwell and Kaufman 2008). The RT goal has been implemented as policy goal in many EMS departments around the world (Roudsari et al. 2007; Cho et al. 2017; Washington D.C. Fire and EMS Department 2018). A vast body of literature has been dedicated to developing models and methods to allocate EMS locations to meet RT requirements, using optimization and location-allocation models (Revelle et al. 1977; Li et al. 2011), GIS mapping and simulation (Peters and Hall 1999; Peleg and Pliskin 2004; Hong et al. 2008), and cost-effective analysis (Savas 1969).

GIS-based hotspot analysis has been widely used to identify where EMS resources would be most needed, but static geospatial approach ignoring temporal variations of EMS demands could be flawed or inaccurate. Emergent events and ambulance calls are not random events but occur in spatial, temporal, and spatiotemporal patterns and trends that can be observed in large-sized historical data. Although numerous studies have attempted to suggest the best arrangement of EMS resources based on the estimated demand incorporating non-random historical event patterns (Ong et al. 2009), most of them assume that the spatial distribution of the demand is static over time. However, numerous articles have reported time-geographic patterns of ambulance calls, indicating that hotspots change in space and over time (Bassil et al. 2009; Ong et al. 2009). Due to the dynamic patterns of EMS demand, a conventional allocation of EMS services based on the fixed-hotspot framework may be limited in its ability to maintain low RTs for all areas, at all times of the day. With a good understanding of spatiotemporal patterns of EMS demand, we may be able to predict future demand and reallocate resources to reduce RTs if the real-time demand modeling and allocation practices are well developed and implemented.

In order to successfully make predictions, the size and complexity of the spatiotemporal data require the sophisticated statistical reasoning and extensive computation made possible with machine learning. Although several articles have recently built a theoretical framework for real-time EMS vehicle dispatching and redistribution (Haghani et al. 2003; Zhou et al. 2013; Chen et al. 2016), tangible evidence and empirical evaluation are still lacking. Machine learning has been applied to monitoring and predicting the demand for public health services (Obermeyer and Emanuel 2016), but to date, there has been little use of spatiotemporal machine learning for demand forecasting of EMS. Zhou (2016) developed three types of machine learning methods – time-varying Gaussian mixture model, spatiotemporal kernel density estimation, and kernel warping – to provide spatiotemporal predictions for ambulance demand in Toronto and Melbourne (Zhou 2016). Additionally, Chen and Lu (2014) applied methods such as moving average, artificial neural network, liner regression, and support vector machine to predict prehospital emergency medical demand using the EMS data in New Taipei City (Chen and Lu 2014). However, they did not demonstrate how predicted demands could be used for staff/fleet management and dynamic deployment. In another study, Levi et al. (2017) used machine learning to build a live dispatch system for the City of Cincinnati based on the predicted incidents, arguing that the system could improve dispatch accuracy and RTs. However, their study lacks detailed information about how the real-time predictive model was utilized in the study works (Levi et al. 2017).

There is no doubt that as spatiotemporal data become more widely available in the fields of health and safety, time-dependent machine learning tools and techniques should rapidly advance. It is also certain that EMS services could become more efficient and responsive with the assistance of dynamic dispatch protocol. To verify usefulness of this approach, this study aims to build empirical evidence for the effectiveness of dynamic resource allocation for EMS dispatch systems using actual data. Therefore, we used long short-term memory (LSTM) method to forecast prehospital emergency medical demands, based on past records of emergency call data in the Gyeongnam Province in Korea between January and December of 2014. This model allowed us to detect time-varying hotspots and allocate the EMS centers and vehicles in a dynamic way. We then calculated the coverage rates by RT standards (5 or 10 minutes) based on the model-recommended allocations and compared them with those for the existing distribution of resources in order to evaluate the effectiveness of the system both methodologically and practically.

## 2 EMS Dispatch Process in Gyeongnam Province, Korea

In South Korea, 119 Safety Centers (SC) are the main locations of EMS vehicles that are routinely dispatched to incident sites when ambulance service is requested. All emergency calls (or "119 calls") within a municipality are received at the Municipal Emergency Dispatch Center and are automatically assigned by the automatic dispatch system to an available EMS team located at the proximate SC. The computer-based automatic dispatch system obtains information about location and the type of incident, either through GPS or verbal information, and sends a dispatch order to the nearest SC, where the requisite personnel and vehicle are available for deployment. If the nearest SC is not able to dispatch the required EMS team, the system automatically contacts the next closest SC. Approximately, 95% of the emergency calls in South Korea requires an EMS dispatch. However, a significant portion of the calls are found to be false alarms which eventually lead to longer response times for true emergencies.

Historically, the locations of South Korean SCs and their jurisdiction boundaries have been determined with an eye to maximizing administrative convenience. Despite the use of the computer-assisted assignment system of emergency calls to the closest available SC, substantial proportion of the calls has failed to be addressed within their target RT goal mostly because the current allocation of SCs in South Korea is suboptimal. Gyeongnam Province, the target area of this study, is 1 of 13 provinces in South Korea and home to approximately 3.4 million people. As of 2016, 85.5% of the provincial population lived in urban areas (Korean Statistical Information Service, 2016). Although the authorities attempt to keep RTs for all EMS services within 5 minutes, in 2014, times ranged between 1 and 30 minutes, showing huge variation within the province. Average RTs were over 20 minutes in rural areas such as Euiryong, Geochang, Hapcheon and Haman-gun.

Recommended redistribution should be based on a systematic investigation of historical demand. However, neither routine nor non-routine reallocation of EMS resources incorporating historical emergency data or estimated demand has been implemented. Moreover, strict enforcement of municipal EMS boundaries may have led to a longer RT in certain regions. It may be unrealistic to relocate SC buildings and large facilities on a regular basis, due to construction and administrative costs. However, it would be realistic to relocate emergency fleets and personnel temporarily based on demand patterns. The dire lack of resources in some SCs could make this approach critically important during high-demand time periods. GIS-based real-time monitoring and proactive allocation of mobile EMS resources based on estimated demand may reduce RTs substantially when and where demand is highest.

Moreover, despite a growing interest in increasing EMS efficiency by using GIS to identify hotspots and allocate resources (including budget, personnel, facilities, equipment, fleets and command structure), GIS has not been incorporated into policy instruments and decision-making tools in South Korea. Major barriers to GIS incorporation include not only the technical limitations of computer-aided dispatch systems (Dean 2008), but also a lack of awareness of GIS's advantages as a decision-making tool (Kim et al. 2016). GIS-based resource allocation and dispatch systems may not be able to take all relevant factors into consideration, but they provide some guidelines for evidence-based solutions.

## 3 Methods

We first looked at the spatial distribution of 323,993 emergency calls received at Gyeongnam SCs between January 1 and December 31 of 2014, and evaluated the appropriateness of their responses, considering the locations of SCs from where EMS services were deployed to each emergency location. Using the spatiotemporal data, we identified three hotspots of emergency calls within the Province based on the Getis-Ord method (Getis and Ord 1992) and calculated an average RT within each hotspot. We then reviewed the temporal trend of the emergency calls within the three hotspots (i.e., zones) over time to understand how RTs fluctuated over a certain time period and varied between hotspots.

We split surface area of the Province into 1367 square grids (3 km by 3 km) for the learning process. We argue that RTs are significantly affected by not only road network distance between emergency and deployment locations but also by various environmental conditions, such as weather and traffic congestion. The real-time weather data were obtained from the API website of the Korea Meteorological Administration and interpolated to each grid point using IDW (Inverse Distance Weighting). The RT to the nearest SC (dependent variable) was predicted using independent variables including month (January to December), day of the week (Monday to Sunday), time of day (midnight to 6 am, 6 am to noon, noon to 6 pm, 6 pm to midnight), weather (clear, cloudy, gray, rain), and the network distance between the emergency call location and the nearest of the 103 SCs within the Province (in kilometers).

The statistical association between each of these factors and RTs can be used to predict future demand patterns. We first ran the OLS (ordinary least squares) regression to estimate the EMS demands on every grid for each of the time periods as baseline for comparison purposes. We then performed two machine learning algorithms, multilayer perceptrons (MLPs) with five neural networks and long short-term memory (LSTM), to estimate the demand for EMS services by grid level over the study area (Orbach 1962; Hochreiter and Schmidhuber 1997) and compared their performances. A cost function was created and used for each machine learning algorithm (Abadi et al. 2016). While the model by MLPs does not reflect the state of previous time, long short-term memory (LSTM) enables the model to incorporate the values from the previous state. The LSTM method was also used to allocate available EMS resources based on their location history during the previous time periods as the input value.

For the experiment, we used the rectifier Relu as an activation function for a hidden layer (Krizhevsky et al. 2012). However, we did not use the activation function for an output layer, since the outcome of the experiment functions as a regression, predicting a numerical value. In order to optimize the machine learning model, we used the Adam algorithm (Kingma and Jimmy 2014), which has been widely used for machine learning processes because it automatically solves the optimal value by adjusting the learning rate. Both MLPs and LSTM models were evaluated by the k-fold cross-validation method (Kohavi 1995). This method provides robust evaluation results on model performance regarding data used for experiments. In other words, it repeats the process of creating a training model using the training dataset with k-1 sets and then validates the model on the remaining dataset until it determines the final model with the lowest mean squared error (MSE) value:

$$\mathrm{MSE} = \frac{1}{n}\sum_{i=1}^{n}\left(Y_i - \hat{Y}_i\right)^2$$

In our experiment, we set k as tenfold and compared MSEs among the models during the learning process. The user tuning for reducing MSE values was performed five times for MLPs, followed by machine tuning.

Once the EMS demands were estimated for a specific time and location (grid) by OLS and machine learning processes, we allocated 103 SCs (or EMS vehicles, whichever feasible) to optimal locations based on the predicted demands for each time period using a dynamic maximal covering location model (MCLM) suggested by Zarandi et al. (2013) as follows:

$$\text{Maximize } Z = \sum_{t=1}^{T}\sum_{i=1}^{I} a_{it} Y_{it}$$

$$\text{subject to } Y_{it} \le \sum_{j \in N_i} X_{jt}, \qquad i \in I, \quad t \in T$$

$$\sum_{t=1}^{T}\sum_{j=1}^{J} X_{jt} = 103$$

$$0 \le Y_{it} \le 1, \qquad i \in I, \quad t \in T$$

$$X_{jt} \in \{0,1\}, \qquad j \in J, \quad t \in T$$

where: index of all EMS demand grid locations $I$;
$j$: index of all candidate EMS locations $J$;
$a_{it}$ = the number of predicted EMS demands present at grid $i$ in period $t$;
$d_{ij}$ = the travel distance (or time) from $i$ to $j$;
$S$ = the RT standard within which coverage is desired (either 5 or 10 minutes)
$N_{ij} = \{j | d_{ij} \le S\}$ = the set of grid locations that are within RT from $i$;

$$x_{jt} = \begin{cases} 1 & \text{if site } j \text{ is selected for EMS in period } t \\ 0 & \text{otherwise} \end{cases}$$

$$Y_{it} = \begin{cases} 1 & \text{if demand at } i \text{ in period } t \text{ is covered by EMS(s) stationed within } S \\ 0 & \text{otherwise} \end{cases}$$

Using the Python's location-allocation package (https://github.com/chrisfilippis/location-allocation-analysis), the optimal solutions were sought in order to maximize the number of predicted EMS demands covered by at least one EMS location within 5- or 10-minute travel distance, respectively. We allocated the EMS locations using not only a conventional linear programming suggested by Church and ReVelle (1974) ignoring the previous EMS location history, but also a nonlinear programming in order to incorporate the past records of EMS deployment at the three previous time points ($t - 1$, $t - 2$, $t - 3$) as an additional constraint into the process of determining the optimal location at current time ($t$), as illustrated in Fig. 1.

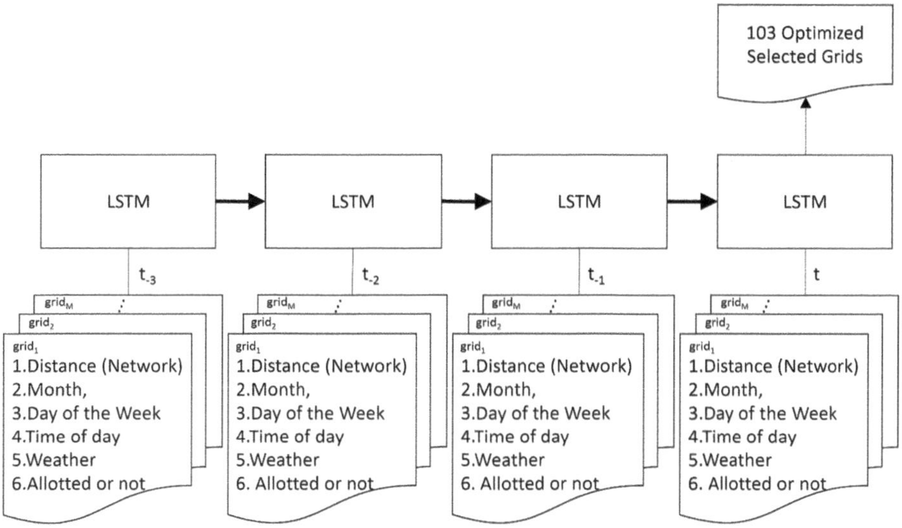


**Fig. 1** LSTM structure

Both demand prediction and location-allocation processes were employed using Python, which is popular as a machine learning language. Once the optimization processes were completed, the coverage rates by 5- and 10-minute standard were calculated for each method and compared to those from the current EMS arrangements to evaluate if the EMS dispatch processes based on machine learning could outperform in terms of covering more demands within their target RT goals in Gyeongnam Province.

## 4 Results and Discussion

Figure 2 shows how emergency calls between January 1 and December 31 of 2014 were spatially distributed in Gyeongnam Province, along with the locations of 103 SCs. The actual deployment of EMS from each emergency event location to the assigned SC (denoted as a red cross) is illustrated as a straight line in the map. However, the road network distance and travel time were used for actual data analysis. It looks that both emergency calls and SCs are spatially clustered, indicating some levels of spatial conformity between demand and supply of EMS services.

However, the actual response times were found to vary. Compared to other areas, the RTs were significantly larger in the three hotspots displayed in Fig. 3. Table 1 summarizes the area size, total number of emergency calls, mean, and standard deviation of RTs in each zone identified as hotspot.

Figure 4 shows the temporal pattern of daily average RTs (in minutes) for all the emergency calls in each of the three hotspot zones over the study period. The fluctuation

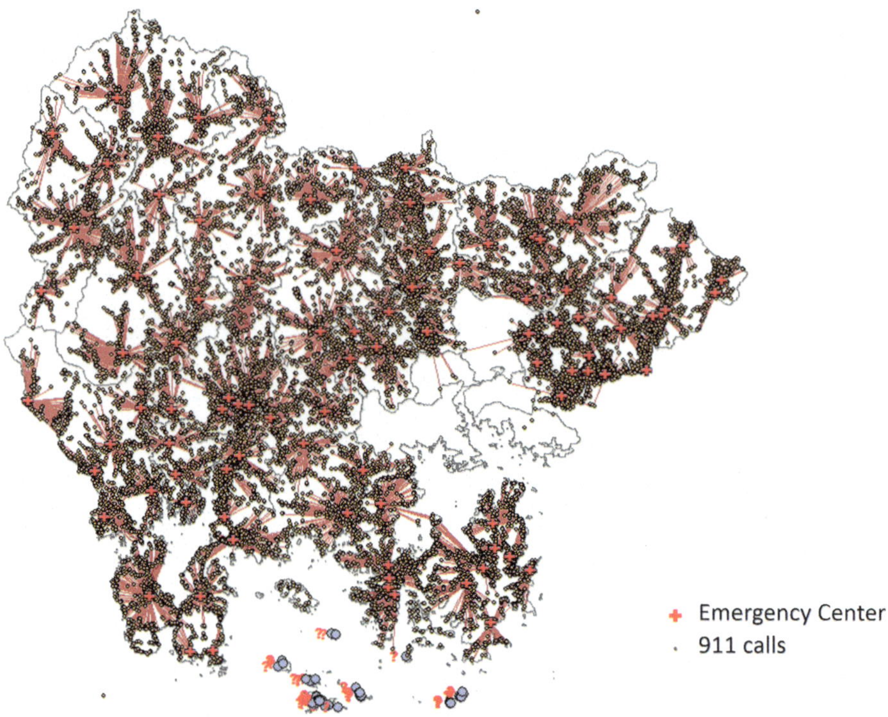


**Fig. 2** Spatial distribution of emergency calls and EMS locations in Gyeongnam Province between January 1 and December 31, 2014

of RTs looks relatively smaller in Zone B than in Zones A or C, except for a few outliers. The average RT ranges between 14 and 18 minutes for all three zones. The hourly variation graph (not shown) looks similar to Fig. 4. This evidence of temporal variation confirms that hotspots with greater RTs move over time instead of staying static in a specific area.

Figure 5 shows a loss function of the model for both training and test datasets during the user and machine tuning processes. The experimental process was iterated 150 times and the parameters were adjusted for both training and validation datasets at each iteration, along with the loss value of each model. Once the training process was complete, various neural network structures were formed to generate the models through a stepwise process until the final model was determined as that with the lowest MSE.

As summarized in Table 2, it is found that both machine learning methods performed better in estimating the EMS demands than OLS, with the MSE values being substantially reduced. In addition, MLPs were outperformed by the LSTM model with a smaller absolute value of MSE (3.98 vs. 5.11). Figure 6 shows how the MSEs have been reduced at each of the five neural networks of MLPs (baseline, deeper, wider#1, wider#2, final). The performance of the pre-tuning MLPs was

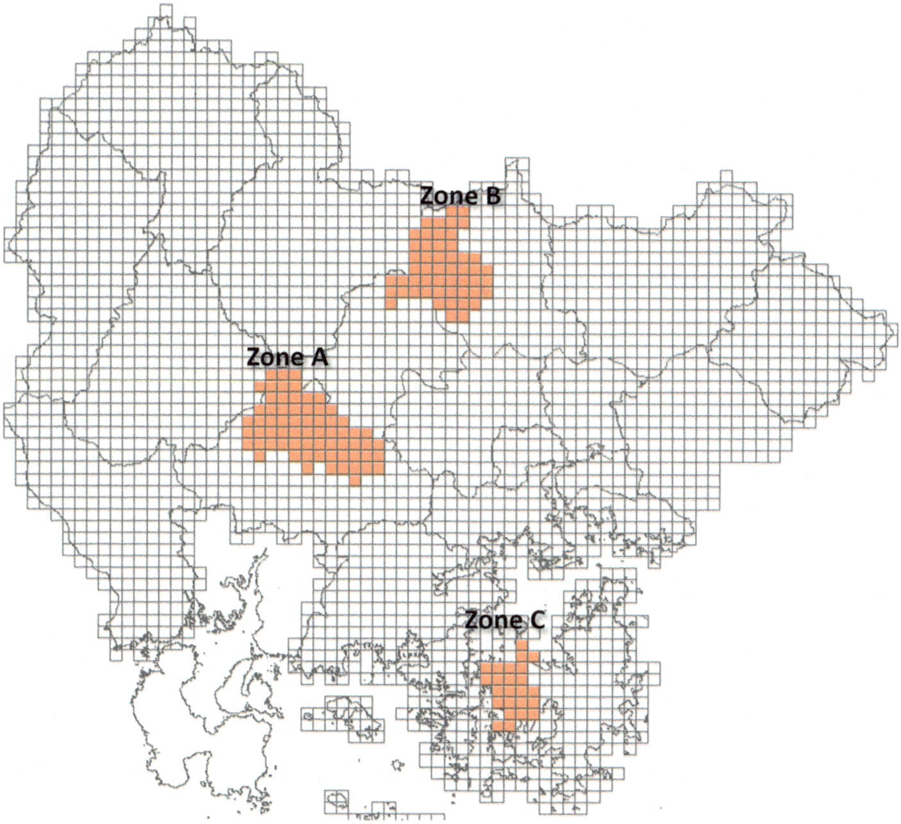


**Fig. 3** Three zones identified by hotspot analysis of RTs

**Table 1** Emergency calls and average RT at each zone

| Zone | Area ($km^2$) | Total number of calls | Average RT (min) Mean (SD) |
|---|---|---|---|
| Zone A | 504 | 712 | 16.86 (5.93) |
| Zone B | 324 | 434 | 14.47 (8.21) |
| Zone C | 160 | 623 | 15.62 (5.15) |

lower than that of OLS, but substantially improved during the tuning process. As LSTM was found best performing model, it was used for all subsequent analyses for predicting the demands for EMS services in each grid and allocating EMS resources via optimization method.

Figure 7 shows the map illustrating the predicted RT at each of 1367 grids, based on each of the four scenarios: (a) Sunday, rainy day, noon–6 pm; (b) Monday, rainy day, noon–6 pm; (c) Thursday, rainy day, 6 pm–midnight; (d) and Saturday, cloudy day, 6 am–noon.

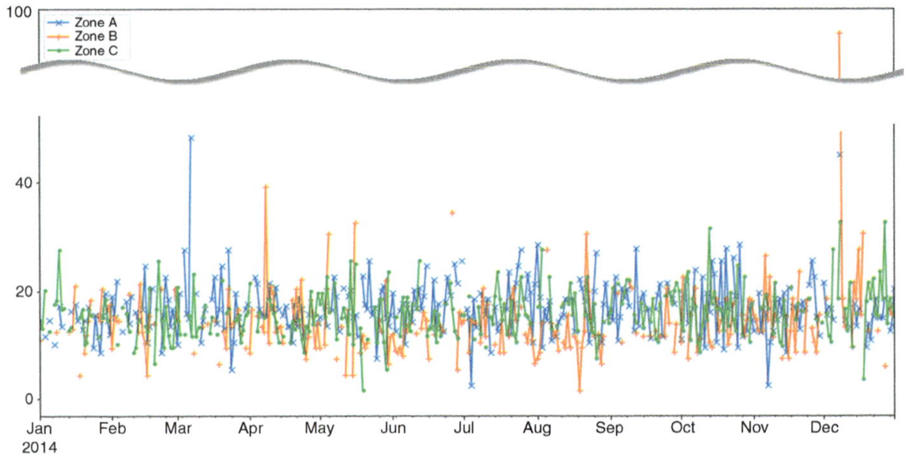


**Fig. 4** Temporal trends of RTs in three hotspots (daily variation)

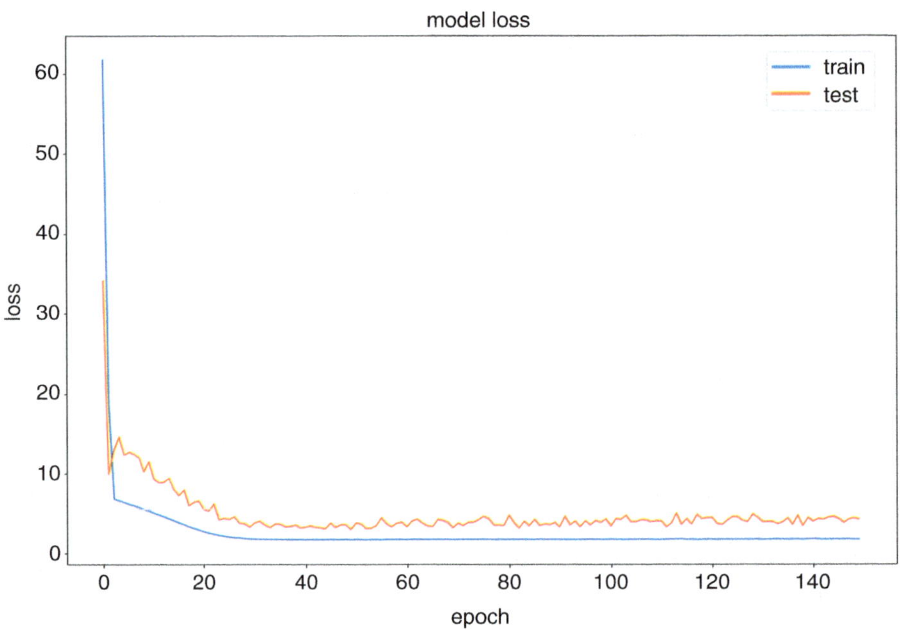


**Fig. 5** Loss function for training and test datasets

Figure 8 compares the RTs of the current distribution of SCs ("Current SC Locations") with the predicted RTs after optimizing the location of SCs for the year ("Optimal SC Locations by Static MCLM") and repeat allocation of EMS resources over multiple time period in each day ("Optimized EMS Locations by Dynamic MCLM"). We also compared the performances of the two different dynamic

**Table 2** Model comparison between OLS and machine learning

| Model | | Main estimation/learning structure | MSE (s.d) |
|---|---|---|---|
| OLS | | Linear | −23.24 (8.37) |
| Machine learning | MLPs | 5 neural networks | −5.11 (2.30) |
| | LSTM | 3 previous time periods | −3.98 (1.79) |

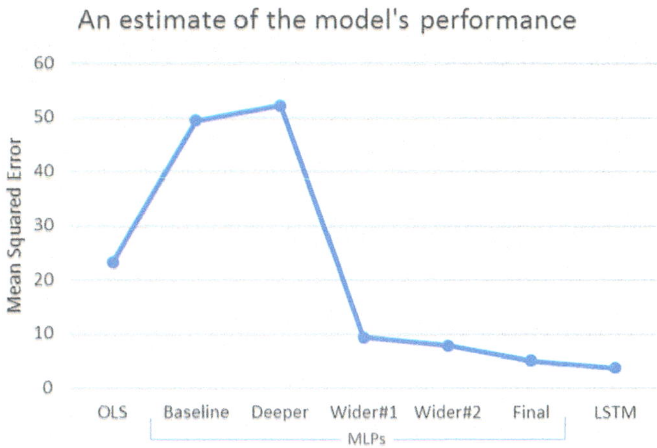


**Fig. 6** Change in model performance during learning process

MCLM, one with a conventional linear programming and the other with a nonlinear programming that incorporates the past records of EMS deployment at the three previous time points ($t - 1$, $t - 2$, $t - 3$) as an additional constraint into the process of determining the optimal location at current time ($t$). We used two target RTs (5 and 10 minutes) and calculated the percentages of emergency calls exceeding the target RT before and after the reallocation of EMS resources.

It is predicted that, at the current SC locations, 41% of the emergency calls will not receive EMS services within 5 minutes, and 16% of those calls will not receive services within 10 minutes. Interestingly, only a slight reduction of these percentages (2% reduction for both 5- and 10-minute cases) is predicted when the locations of the 103 SCs was optimized over the entire study period (i.e., 1 year) without regard to temporal variations throughout the day. However, if temporal patterns of EMS demand are fully incorporated into the model across the four daily time periods, a significant reduction was found for both above-five-minute RTs (14% reduction) and above-10-minute RTs (7% reduction), compared to the current SC allocation. Moreover, an additional 6–9% reduction was observed when the location history at the past time periods was incorporated into optimization process by the "memory-based dynamic MCLM."

This result highlights a potential improvement of RTs in the case of dynamic redistribution of EMS locations. This type of distribution would be made possible by allocating EMS personnel and vehicles (mobile resources) to the optimal locations

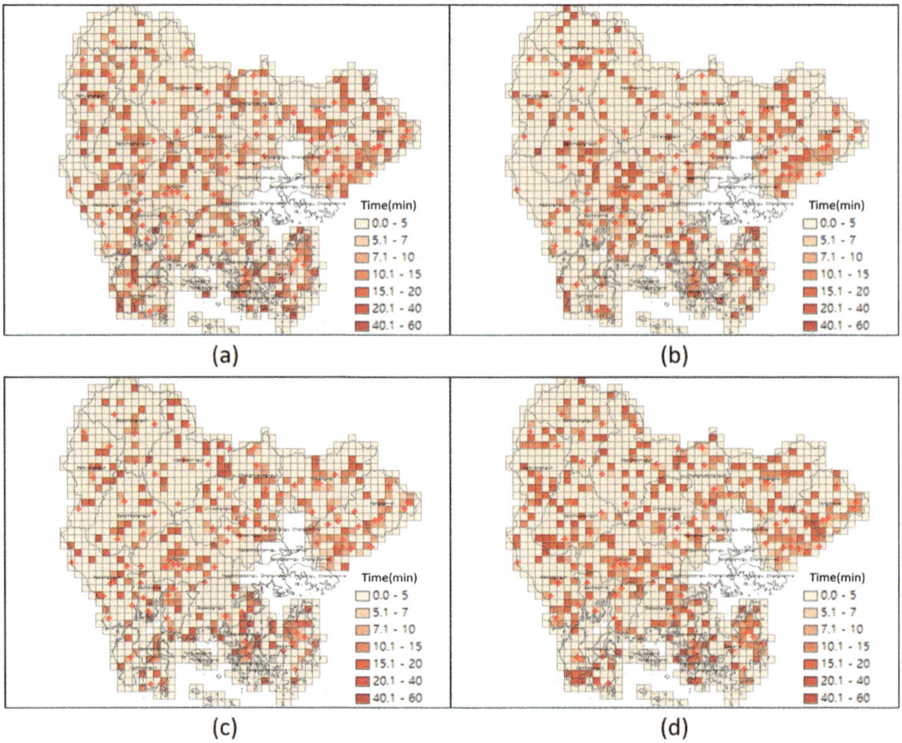


**Fig. 7** Spatial distribution of the predicted RTs at each grid for four scenarios

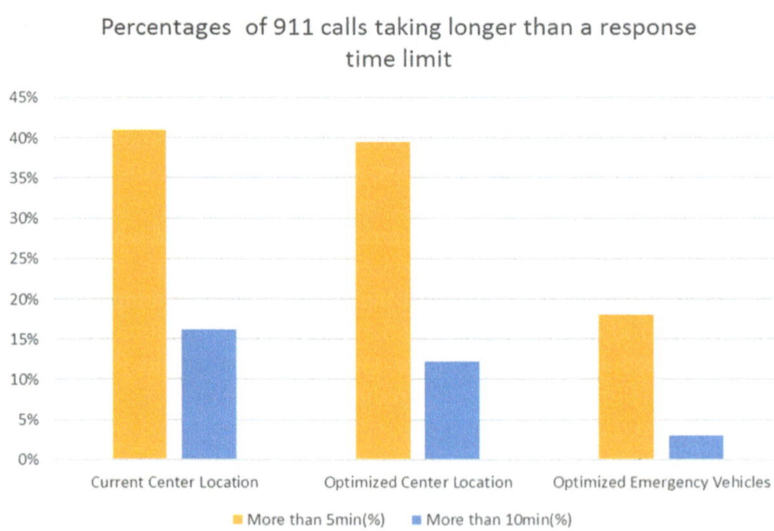


**Fig. 8** Comparison in percentages of emergency calls with RT over the two RT standards: before vs. after reallocation of EMS resources

**Table 3** EMS performance by time period and model: before vs. after reallocation

| | Dispatched from existing SC locations | | Dispatched from optimal location of EMS vehicles/outposts by | | | |
|---|---|---|---|---|---|---|
| | | | Conventional/linear MCLM | | Memory-based/nonlinear MCLM | |
| | Within 5 minutes (%) | Within 10 minutes (%) | Within 5 minutes (%) | Within 10 minutes (%) | Within 5 minutes (%) | Within 10 minutes (%) |
| 6 am–noon | 61.8 | 84.1 | 72.2 | 92.5 | 83.2 | 98.7 |
| Noon–6 pm | 57.1 | 79.9 | 73.6 | 89.9 | 81.7 | 96.4 |
| 6 pm–midnight | 62.6 | 84.2 | 72.3 | 91.3 | 85.4 | 97.3 |

temporarily rather keeping them at permanent locations as static resources. While full-scale center redistribution may seem impractical and ineffective, the dynamic real-time reallocation of EMS resources based on the predicted demand could guarantee a significant improvement in arriving on site within the target RTs.

Table 3 confirms that the optimization-based dynamic allocation of EMS resources improves the chances of EMS arriving on site within 5 or 10 minutes of a call over multiple time periods, without any additional resources. If the same number of EMS resources (fleet or any other mobile outposts) is dispatched from the time-varying optimal locations of EMS outposts suggested by the machine learning method on this specific date (August 24, 2014), over 70% of the emergency calls would have an RT of 5 minutes or less, which is a significant improvement from the dispatch based on the current SC locations (10–16% increase for 5-minute RT and 8–10% increase for 10-minute RT). This method of optimization particularly improves RTs during the afternoon.

Furthermore, we found an additional improvement in EMS coverage within a desired RT when MCLM was solved using a nonlinear programming based on the memory of SC locations during the past time periods. Additional 8–13% of the 911 calls were covered within 5-minute RT, while additional 6–7% of the demands met the 10-minute RT standard. Thus, using machine learning to optimize EMS resources according to spatiotemporal patterns of demand can substantially increase the number of calls that can be reached within the "platinum 5 (or 10)" minute periods.

Figure 9 illustrates the performance of time-varying memory-based dynamic allocation of EMS outposts. Using August 24, 2014, as an example, Fig. 9 shows the optimal locations of emergency outposts (or vehicles) suggested by machine learning for maximizing the number of EMS demands covered by EMS within 5/10 minutes of RT for each of the three time zones (6 am–noon, noon–6 pm, 6 pm–midnight), along with the coverage area of each EMS location by 5 minutes (light gray) and 10 minutes (dark gray). The average RT ranges between 3.2 minutes (6 pm–midnight) and 3.8 minutes (noon–6 pm), when incorporating the temporal changes of the predicted EMS demand in the process of location optimization.

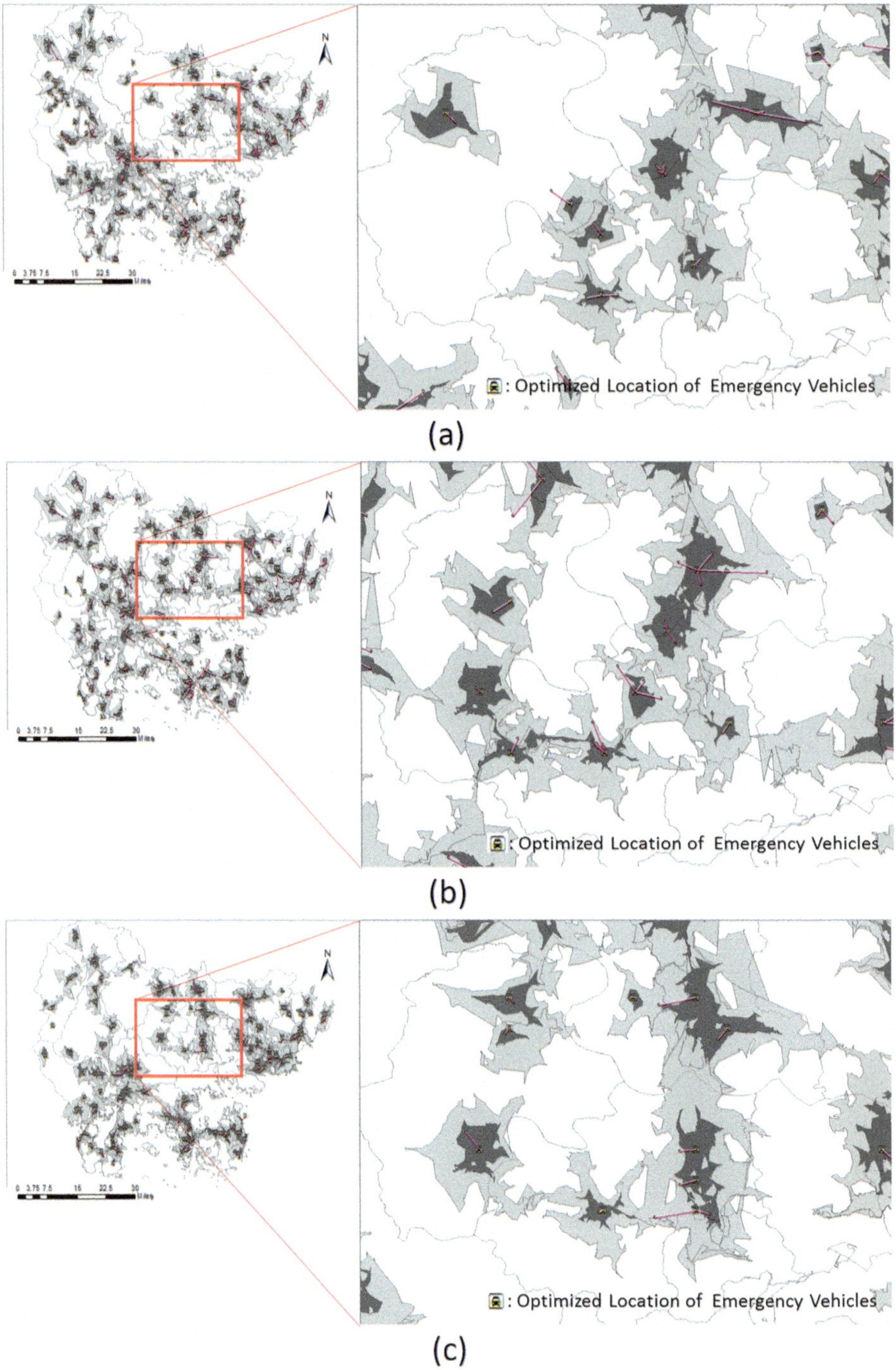


**Fig. 9** Illustration of optimal EMS locations: (**a**) 6 am–noon (3.5 minutes), (**b**) noon–6 pm (3.8 minutes), (**c**) 6 pm–midnight (3.2 minutes)

## 5 Conclusions

Rapid urbanization and population growth in many developed and developing countries have increased the need for emergency medical services and made it highly important to create effective deployment systems for the prompt care of injured patients. Understanding the spatiotemporal patterns of emergency events is key in predicting the future trends of EMS demand and allocating relevant resources according to demand patterns. Despite the growing availability of spatiotemporal data in the field of emergency medicine and rescue, there have been few attempts to maximize the potential of these data to improve evidence-based policymaking tools. GIS has been used to manage and visualize the spatial distribution of demand data. However, in the presence of substantial temporal variations of service demands, conventional hotspot or clustering approach overlooking temporal trends could be inappropriate or misleading. Despite the recent methodological development of spatiotemporal machine learning techniques, there is still a need for more evidence which could strengthen the practicality of the tools in improving RT coverage by allocating EMS based on time-varying predicted demands. Our research confirms that one approach to enhancing response times for emergency dispatches is to use real-time dynamic deployment based on the spatiotemporal demand forecasting done through machine learning.

Machine learning has been used to predict mortality or deterioration of patients due to a specific disease or risk factors. But, to date, there has been no attempt to develop a GIS-based machine learning framework for demand forecasting of emergency medical services. In this research, we used long short-term memory (LSTM) method to forecast prehospital emergency medical demand, based on actual emergency call data in the Gyeongnam Province in Korea, along with the training and validation of the models. The predicted demands were then used to reallocate existing EMS resources. We compared the RTs of the original arrangements with those of the optimized arrangements via multiple MCLMs. We found that when previous spatiotemporal patterns of EMS demands and resources were fully incorporated into the model across the four time periods, there was significant improvement in meeting the policy targets of 5-minute RT (23% reduction in calls with RTs > 5 minutes) and 10-minute RT (13% reduction in calls with RTs > 10 minutes), compared to the current SC allocation. This study provides empirical evidence of the potential benefits of dynamic redistribution of EMS resources, as opposed to permanent reallocation of SCs or the creation of EMS centers/facilities.

We believe that a demand-based dynamic emergency dispatch system has the potential to become more effective when machine-learning-based spatiotemporal demand forecasts provide supportive evidence for allocation decisions, especially when public health resources are limited. Further research should follow in order to ensure successful implementation of this system. First, administrative costs and other practical barriers of reallocating EMS facilities or vehicles need to be fully taken into consideration in the context of South Korea. Second, more accurate road traffic patterns and driving condition data, in addition to weather and historical trends, should be maintained and routinely inputted into the real-time machine learning algorithms to fine-tune the accuracy and robustness of the learning and prediction processes. Lastly, machine learning researchers and developers should maintain open channels of communication with policymakers and attempt to make highly technical computer-based tools more accessible and user-friendly.

## References


Abadi, M., Barham, P., Chen, J., Chen, Z., & Davis, A. (2016). Tensorflow: A system for large-scale machine learning. OSDI, Savannah, GA, USENIX.

Bassil, K., Cole, D. C., Moineddin, R., Craig, A. M., Lou, W. Y., Schwartz, B., & Rea, E. (2009). Temporal and spatial variation of heat-related illness using 911 medical dispatch data. *Environmental Research, 109*(5), 600–606.

Blackwell, T. H., & Kaufman, J. S. (2008). Response time effectiveness: Comparison of response time and survival in an urban emergency medical services system. *Academic Emergency Medicine, 9*(4), 288–295.

Chen, A., & Lu T. (2014). A GIS-based demand forecast using machine learning for emergency medical services. 2014 international conference on computing in civil and building engineering. Orlando, FL, USA.

Chen, A. Y., Lu, T., Ma, M. H., & Sun, W. (2016). Demand forecast using data analytics for the preallocation of ambulances. *IEEE Journal of Biomedical and Health Informatics, 20*(4), 1178–1187.

Cho, J., You, M., & Yoon, Y. (2017). Characterizing the influence of transportation infrastructure on emergency medical services (EMS) in urban area—A case study of Seoul, South Korea. *PLoS One, 12*(8), e0183241.

Church, R., & ReVelle, C. (1974). The maximal covering location problem. *Papers of the Regional Science Association, 32*, 101–118.

Dean, S. F. (2008). Why the closest ambulance cannot be dispatched in an urban emergency medical services system. *Prehospital and Disaster Medicine, 23*(2), 161–165.

Getis, A., & Ord, J. K. (1992). The analysis of spatial association by use of distance statistics. *Geographical Analysis, 24*, 188–205.

Haghani, A., Hu, H., & Tian, Q. (2003). *An optimization model for real-time emergency vehicle dispatching and routing*. Washington, DC: Transportation Research Board.

Hochreiter, S., & Schmidhuber, J. (1997). Long short-term memory. *Neural Computation, 9*(8), 1735–1780.

Hong, K. H., Lee, K. J., Kim, J. T., & Lee, D. H. (2008). Severity-based analysis of prehospital transportation time using the geographic information system (GIS). *Journal of the Korean Society of Emergency Medicine, 19*(2), 153–160.

Kim, D., Sarker, M., & Vyas, P. (2016). Role of spatial tools in public health policymaking of Bangladesh: Opportunities and challenges. *Journal of Health, Population and Nutrition, 35*(8), 1–5.

Kingma, D., & Jimmy B. (2014). Adam: A method for stochastic optimization. *arXiv Preprint arXiv* 1412(6980).

Kohavi, R. (1995). A study of cross-validation and bootstrap for accuracy estimation and model selection. *International Joint Conference on Artificial Intelligence*. Montreal, QB, Canada.

Krizhevsky, A., Sutskever, I., & Hinton, G. (2012). Imagenet classification with deep convolutional neural networks. In Advances in neural information processing systems 25 (NIPS Proceedings 2012), pp. 1097–1105.

Levi, K., Kharkar, R., Kiang, M., & Hartmann, C. (2017). Using machine learning to improve emergency medical dispatch decisions. 23rd ACM SIGKDD conference on knowledge discovery and data mining. Halifax, NS, Canada.

Li, X., Zhao, Z., Zhu, X., & Wyatt, T. (2011). Covering models and optimization techniques for emergency response facility location and planning: A review. *Mathematical Methods of Operations Research, 74*(3), 281–310.

Obermeyer, Z., & Emanuel, E. J. (2016). Predicting the future—Big data, machine learning, and clinical medicine. *New England Journal of Medicine, 375*(13), 1216–1219.

Ong, M. E., Ng, F. S., Overton, J., Yap, S., Anderson, D., Yong, D. K., Lim, S. H., & Anantharaman, V. (2009). Geographic-time distribution of ambulance calls in Singapore: Utility of geographic information system in ambulance deployment (CARE 3). *Annals Academy of Medicine, 38*(3), 184–191.

Orbach, J. (1962). Principles of neurodynamics. Perceptrons and the theory of brain mechanisms. *Archives of General Psychiatry, 7*(3), 218–219.

Peleg, K., & Pliskin, J. S. (2004). A geographic information system simulation model of EMS: Reducing ambulance response time. *The American Journal of Emergency Medicine, 22*(3), 164–170.

Pell, J. P., Sirel, J. M., Marsden, A. K., Ford, I., & Cobbe, S. M. (2001). Effect of reducing ambulance response times on deaths from out of hospital cardiac arrest: Cohort study. *BMJ, 322*, 1385–1388.

Peters, J., & Hall, G. B. (1999). Assessment of ambulance response performance using a geographic information system. *Social Science & Medicine, 49*(11), 1551–1556.

Pons, P. T., Haukoos, J. S., Bludworth, W., Cribley, T., Pons, K. A., & Markovchick, V. J. (2005). Paramedic response time: Does it affect patient survival? *Academic Emergency Medicine, 12*(7), 594–598.

Revelle, C., Bigman, D., Schilling, D., Cohon, J., & Church, R. (1977). Facility location: A review of context-free and EMS models. *Health Services Research, 12*(2), 129–146.

Rogers, F. B., Rittenhouse, K., & Gross, B. W. (2015). The golden hour in trauma: Dogma or medical folklore? *Injury, 46*, 525–527.

Roudsari, B. S., Nathens, A. B., Arreola-Risa, C., Cameron, P., Civil, I., Grigoriou, G., Gruen, R. L., Koepsell, T. D., Lecky, F. E., Lefering, R. L., Liberman, M., Mock, C. N., Oestern, H. J., Petridou, E., Schildhauer, T. A., Waydhas, C., Zargar, M., & Rivara, F. P. (2007). Emergency medical service (EMS) systems in developed and developing countries. *Injury, 38*(9), 1001–1013.

Savas, E. S. (1969). Simulation and cost-effectiveness analysis of New York's emergency ambulance service. *Management Science, 15*(12), 608–627.

Washington D.C. Fire and EMS Department. (2018). EMS response time. Retrieved 20 May 2018, from https://fems.dc.gov/page/ems-response-time.

Zarandi, M., Davari, S., & Sisakht, S. (2013). The large-scale dynamic maximal covering location problem. *Mathematical and Computer Modeling, 57*, 710–719.

Zhou, Z. (2016). Predicting ambulance demand: Challenges and methods. 2016 ICML workshop. New York, NY.

Zhou, Z., Matteson, D. S., Woodard, D. B., Henderson, S. G., & Micheas, A. C. (2013). A spatio-temporal point process model for ambulance demand. *Journal of the American Statistical Association, 110*(509), 6–15.

**Sunghwan Cho** is a researcher at the Fusion & Convergence Division in the Korea Land and Geospatial Information Institute. His research focus on spatial big data analysis for crime and transportation applications.

**Dohyeong Kim, Ph.D.** is an Associate Professor of Public Policy and Geospatial Information Sciences at the University of Texas at Dallas. He received a Ph.D. on spatial health planning from the University of North Carolina at Chapel Hill and postdoctoral training at Duke University. His research efforts have been dedicated to develop statistical, economic, geospatial, and decision-analytic approaches to address a variety of health, environmental, and safety concerns both in the USA and internationally.

# Part III
# Healthy Behavior and Urban Lifestyle

# Incorporating Online Survey and Social Media Data into a GIS Analysis for Measuring Walkability

**Xuan Zhang and Lan Mu**

**Abstract** Existing walkability measurements have not considered some important components of the built environment, pedestrians' preferences, or all walking purposes. As area-based measurements, they may overlook some detailed walkability changes. We propose a **P**erceived importance and **O**bjective measure of **W**alkability in the built **E**nvironment **R**ating (POWER) method, which is a line-based approach considering both the perception of pedestrians and subjective characterizing of the urban built environment. Incorporating online survey and social media data, we present a built environment walkability study in a specific environment and the potential for more general scenarios. The survey can be customized for the particular urban environment and capture the preferences of a local population. The social media obtain general opinions from a broader audience. Although focusing on the specific setting at a university campus, we also included the general social media results to supplement the POWER structure and survey findings. Using social media and survey results can bring two scales together to provide a more complete understanding of walkability.

## 1 Introduction

The urban environment, along with factors such as dietary patterns and genetics, has a great impact on physical activities and potential to benefit the community at large. Spurred by health awareness and the popularity of the sedentary lifestyle, researchers in many fields share increasing interests in the effects of the urban built environment on physical activities, especially walking.

Walkability is an essential measurement of how the urban built environment supports walking. Existing walkability studies are often coupled with health concerns such as sedentary lifestyle and obesity. The concept of obesogenic environment

X. Zhang · L. Mu (✉)
Department of Geography, University of Georgia, Athens, GA, USA
e-mail: xuan.zhang@uga.edu; mulan@uga.edu



Y. Lu, E. Delmelle (eds.), *Geospatial Technologies for Urban Health*,
Global Perspectives on Health Geography, https://doi.org/10.1007/978-3-030-19573-1_8

explains why and how certain neighborhood environment discourages people from physical activities and leads to obesity. Previous studies have shown that an obesogenic, or walking-unfriendly, built environment negatively influences people's travel behaviors and lead to some health issues (Powell et al. 2010). A better understanding of the impact of the built environment on walkability can help make possible changes, shape people's behavior, promote a healthier lifestyle, and improve population health in the long run.

We introduce a hybrid objective-subjective walkability measurement: The **P**erceived importance and **O**bjective measure of **W**alkability in the built **E**nvironment **R**ating (POWER). The POWER considers both the perception of pedestrians and subjective characterizing of the urban built environment. In order to understand and evaluate the built environment walkability from multiple scales, we investigated two scenarios on different scales using data from both an online survey and social media. On a local scale, the online survey can be a platform to gather pedestrians' walking considerations and concerns in a particular urban environment, such as a university campus, and it can further quantify walkability using the POWER. On a larger geographical scale, such as a region or a nation, social media is a quick and accessible data source to obtain opinions from more general settings and potentially qualify the factors influencing walkability. Instead of being directly applied in the POWER calculation, social media data identify other factors that people expect for walkability, and further help shape the structure of the POWER in terms of what factors of the built environment should be considered. In local and general scopes, survey and social media results complement each other. With the geospatial techniques to analyze and visualize the survey results and social media data of walkability, this study can offer new perspectives for understanding walkability and urban health issues.

## 2 Literature Review

### *2.1 Physical Activities, Obesogenic Environment, and Walkability*

Beyond benefiting people's health by preventing obesity and other diseases among all ages, physical activities such as walking, can also reduce traffic congestion, energy consumption, air pollution, as well as economically benefiting the local business and real estate markets (Warburton et al. 2006; Loo and Lam 2012; Slater et al. 2013; Duncan et al. 2014; Litman 2018). Compared with other transportation methods, physical activities are more affordable for the economically or socially disadvantaged (Litman 2018). However, some aspects of the built environment can discourage people from being physically active. For example, an obesogenic environment describes "an environment that promotes gaining weight" (Swinburn et al. 1999; Powell et al. 2010). Automobile-oriented planning, the dominant planning strategy in the U.S., is one of the causing factors of an obesogenic environment. The automobile-oriented strategy has been criticized for emphasizing the needs of automobiles while neglecting the demands of pedestrians, bikers, and bus riders (Jackson and Kochtitzky 2001; Forsyth and Southworth 2008). Litman (2014) has emphasized this issue by asking whether the land should be used for people or vehicles. Failing to provide efficient and sufficient walking support, the walking-unfriendly environment can influence people's daily travel-mode choices (Duncan et al. 2013), resulting in automobile dependency and related health issues (Rundle et al. 2009; Slater et al. 2013).

Scholars have scrutinized ways of understanding walkability and constructing walkability measurements (Loo and Lam 2012; Vargo et al. 2012; Yin 2017). Walkability can be conceptualized as the overall support that pedestrians can receive for walking in a certain area (Hung et al. 2010). It also describes the extent to which characteristics of the built environment are conducive to different walking purposes, such as leisure, exercise, recreation, accessing services, or traveling to work (Leslie et al. 2007).

There are two types of commonly used walkability measurements. The first is the Walkability Index, which comprises intersection density, net residential density, retail floor-area ratios, and entropy scores (Frank et al. 2010). Using Walkability Index, a place is walkable if it has high-density intersections, high residential density, diverse land-use types, and less surface parking (Dobesova and Krivka 2012; Jun and Hur 2015). Using the Walkability Index, Slater et al. (2013) found that living in more walkable communities, American adolescents have less chance to be overweight or obese. After developing a walkability index calculation tool, Dobesova and Krivka (2012) applied the tool to a European city and concluded that the Walkability Index might have a lot of applications and potential in urban design and planning. The U.S. Environmental Protection Agency (EPA) followed the same idea, and calculated the National Walkability Index by changing the factors to intersection density, occupied housing, mix of employment types, and predicted commute mode split (Walkability Index 2017). The Walkability Index does an excellent job capturing the general environmental walkability with easily accessible land-use data, and it can be used to compare different locations due to the standardized calculation. However, this method does not consider people's preference and perception of walkability or detailed built environment design (Gu et al. 2018). Jun and Hur (2015) advocate a reconfiguration of the dimensions of the Walkability Index to include other factors and perceived walkability.

The second type is a group of similar measures, in which a place is more walkable if more amenities are available within a certain distance or travel time. The most widely-used one is Walk Score, a public website estimating walkability to nearby amenities (Carr et al. 2010). Although Walk Score was initially for real estate purposes, several groups of researchers have validated the methodology behind Walk Score using very similar methods in populous areas, such as Boston and U.S. metropolitan areas (Carr et al. 2010, 2011, Duncan et al. 2011, 2013). Hirsch et al. (2014) illustrated that a location with a higher Walk Score was associated with more walking, as a transportation choice, and reduction in body mass index (BMI). Hall and Ram (2018) used Walk Score in tourism research and found only a weak relationship between Walk Score with the number of visitors and TripAdvisor ratings for top London attractions. The popularity of Walk Score definitely has facilitated walkability-related research with its ready-made dataset. Nevertheless, this group of walkability measures is more like a proxy for amenity availability, which examines the distribution of possible destinations while overlooking other perspectives of the built environment and various walking purposes (Forsyth and Southworth 2008; Duncan et al. 2013; Fan et al. 2014; Gu et al. 2018).

The built environment consists of three components: land-use patterns, the transportation system, and urban design (Handy et al. 2002; Saelens and Handy 2008). However, existing walkability measurements usually lack the urban design part, and some critical transportation system elements, such as sidewalks and bike paths, which can provide a better and safer walking experience. Additionally, walkability should consider not only the objective condition of the built environment components, but also how pedestrians value some parts of the built environment differently (Park 2008; Jun and Hur 2015). Various groups of people at diverse locations may consider walkability differently, and some aspects of the built environment may be weighted dissimilarly for commercial and residential areas. If the local pedestrians conceive certain factors of the built environment are more important, these factors should be weighted more in the walkability calculation. Furthermore, compared with area-based measures, which are used in both Walkability Index and Walk Score, the line-based method (Park 2008), using road features for example, is more intuitive for representing walking activities. It can capture more detailed variation with small road segments and avoid the risk of ecological fallacy, i.e., applying a homogeneous value of an area to all locations within that area (Robinson 1950; Selvin 1958).

Existing walkability measurements have rarely been conducted from both the viewpoint of pedestrians and the condition of the built environment. Moreover, most walkability studies concentrate on walking as a mode of transportation with predetermined destinations, and leave out other purposes such as workouts, social choices, entertainment, or aimless activity (Lo 2009). In addition, walkability may be better represented as high-resolution line-based features to notice the nuanced difference. It is necessary to construct walkability with all of the perspectives beyond those commonly used.

## 2.2 Surveys for Walkability Studies

Surveys have been widely used in social science research, especially human research, to collect quantitative and qualitative data. By asking questions targeting a particular group of people, a survey can obtain their opinions and feelings (Anderson et al. 2010). Surveys are designed for various purposes, such as public safety and public health, and they can be specific as well as general based on the goals (Kilpatrick et al. 1985; Gravel and Béland 2005; Anderson et al. 2010).

Researchers have applied surveys to physical activity-related studies. For example, a web-based survey was used to study the relationship between social travel and other factors (e Silva et al. 2017). Travel behavior surveys have also been used with GIS technologies to understand how neighborhood design influences travel behaviors (Crane and Crepeau 1998). Similarly, walkability surveys have been used to assess certain locations or their environmental attributes (Livi and Clifton 2004; Gota et al. 2010). Using surveys is a straightforward and intuitive method for obtaining first-hand data and assessing the built environment from the pedestrians' perspective.

## 2.3 Social Media for Walkability Studies

Social media, as an accessible real-time data source, can be used to acquire, react to, communicate, and participate in what is happening. Sui and Goodchild (2003, 2011) believe the media theories cast light on the social applications of GIS, and they agree with McLuhan's law of the media, that modern media are modifiable perceptive extensions of human thought (McLuhan 1975). Social media provides user-generated data about daily activities or feelings (Brooker et al. 2016), generates valuable information regarding people's behaviors and health outcomes, and even predicts infectious disease outbreaks (Liu and Young 2018). There is big potential for understanding phenomena through social media data mining and analysis (Felt 2016).

Recently, social media has also been used in research about human mobility and physical activities. Hasan et al. (2013) used location-based social media data to understand urban human activity and mobility patterns for different activity categories, and further designed the purpose-specific activity distribution maps. Individual geotaged Twitter data were used to study human mobility patterns in Australia (Jurdak et al. 2015). Shen and Karimi (2016) state that social media can enrich the current description and understanding of urban network systems and network accessibility. For walkability in particular, researchers have used social media data, such as pictures, to automatically identify safe and walkable streets (Quercia et al. 2015). Berzi et al. (2017) used Foursquare and Flickr for a bottom-up assessment of the walkability in Milano.

Among multiple social media platforms including Facebook, WhatsApp, Foursquare, Tumblr, and more, Twitter is widely used for academia because of its popularity and free access. Twitter users share their opinions through tweets, whose word limit increased from 140 to 280 characters in late 2017. According to the earnings report of Twitter, it has 336 million monthly active users worldwide, and the United States has 69 million users as of the beginning of 2018 (Walker 2018). There are about 340 million tweets per day. The big number of users and tweets can provide ingredients for us to gather insights about people's attitudes on certain topics. For example, researchers have used tweet content to analyze political deliberation via Twitter users' attitudes (Tumasjan et al. 2010; Larsson and Moe 2011; Diehl 2017), and unearthed sentiments, opinions, and even detected some mental disorders (Pak and Paroubek 2010; Kouloumpis et al. 2011; Yang and Mu 2015; Yang et al. 2015).

Although social media data have been used in walkability research, it is still new to use social data to qualify people's considerations in assessing walkability. We used Twitter data and data mining to expose the walking considerations when people comment about walkability or walkable places, and further to structure the POWER main factors.

## 3 Data and Methods

To conduct the walkability research, we designed and customized the POWER, online survey, and social media methods. Using the Customized Analytic Hierarchy Process (CAHP), the online survey data was applied in the POWER method to quantify walkability for our study area. Social media results, although not used directly in the POWER presented in this study, provided a broader idea of walkability factors from the general settings.

### *3.1 The POWER*

We propose the POWER method, which conceptualizes walkability with perceived importance and objective measurement. Building upon existing literature, we include sidewalk availability, bike lane availability, traffic speed, and more, as the built environment factors in the POWER structure. As Fig. 1 shows, there are four categories in the POWER structure: (1) sidewalk condition (availability, width, and slope); (2) connectivity (route density, pedestrian-crossing connectivity, and bus-stop connectivity); (3) amenities (variability and density); and (4) walking environment (traffic speed, buffer, and share situation). Under these four categories, other potential factors, such as road width, sidewalk evenness, presence of trees, and more, can be added to the structure based on the localized settings in order to be better applied in different scenarios.

This structure focuses on all three parts of the built environment: land-use patterns are reflected in the amenities; the transportation aspect is represented by connectivity; and the urban design part is sampled by the sidewalk condition and walking environment. In our study, ten factors (check marked in Fig. 1) were considered after assessing the local built environment condition, and were included in both the perceived importance and objective measures. Under amenities, the variability and density category includes different types of amenities to show nuanced differences between various destinations.

The perceived importance ($p_j$) indicates the preference for the built environment factor *j*. Using survey results (discussed later in this chapter), $p_j$ is calculated with the CAHP (Zhang and Mu 2019; Zhang 2016). CAHP is a modified version of the classic Analytic Hierarchy Process (AHP), a structured technique for analyzing complex decisions by pairwise comparisons (Saaty 1980, 1987). A paired comparison, say factors A and B, uses a scale from 1 to 9, from equally important to

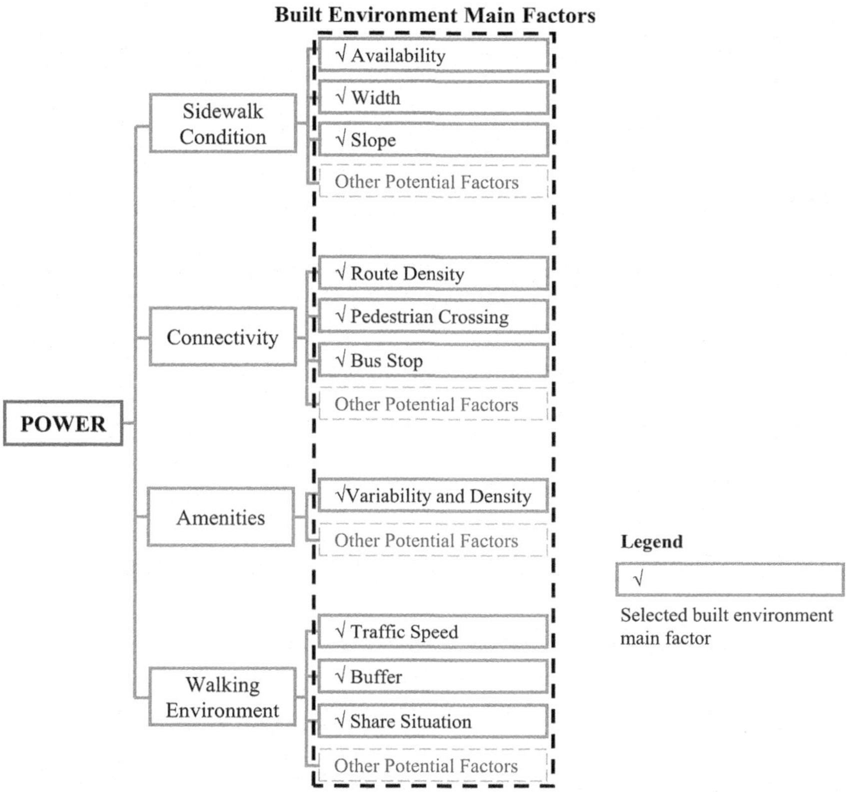


**Fig. 1** The Structure and Built Environment Main Factors of the POWER

extremely more important, to indicate the relative importance of A over B, and uses the reciprocal value for B over A. For n factors, the AHP uses a set of paired comparisons to form the $n * n$ relative importance matrix (Saaty 2008). Then relative importance in the matrix is standardized by dividing the summed relative importance of the column factor. Finally, $p_j$ is calculated by averaging the standardized relative importance of each row factor, and $\sum_{n}^{j-1} p_j = 1$ (Saaty 2004). For n factors, the classic AHP asks $C(n, 2) = \frac{n!}{2(n-2)!}$ pairwise comparison questions which fall into two categories: (1) how much more important is A than B, for example, how much more important is breakfast than snacks, and (2) how many times is A more than B, for example, how many times more he/she drinks coffee than milk. Compared with the AHP, the CAHP only asks $n$ questions directly on n factors; the answers, expressed as values, are used in pairwise comparisons to calculate the perceived importance as in AHP. For example, on the importance scale of 1 to 9, factor A receives 3 from the direct question, while B receives 5. Then the relative importance of A over B is 3/5, and B over A is 5/3. The calculated relative importance can be used in original AHP to form the relative importance matrix.

For individual road segments, the objective measures ($o_{ij}$) are determined based on the physical condition of the road segment $i$ and the built environment factor $j$, such as sidewalk availability. For example, ($o_{ij}$) could be 0, 0.5, or 1 for sidewalk availability: unavailable, available on one side, or on both sides. Details of the schemes for objective measures can be found in related references (Zhang and Mu 2019; Zhang 2016). To obtain the objective condition of the factors, we used various GIS data, including buildings, parking, roads, elevation, and more, provided by the Office of University Architects, the University of Georgia. Auxiliary data, such as speed limits, were collected or updated, and validated via fieldwork.

Both $p_j$ and $o_{ij}$ range between 0 and 1. We calculated the POWER for each 14-meter road segment, which is a 10-second walking distance at a preferred pace (Browning et al. 2006). POWER$_i$, the POWER value of road segment $i$, is the sum product of $p_j$ and $o_{ij}$ times 100 (Eq. 1).

$$\mathrm{POWER}_i = \sum_j \left(p_j \times o_{ij}\right) \times 100 \tag{1}$$

With the range from 0 to 100, the POWER reflects the walkable level which describes how the built environment can support walking activities considering pedestrians' preferences.

## 3.2 Conducting a Specific Online Survey on a University Campus

We chose a university campus as the study area to survey affiliated people, mostly students, about their walking preferences. Universities and colleges are a vital sector of society. In 2016, there are approximately 17 million undergraduates enrolled in the U.S. (National Center for Education Statistics 2018). It was reported that about 40%–50% of college students were physically inactive (Keating et al. 2005). Another study surveyed over 700 college students and most of them did not meet the physical activity guideline (Huang et al. 2003). A typical campus size ranges from a few hundred to several thousand acres, and many U.S. universities locate in urban areas or campus towns. In both scenarios, the university campus is usually urbanized. Although some universities, such as Princeton University, have the idea of a walkable campus in their campus plan (Princeton University 2008), more efforts are needed to understand how the campus, a specific location serving specific groups of people, can be walkable.

Our study area is the main campus of the University of Georgia, which is about 70 miles east of Atlanta, Georgia. The main campus sits in an urban environment with around 200 city blocks and over 300 buildings. It has good vegetation coverage, a compact design, and attractive scenery with mild weather. These factors make it suitable for walking all year round. Students and other affiliated people go across campus for classes, talks, and other activities, and many of them live on campus or nearby for easier commutes.

To collect pedestrian walking preferences and personal information, we designed an online survey asking questions regarding built environment main factors (Fig. 1) and more. For example, we asked, "I would choose a route with a sidewalk over a route without a sidewalk" regarding the sidewalk availability factor, and "I prefer walking on a sidewalk with a buffer zone (grass, trees or parking) from the road" for the buffer factor. The questions used a Likert scale ranging from 1 (strongly disagree) to 9 (strongly agree) to capture the slight differences of opinions, because the values can be applied easily in multiscale analysis methods, such as the CAHP. For each main factor, the median agreement value of all survey responses represented that factor's importance level. Following the CAHP method, a pairwise comparison was made between every two factors to calculate relative importance, and then to calculate the perceived importance. For this specific campus setting, the survey covers the nine groups of typical amenities: (1) food services/coffee shops/restaurants/bars, (2) multi-functional centers, (3) athletic fields/recreational facilities, (4) teaching/lab buildings, (5) administration buildings, (6) green space, (7) libraries/bookstores, (8) residence halls/apartments, and (9) parking spaces on and near campus. For each amenity group, participants also selected their preferences from 1 to 9 (extremely important to extremely unimportant), and the perceived importance for the amenity groups can be calculated in the same way as the main factors. Each amenity group represents one aspect of the amenities variability and density main factor in the POWER.

In addition, the survey collected the participants' gender, year of birth, and occupation. It also investigated other walking considerations, motivations, and the most walkable or unwalkable places. More survey questions can be found in the related literature (Zhang and Mu 2019). After the sample interview and revision of the questions, we distributed the walking preference survey to all potential sources for the participant recruitment. Project recruitment flyers were posted on noticeboards and at bus stops. Additional emails were sent out to mailing lists. This study met the Human Subjects requirements and received approval from the University Institutional Review Board (IRB).

In the POWER, the quantitative survey result was processed into perceived importance in order to integrate pedestrians' preferences toward the built environment design. For the open questions, we summarized the text answers to understand other perspectives which were also important from the view of pedestrians.

To visualize the benefits of walking and to better communicate the results to the public (Trumbo 2000), we designed calorie maps to estimate the energy burned when walking from different campus landmarks to possible destinations via the shortest walking path. Based on the average male weight (164 lbs.) of college students and average slope on campus (5 degrees), we calculated the calories burned by walking (Fellrnr.com 2017), depicting them as fictitious coke cans (100 calories per can). The same method was also used to create a calorie matrix between popular campus destinations.

### 3.3 Collecting Social Media Data in General

For a general setting, social media data were collected to understand the context when people talk about walkability or walkable places. Using the related keywords walkable and walkability, we can understand which aspects are associated with walkability. We used R programming (rtweet package) (Kearney 2018) and the Twitter Streaming API to collect real-time tweets for two weeks, without specifying a geographical extent. English is the most-used language on Twitter (Statista 2013), and the U.S. is the country with the most users (49.35 million) while the U.K. is the third at 13.7 million. Moreover, the U.S. is the country with the most active Twitter users (Statista 2018). For this sake, we did the analysis based on English tweets. The data was processed to filter out the website hyperlinks, punctuations, the retweet mark, and more. We used the natural-language processing techniques and R packages, including tm, textstem, stringr, and qdapRegex, to lemmatize the tweet words in order to reduce inflectional forms of a word to a common base form, which is useful for late tasks including frequency and co-occurrence analysis (Feinerer et al. 2008; R Development Core Team 2008; Rinker 2017, 2018; Kearney 2018; Wickham 2018; Wikipedia Contributors 2018). For example, lemmatization transfers "am," "are," and "is" into "be," and "car," "cars," "car's," and "cars" into "car." The process transformed the original tweets into the vocabulary rather than the changeable forms (Manning et al. 2008). To understand the factors that people tweet about walkability-related topics, we calculated the word frequency and co-occurrence, to gain the insights out of context. The co-occurrence is the frequency of two words appear in the same tweet. As Matsuo and Ishizuka (2004) state, the co-occurrence is likely to have an important meaning if some words appear selectively with each other.

To obtain a comprehensive understanding of walkability considerations on different scales, we incorporated the online survey into the POWER method and used social media data to extend the research for future use. The online survey was for a location and audience-specific setting, under the recommendation that specific behaviors, in this case walking, should be studied in particular environments (Saelens and Handy 2008). The social media data was applied for a more general context to understand people's walking concerns which can extend the picture of walkability and the structure of the POWER measurement. The combination of working in specific and general settings via survey and social media data complements the study of walkability using the POWER.

## 4 Results

### 4.1 Results of the Location and Audience-Specific Survey

In total, 413 people (undergraduates, graduate students, faculty members, and staff) had participated within five months (from November 20, 2015, to April 20, 2016). The following results were derived from the data collected during the first 4 months

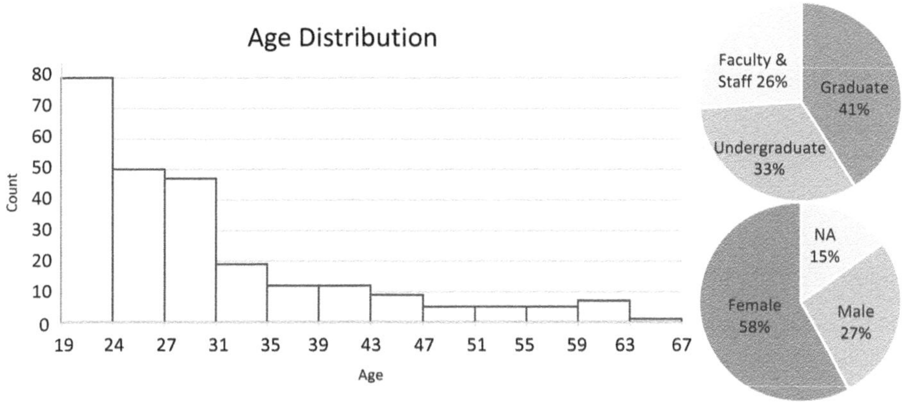


**Fig. 2** The Age Distribution, Group, and Gender of Survey Participants

from 307 participants with the fifth-month data for validation. Out of the initial responses, 23 were incomplete and excluded from further analysis. Mirroring the campus population, the remaining 284 valid survey responses (92.51%) covered an age range from 19 to 65, with more young participants (48% participants were younger than 27) (Fig. 2). Undergraduate (33%), graduate (41%), and faculty/staff (26%) groups all had pretty good representation, and more self-reported female (58%) participated in the survey.

Zooming into the survey result, we processed it with the CAHP and calculated the perceived importance of individual built environment factors. Sidewalk availability and width received the highest perceived importance, and the bus stop connectivity was the lowest. For various amenities, people have the most preferences to green space, then food amenities and book amenities, and all the others have relatively lower perceived importance. With all perceived importance and objective measurements, we calculated the POWER based on Eq. 1. The campus POWER ranges from 25.3 to 88.94 on a scale from 0 to 100 (Fig. 3A). We also included the EPA Walkability Index map (Fig. 3B) to see the difference from the POWER result. In the EPA Walkability Index, large parts of the campus are below the average walkable (in yellow) or least walkable (in orange). Compared to that, the POWER result has more variations at a finer scale when the area is in the same categories of Walkability Index. Specific locations with low POWER value stand out by using a two-color tone cartographic rendering. The POWER map can provide planners a visual reference about specific locations which needs to be improved. For example, the east and west side of the campus, as well as some boundary parts, are in orange or red, indicating low POWER scores.

The POWER result was validated with the fifth-month survey data using the information from the question asking about the most unwalkable places. Participants mentioned particular areas, roads, and buildings/intersections. The result matched up with the POWER calculation. The 11 roads mentioned more than twice received relatively low POWER scores in Fig. 3A, and most of the roads were at the edge of the campus. The people' choice of the most unwalkable road (Fig. 3A, the road

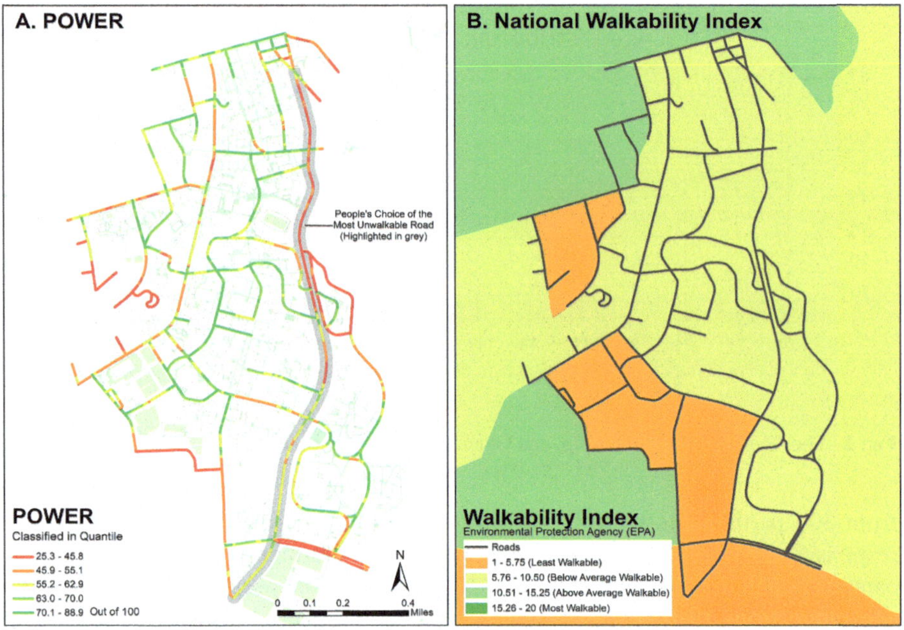


**Fig. 3** The POWER Map and National Walkability Index for the Study Area

shaded with gray background) only received POWER score of 49.36 on average, and it has 80% road segments scoring less than 60.

Although the survey included most amenity types surrounding the campus, there may be other amenities that people consider. Intended to cover more situations for future research, the survey asked a question about other important amenities that may influence walking decisions. The results (Fig. 4, left), regenerated by WordArt.com (WordArt.com 2016) based on the same data (Zhang and Mu 2019), are interesting and consistent with the literature (Jun and Hur 2015). People consider bathrooms and water fountains as most important (other) amenities when making walking decisions. Parks and gardens, which were mentioned a lot, have been considered as green space under amenities in the POWER structure and calculation.

We also asked about the factors that make people avoid walking on campus. The word cloud (Fig. 4, right) highlights factors such as dark, crowds, traffic, and the lack of safety. It offers constructive ideas for campus administrators and planners to promote walking and provide better walking experiences. Safety and darkness issues have been mentioned in the previous research (Foster et al. 2014; Hall and Ram 2018), and they can be future main factors included in the POWER under the category of walking environment. Participants mentioned pedestrian crossing, sidewalk condition, parking lot, bus stop, sidewalk width, and many others, which have all been included as the built environment factors in the POWER method.

To summarize, the survey revealed that participants prefer to walk when a sidewalk is available, sidewalk width is sufficient, a buffer is available from the traffic lanes, and more. It also shows that people would avoid walking if it is crowded, dark,

**Fig. 4** Other Amenities that Influence Walking Decisions and Factors Discourage People for Walking

or unsafe. Pedestrians also consider other amenities such as water fountain, bathroom, and shade while walking. We calculated and visualized the POWER, a line-based measurement which captures subtle variations of walkability better than the widely used area-based methods. Using survey results in the POWER, this study helps us to better understand pedestrians' perceived importance upon the built environment for a specific location and quantify walkability.

The calorie map (Fig. 5) illustrates the benefits that people can gain from walking, beyond exercising in long time slots at the gym. Using the same method, we also created a calorie matrix including 11 popular places, such as landmarks, classroom buildings, undergraduate dorms, graduate housing, libraries, and other destinations (Fig. 6). We designed the calorie map and matrix to promote walking, raise people's awareness of health, and broaden the impact of this study. These are ballpark estimates using average male weight and slope, and other variations can be generated for different target groups, such as female students, older adults, children, and more. For other places, such as large areas with dense points of interests, the idea of encouraging walking can be achieved by using an interactive map to show the calories with a user-defined route and with local food or drink.

The survey result showed different walking preferences among varied groups of survey participants. Toward the preference on bus stop connectivity, the difference was statistically significant between faculty-and-staff group and the undergraduate students ($p$-value = 0.01), or the graduates ($p$-values = 0.09). However, it was not significant between the two student groups. This result may be explained that faculty and staff have higher priority to purchase parking permits, and they generally commute by car. In this case, the students, who often use public transportation coupled with biking or walking, care more about bus stops. Meanwhile, the undergraduates have relatively tighter schedule than the graduates regarding on-campus classes and activities, so they may have slightly more preference on more available bus stops.

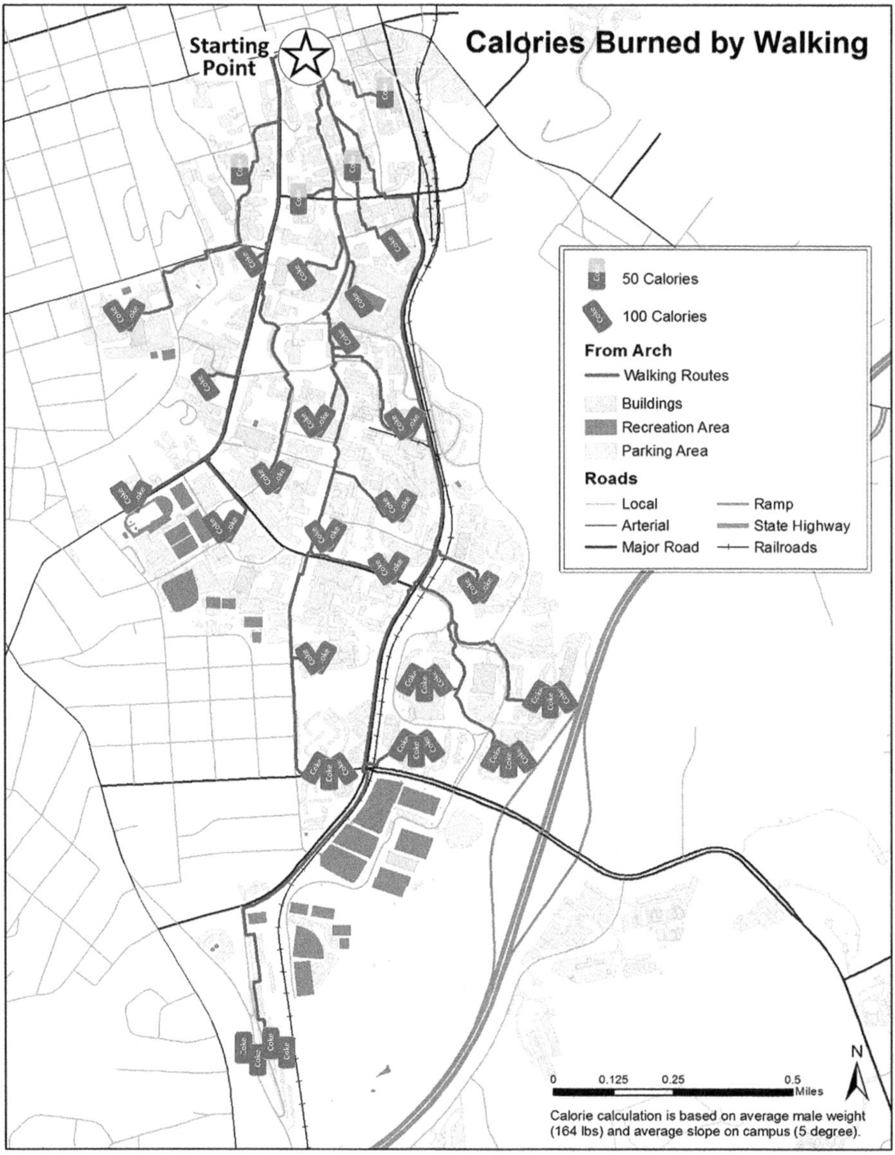


**Fig. 5** Calorie Map

## 4.2 Results from the Social Media Data in a General Setting

Using predefined keywords (walkable, walkability), we collected more than 14,500 tweets, with 16.88% original tweets and 83.12% retweets during two weeks, winter 2018. Those tweets were with hashtags like #liveablecities18, #smartcities,

| Approximate Calorie Burned | Calorie calculation is based on the average male weight (164 lbs.) and average slope on campus (5 degree) without considering the downhill situation. | | | | | | | | | |
|---|---|---|---|---|---|---|---|---|---|---|
| 1 | | | | | | | | | | |
| 2 | 224 | | | | | | | | | |
| 3 | 75 | 177 | | | | | | | | |
| 4 | 196 | 45 | 152 | | | | | | | |
| 5 | 69 | 271 | 141 | 239 | | | | | | |
| 6 | 207 | 414 | 274 | 382 | 176 | | | | | |
| 7 | 168 | 142 | 107 | 126 | 232 | 366 | | | | |
| 8 | 81 | 154 | 35 | 126 | 139 | 282 | 104 | | | |
| 9 | 155 | 104 | 91 | 76 | 211 | 354 | 86 | 85 | | |
| 10 | 96 | 303 | 164 | 271 | 40 | 138 | 256 | 171 | 243 | |
| 11 | 103 | 275 | 121 | 246 | 87 | 160 | 212 | 131 | 205 | 50 |
| | 1 | 2 | 3 | 4 | 5 | 6 | 7 | 8 | 9 | 10 | 11 |

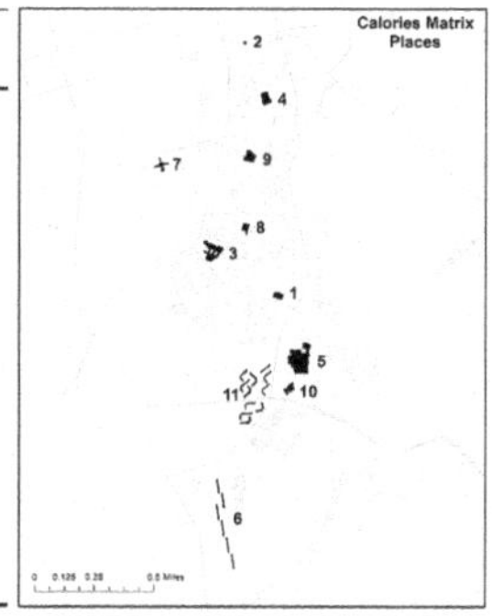


**Fig. 6** Calorie Matrix

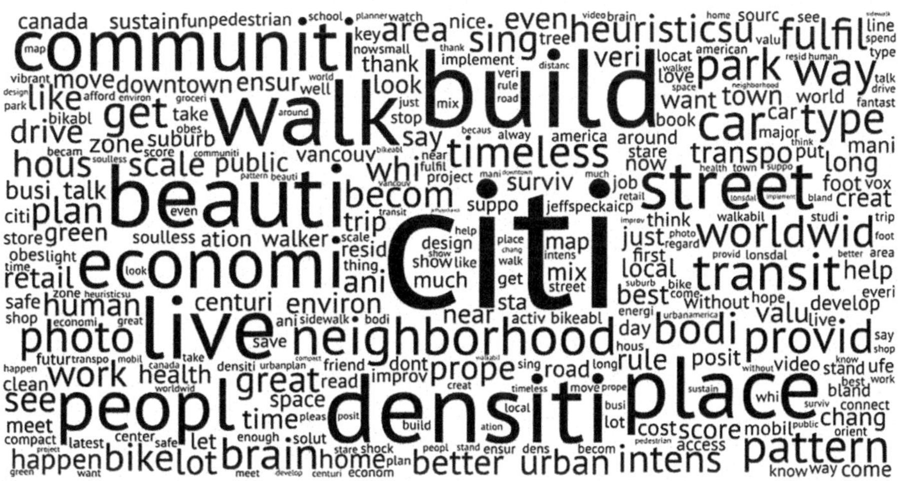


**Fig. 7** Word Cloud of the Tweets excluding Keywords (Walkable, Walkability)

#StreetsAreForPeople, and more. Only 14 (0.09%) of our collected tweets were geocoded so we did not consider geolocational information in our analysis. According to the Tweet data dictionary (Twitter Inc. 2018), our tweets were in 17 languages, and more than half of them (64.64%) were in Thai[1], and the second most language was English (33.91%). Since overall English is the most-used language (Statista 2013), we focused only on the 4919 English-language tweets for this study.

For the word frequency, there are 686 words mentioned more than 20 times. We made a word cloud (Fig. 7), without considering the searching keywords walkability and walkable. The word *citi*, as the lemmatized word for city, cities, and

[1] We did some digging and found one particular Thai tweet, posted on the first day of our data collection. It was posted by a user with 18.8 k followers, and was retweeted 8352 times during our collection, eventually with more than 13,000 times with some follow-up replies and retweets.

more, has been mentioned the most for 2375 times. Other lemmatized words such as *densit* (with a frequency of 691), *walk* (634), *beauti* (569), *place* (541), *communiti* (517), *economi* (490), *street* (456), *transit* (411) and more, stand out as the top most mentioned words. The walkability issue is in different scales and settings, from the street, neighborhood, community, to the city scope. Specific place names, such as *Vancouver*, *Canada* were also in the word cloud.

The Twitter result from a general setting echoes with the campus-focused study and the POWER method. The word *people*, as well as *pedestrian*, was mentioned a lot, and it emphasizes the importance of people' needs and considerations, which have been considered as the perceived importance of the POWER method. Moreover, specific built environment aspects were called out in the tweet contexts. For example, some words, such as *densi* and *transit*, are corresponding to the connectivity. Amenities are in the forms of *park*, *shop*, *retail*, and more, which are mentioned more than 140 times each in our collected tweets. The walking environment factors appeared in the tweets as *bike* (286 times), *bikeable* (101 times), among others.

For the top 10 co-occurrence word combinations, *walkabl* (walkable/walkability) is in nine of them as they were used as search keywords. The lemmatized words *citi* and *walkabl* have the highest co-occurrence frequency as 1548 times. With the top 12 co-occurrence word pairs, we interpret the results that people desire to *build* (mentioned with *walkabl* for 664 times, and with *beauti* for 471 times), *provide* (337 times with *walkabl*), and *live* (523) in a walkable environment in the form of *beauti* (551), *dense* (378) urban design at the *place* (493), *community* (415), and *city* levels.

There are some interesting findings in the tweet content. Over 80 users expressed that "the more walkable a country is, the more it saves on healthcare costs." A few tweets mentioned that "unfortunately car-dependence is built into most US communities" which was the issue that we discussed earlier in this book chapter. Twitter users also spread their opinions that different age groups, older adults, millennials, and kids, all desire walkable neighborhood. Some of them provide information that walkable communities can promote more exercises, and further benefit people's health and social life. People also use Twitter as a platform to share related conferences, research results of walkability and show their support to certain plan, campaign, or change. One tweet, "density can be intense, beautiful & walkable," which has been retweeted over 600 times, refers to the low-density planning issue and shows the desire of compact cities with the walkable environment.

## 5 Conclusion and Discussion

We present a walkability study incorporating data from both a location-specific online survey and social media in a more general setting. Our research shows that survey is efficient in customizing for specific urban sites and local people, and social media can reach to a broader audience to obtain general opinions toward the same topic. The survey result and the social media data are mutually supplementary for understanding the whole picture of walkability. On different local settings, such as commercial areas, residential areas, or mixed land-use areas, local residents may have various preferences toward different aspects of the built environment because of their commute mode, occupation, income, age, and characteristics of the area, among others. For example, older adults may consider flat walking space as a must while others not, or people of hot regions may consider tree shade vital in their walking environment. On a general setting, the social media result demonstrates that people, as pedestrians, desire walkability at different geographical levels, and have their emphasis on the built environment design for perceiving walkable places. Using these two approaches in combination can provide a more complete understanding of walkability.

This research contributes to the existing walkability literature. Beyond the traditional factors of land use and transportation, the POWER method embraces more urban design perspectives, including sidewalk conditions, traffic speed, and more. Moreover, the walking-preference survey captures the preferences and considerations of local pedestrians. The POWER incorporates people's walking preferences and the built environment conditions, making the concept of walkability more human-oriented. Social media is an efficient approach to obtain people's feeling and attitudes. Compared with other walkability measures, the POWER takes more efforts in collecting various data and people's input, and therefore can capture the local people's demands and understand the perceived walkability. The same built environment may be very walkable or not based on individual needs.

Incorporating survey and social media data for specific and general settings further complements the walkability study. In this study, we used social media data as an add-on part to supplement some the structure of the POWER and survey findings. Following the revised version of the method (Fig. 8), future studies can be more efficient to understand the walkability on different scales. The major changes between the current and future practices are shaded in dark gray, and involved individual processes are dashed outlined. Instead of starting with a local focus, future studies can take advantage of social media to acquire understandings from the general population. Based on the results from the social media data and literature review, more local elements are taken into consideration, and the sample interview can validate the choices. Using social media and survey can bring two scales together to understand the specific environment.

Using social media data added more flavors to our walkability research. However, it has limitations. First, our study focuses on English-language Tweets. Although other language tweets were collected, we can do very limited analysis without understanding the meaning. This limited the extent to the English-spoken people and places. Second, we have a relatively short time to collect tweets, and it was the winter time of the North Hemisphere, where the USA and many other English-speaking countries locate. The seasonality may influence people's experience toward walkability and further influence the quantity and content of the related tweets. Third, although the word frequency can capture the count of the word even in different forms, it is hard to automatically group the words with the same meaning (cyclable and bikeable). Last, the representativeness of the Twitter user is not clear.

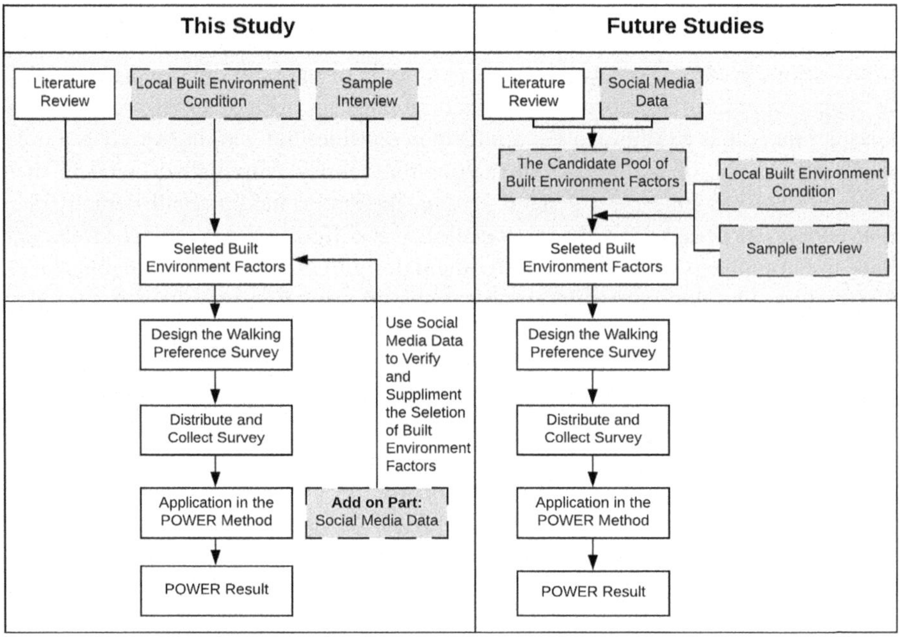


**Fig. 8** The Flowcharts of This and Future Walkability Studies Using Social Media (Gray shading: major changes. Dashed outlines: changed processes)

Only 1% of tweets can be collected freely via the Streaming API, and that can mask some parts of the whole picture (Morstatter et al. 2014). Meanwhile, different age groups have various proportions in Twitter users, and the situation may vary dramatically by region.

For future endeavors, we would like to explore more approaches. There are some new techniques for measuring walkability. For example, some researchers are using Wi-Fi connections to predict how many pedestrians are there as a proxy of walkability. Others apply high-resolution street view imagery to evaluate the neighborhood. With the development of geospatial technologies, more possibilities will be available for urban health measurements.

**Acknowledgment** Thanks for the support received from the UGA Sustainability Grant.

## References

Anderson, D., Al-Tarawneh, H. A., Amorose, A. J., & Horn, T. S. (2010). *Research methods in psychology*. http://search.ebscohost.com/login.aspx?direct=true&db=psyh&AN=2000-08059-004&lang=pt-br&site=ehost-live%0Ahttp://search.ebscohost.com/login.aspx?direct=true&db=psyh&AN=2011-20515-022&lang=pt-br&site=ehost-live%0Ahttp://search.ebscohost.com/login.aspx?dire

Berzi, C., Gorrini, A., & Vizzari, G. (2017). Mining the social media data for a bottom-up evaluation of walkability. arXiv preprint arXiv:1712.04309.
Brooker, P., Barnett, J., & Cribbin, T. (2016). Doing social media analytics. *Big Data & Society, 3*(2), 2053951716658060.
Browning, R. C., Baker, E. A., Herron, J. A., & Kram, R. (2006). Effects of obesity and sex on the energetic cost and preferred speed of walking. *Journal of Applied Physiology, 100*(2), 390–398.
Carr, L. J., Dunsiger, S. I., & Marcus, B. H. (2010). Walk Score™ as a global estimate of neighborhood walkability. *American Journal of Preventive Medicine, 39*(5), 460–463.
Carr, L. J., Dunsiger, S. I., & Marcus, B. H. (2011). Validation of Walk Score for estimating access to walkable amenities. *British Journal of Sports Medicine, 45*(14), 1144–1148.
Crane, R., & Crepeau, R. (1998). Does neighborhood design influence travel? A behavioral analysis of travel diary and GIS data. *Transportation Research Part D: Transport and Environment, 3*(4), 225–238.
Diehl, T. (2017). Citizenship, social media, and big data: Current and future research in the social sciences. *Social Science Computer Review, 35*(1), 3–9.
Dobesova, Z., & Krivka, T. (2012). Walkability index in the urban planning: A case study in Olomouc City. In J. Burian (Ed.), *Advances in spatial planning* (pp. 179–196). InTech.
Duncan, D. T., Aldstadt, J., Whalen, J., & Melly, S. J. (2013). Validation of Walk Scores and Transit Scores for estimating neighborhood walkability and transit availability: A small-area analysis. *GeoJournal, 78*(2), 407–416.
Duncan, D. T., Aldstadt, J., Whalen, J., Melly, S. J., & Gortmaker, S. L. (2011). Validation of Walk Score® for estimating neighborhood walkability: An analysis of four US metropolitan areas. *International Journal of Environmental Research and Public Health, 8*(12), 4160–4179.
Duncan, D. T., Sharifi, M., Melly, S. J., Marshall, R., Sequist, T. D., Rifas-Shiman, S. L., & Taveras, E. M. (2014). Characteristics of walkable built environments and BMI z-scores in children: Evidence from a large electronic health record database. *Environmental Health Perspectives, 122*(12), 1359–1365. http://www.pubmedcentral.nih.gov/articlerender.fcgi?artid=4256697&tool=pmcentrez&rendertype=abstract
Fan, J. X., Wen, M., & Kowaleski-Jones, L. (2014). An ecological analysis of environmental correlates of active commuting in urban U.S. *Health & Place, 30*, 242–250.
Feinerer, I., Hornik, K., & Meyer, D. (2008). Text mining infrastructure in R. *Journal of Statistical Software, 25*(5), 1–54. http://www.jstatsoft.org/v25/i05/
Fellrnr.com. (2017). *Calories burned running and walking*. http://fellrnr.com/wiki/Calories_burned_running_and_walking?Weight=164&WeightUnits=Pounds. Last accessed 20 June 2017
Felt, M. (2016). Social media and the social sciences: How researchers employ Big Data analytics. *Big Data & Society, 3*(1), 2053951716645828.
Forsyth, A., & Southworth, M. (2008). Cities Afoot—Pedestrians, walkability and urban design. *Journal of Urban Design, 13*(1), 1–3.
Foster, S., Knuiman, M., Villanueva, K., Wood, L., Christian, H., & Giles-Corti, B. (2014). Does walkable neighbourhood design influence the association between objective crime and walking? *International Journal of Behavioral Nutrition and Physical Activity, 11* (1), 100. http://www.ijbnpa.org/content/11/1/100
Frank, L. D., Sallis, J. F., Saelens, B. E., Leary, L., Cain, K., Conway, T. L., & Hess, P. M. (2010). The development of a walkability index: application to the Neighborhood Quality of Life Study. *British Journal of Sports Medicine, 44*(13), 924–933.
Gota, S., Fabian, H. G., Mejia, A. A., & Punte, S. S. (2010). Walkability surveys in Asian cities. Clean Air Initiative for Asian Cities (CAI- Asia), 20. https://www.ictct.net/migrated_2014/ictct_document_nr_663_102A%20Sophie%20Sabine%20Punte%20Walkability%20Surveys%20in%20Asian%20Cities.pdf
Gravel, R., & Béland, Y. (2005). The Canadian Community Health Survey: Mental health and well-being. *The Canadian Journal of Psychiatry, 50*(10), 573–579.
Gu, P., Han, Z., Cao, Z., Chen, Y., & Jiang, Y. (2018). Using open source data to measure street walkability and bikeability in China: A case of four cities. *Transportation Research Record*. https://doi.org/10.1177/0361198118758652.

Hall, C. M., & Ram, Y. (2018). Measuring the relationship between tourism and walkability? Walk Score and English tourist attractions. *Journal of Sustainable Tourism, 9582*, 1–18. https://www.tandfonline.com/doi/full/10.1080/09669582.2017.1404607

Handy, S. L., Boarnet, M. G., Ewing, R., & Killingsworth, R. E. (2002). How the built environment affects physical activity: Views from urban planning. *American Journal of Preventive Medicine, 23*(2 Suppl 1), 64–73.

Hasan, S., Zhan, X., & Ukkusuri, S. V. (2013). Understanding urban human activity and mobility patterns using large-scale location-based data from online social media. In *Proceedings of the 2nd ACM SIGKDD international workshop on urban computing* (p. 6). Chicago, Illinois. ACM.

Hirsch, J. A., Roux, A. V. D., Moore, K. A., Evenson, K. R., & Rodriguez, D. A. (2014). Change in walking and body mass index following residential relocation: The multi-ethnic study of atherosclerosis. *American Journal of Public Health, 104*(3), 49–56.

Huang, T. T.-K., Harris, K. J., Lee, R. E., Nazir, N., Born, W., & Kaur, H. (2003). Assessing overweight, obesity, diet, and physical activity in college students. *Journal of American College Health, 52*(2), 83–86. http://www.tandfonline.com/doi/abs/10.1080/07448480309595728

Hung, W. T., Manandhar, A., & Ranasinghege, S. A. (2010). A walkability survey in Hong Kong. In *The 12th international conference on mobility and transport for elderly and disabled persons (TRANSED)*. Hong Kong, China.

Jackson, R. J., & Kochtitzky, C. (2001). *Creating a Healthy Environment: The impact of the built environment on public health.* Sprawl Watch Clearinghouse Monograph Series. Washington, DC: Public Health and Land Use Planning & Community Design Professionals.

Jun, H.-J., & Hur, M. (2015). The relationship between walkability and neighborhood social environment: The importance of physical and perceived walkability. *Applied Geography, 62*, 115–124.

Jurdak, R., Zhao, K., Liu, J., Aboujaoude, M., Cameron, M., & Newth, D. (2015). Understanding human mobility from Twitter. *PLoS One*, 1–16. https://doi.org/10.1371/journal.pone.0131469.

Kearney, M. W. (2018). *rtweet: Collecting Twitter Data*. https://cran.r-project.org/package=rtweet

Keating, X. D., Guan, J., Piñero, J. C., & Bridges, D. M. (2005). A meta-analysis of college students' physical activity behaviors. *Journal of American College Health, 54*(2), 116–125.

Kilpatrick, D. G., Best, C. L., Veronen, L. J., Amick, A. E., Villeponteaux, L. A., & Ruff, G. A. (1985). Mental health correlates of criminal victimization: A random community survey. *Journal of Consulting and Clinical Psychology, 53*(6), 866–873.

Kouloumpis, E., Wilson, T., & Moore, J. (2011). Twitter sentiment analysis: The good the bad and the omg! In *Proceedings of the fifth international AAAI conference on Weblogs and Social Media (ICWSM 11)* (pp. 538–541). http://www.aaai.org/ocs/index.php/ICWSM/ICWSM11/paper/download/2857/3251?iframe=true&width=90%25&height=90%25

Larsson, A. O., and H. Moe. 2011. Studying political microblogging: Twitter users in the 2010 Swedish election campaign. *New Media & Society,* 14 (5), 729–747.

Leslie, E., Coffee, N., Frank, L., Owen, N., Bauman, A., & Hugo, G. (2007). Walkability of local communities: Using geographic information systems to objectively assess relevant environmental attributes. *Health and Place, 13*(1), 111–122.

Litman, T. (2014). Land for vehicles or people? *Planetizen.* http://www.planetizen.com/node/72454/land-vehicles-or-people. Last accessed 10 Jan 2018.

Litman, T (2018). *Evaluating Active Transport Benefits and Costs.* Victoria, Canada: Victoria Transport Policy Institute.

Liu, S., & Young, S. D. (2018). A survey of social media data analysis for physical activity surveillance. *Journal of Forensic and Legal Medicine, 57*, 33–36. https://doi.org/10.1016/j.jflm.2016.10.019.

Livi, A. D., & Clifton, K. J. (2004). Issues and methods in capturing pedestrian behaviors, attitudes and perceptions: experiences with a community-based walkability survey. In *Transportation research board annual meeting* (17pp). Washington, DC.

Lo, R. H. (2009). Walkability: What is it. *Journal of Urbanism, 2*(2), 145–166.

Loo, B. P. Y., & Lam, W. W. Y. (2012). Geographic accessibility around health care facilities for elderly residents in Hong Kong: A microscale walkability assessment. *Environment and Planning B: Planning and Design, 39*(4), 629–646.
Manning, C. D., Raghavan, P., & Schütze, H. (2008). *Introduction to information retrieval.* Cambridge University Press. https://nlp.stanford.edu/IR-book/
Matsuo, Y., & Ishizuka, M. (2004). Keyword extraction from a single document using word co-occurrence statistical information. *International Journal on Artificial Intelligence Tools, 13*(01), 157–169. http://www.worldscientific.com/doi/abs/10.1142/S0218213004001466
McLuhan, M. (1975). McLuhan' s laws of the media. *Technology and Culture, 16*(1), 74–78. Published by: The Johns Hopkins University Press and the Society for the History of Technology Stable URL: https://www.jstor.org/stable/3102368
Morstatter, F., Pfeffer, J., & Liu, H. (2014). When is it biased?: assessing the representativeness of twitter's streaming API. In *Proceedings of the 23rd international conference on world wide web* (pp. 555–556). ACM.
National Center for Education Statistics. (2018). *Undergraduate enrollment.* https://nces.ed.gov/programs/coe/indicator_cha.asp. Last accessed 23 May 2018.
Pak, A., & Paroubek, P. (2010). Twitter as a Corpus for sentiment analysis and opinion mining. In *Seventh conference on international language resources and evaluation* (pp. 1320–1326).
Park, S. (2008). *Defining, measuring, and evaluating path walkability, and testing its impacts on transit users' mode choice and walking distance to the station.* Berkeley: University of California.
Powell, P., Spears, K., & Rebori, M. (2010). *What is obesogenic environment?* (pp. 1–2). University of Nevada Cooperative Extension (fact sheet 10–11). Reno, NV: University of Nevada Cooperative Extension.
Princeton University. (2008). *2016 campus plan.* http://www.princeton.edu/pr/doc/2006-campus-plan.pdf. Last accessed 1 Dec 2017.
Quercia, D., Aiello, L. M., Schifanella, R., & Davies, A. (2015). The digital life of walkable streets. In *Proceedings of the 24th international conference on World Wide Web* (pp. 875-884). International World Wide Web Conferences Steering Committee.
R Development Core Team. (2008). *R: A language and environment for statistical computing.* http://www.r-project.org
Rinker, T. W. (2017). *{qdapRegex}: Regular expression removal, extraction, and replacement tools.* http://github.com/trinker/qdapRegex
Rinker, T. W. (2018). *{textstem}: Tools for stemming and lemmatizing text.* http://github.com/trinker/textstem
Robinson, W. S. (1950). Ecological correlations and the behavior of individuals. *American Sociological Review, 15*(3), 351–357.
Rundle, A., Neckerman, K. M., Freeman, L., Lovasi, G. S., Purciel, M., Quinn, J., Richards, C., Sircar, N., & Weiss, C. (2009). Neighborhood food environment and walkability predict obesity in New York City. *Environmental Health Perspectives, 117*(3), 442–447.
Saaty, R. W. (1987). The analytic hierarchy process-what it is and how it is used. *Mathematical Modelling, 9*(3–5), 161–176.
Saaty, T. (1980). *The analytic hierarchy process: Planning, priority setting, resources allocation.* New York: McGraw-Hill.
Saaty, T. L. (2004). Decision making — the Analytic Hierarchy and Network Processes (AHP/ANP). *Journal of Systems Science and Systems Engineering, 13*(1), 1–35.
Saaty, T. L. (2008). Decision making with the analytic hierarchy process. *International Journal of Services Sciences, 1*(1), 83–98.
Saelens, B. E., & Handy, S. L. (2008). Built environment correlates of walking: A review. *Medicine and Science in Sports and Exercise, 40*(7 Suppl), S550–S566.
Selvin, H. C. (1958). Durkheim's suicide and problems of empirical research. *American Journal of Sociology, 63*(6), 607–619.
Shen, Y., & Karimi, K. (2016). Urban function connectivity: Characterisation of functional urban streets with social media check-in data. *Cities, 55,* 9–21. https://doi.org/10.1016/j.cities.2016.03.013.

e Silva, J. D. A., De Oña, J., & Gasparovic, S. (2017). The relation between travel behaviour, ICT usage and social networks. The design of a web based survey. *Transportation Research Procedia, 24*, 515–522. https://doi.org/10.1016/j.trpro.2017.05.482.

Slater, S. J., Nicholson, L., Chriqui, J., Barker, D. C., Chaloupka, F. J., & Johnston, L. D. (2013). Walkable communities and adolescent weight. *American Journal of Preventive Medicine, 44*(2), 164–168.

Statista. (2013). Most-used languages on Twitter as of September 2013. *Statista*. https://www.statista.com/statistics/267129/most-used-languages-on-twitter/. Last accessed 4 Dec 2018.

Statista. (2018). Leading countries based on number of Twitter users as of October 2018 (in millions). *Statista*.

Sui, D., & Goodchild, M. (2011). The convergence of GIS and social media: Challenges for GIScience. *International Journal of Geographical Information Science, 25*(11), 1737–1748.

Sui, D. Z., & Goodchild, M. F. (2003). A tetradic analysis of GIS and society using McLuhan's law of the media. *The Canadian Geographer, 1*(1), 5–17.

Swinburn, B., Egger, G., & Raza, F. (1999). Dissecting obesogenic environments: the development and application of a framework for identifying and prioritizing environmental interventions for obesity. *Preventive Medicine, 29*(6), 563–570.

Trumbo, J. (2000). Essay: seeing science: Research opportunities in the visual communication of science. *Science Communication, 21*(4), 379–391.

Tumasjan, A., Sprenger, T., Sandner, P., Welpe, I. (2010). Predicting elections with Twitter: What 140 characters reveal about political sentiment. In *Proceedings of the fourth international AAAI conference on Weblogs and Social Media* (pp. 178–185). http://www.aaai.org/ocs/index.php/ICWSM/ICWSM10/paper/viewFile/1441/1852

Twitter Inc. (2018). *Tweet objects*. https://developer.twitter.com/en/docs/tweets/data-dictionary/overview/tweet-object. Last accessed 23 May 2018.

Vargo, J., Stone, B., & Glanz, K. (2012). Google walkability: A new tool for local planning and public health research? *Journal of Physical Activity & Health, 9*(5), 689–697.

Walkability Index. (2017). *United States environmental protection agency*. https://edg.epa.gov/metadata/catalog/search/resource/details.page?uuid=%7B251AFDD9-23A7-4068-9B27-A3048A7E6012%7D. Last accessed 2 Dec 2018.

Walker, A. (2018). Q1 2018: Twitter now has 336m monthly active users. *Memeburn*. https://memeburn.com/2018/04/twitter-users-q1-2018/. Last accessed 20 May 2018.

Warburton, D. E. R., Nicol, C. W., & Bredin, S. S. D. (2006). Health benefits of physical activity: the evidence. *Canadian Medical Association Journal, 174*(6), 801–809.

Wickham, H. (2018). *stringr: Simple, consistent wrappers for common string operations*. https://cran.r-project.org/package=stringr

Wikipedia Contributors. (2018). *Natural-language processing*. https://en.wikipedia.org/w/index.php?title=Natural-language_processing&oldid=843426453

WordArt.com. (2016). https://wordart.com/. Last accessed 20 July 2016.

Yang, W., & Mu, L. (2015). GIS analysis of depression among Twitter users. *Applied Geography, 60*, 217–223. https://doi.org/10.1016/j.apgeog.2014.10.016.

Yang, W., Mu, L., & Shen, Y. (2015). Effect of climate and seasonality on depressed mood among twitter users. *Applied Geography, 63*, 184–191. https://doi.org/10.1016/j.apgeog.2015.06.017.

Yin, L. (2017). Street level urban design qualities for walkability: Combining 2D and 3D GIS measures. *Computers, Environment and Urban Systems, 64*, 288–296.

Zhang, X. (2016). *Perceived importance and objective measures of built environment walkability of a university campus*. https://athenaeum.libs.uga.edu/handle/10724/36572

Zhang, X., & Mu, L. (2019). The perceived importance and objective measurement of walkability in the built environment rating. *Environment and Planning B: Urban Analytics and City Science*. Advance online publication. https://doi.org/10.1177/2399808319832305

**Xuan Zhang** is a Geography Ph.D. student from the University of Georgia (UGA). She received her M.S. in Geography from UGA and B.S. in GIS from Wuhan University. Her primary interest is GIS application for health and planning. She has worked on projects such as assessing walkability from both perceived importance and objective measurement and examining disparities of the long-term care facilities for the older population.

**Dr. Lan Mu** is Professor of Geography at UGA. Her research interests include GIScience for health and the environment, spatial analysis and modeling, computational geometry, cartography, and geovisualization. She also directs UGA's undergraduate and graduate GIScience Certificate Programs.

# Leveraging Social Media to Track Urban Park Quality for Improved Citizen Health

**Coline C. Dony and Emily Fekete**

**Abstract** In this chapter, we showcase the use of qualitative data available on two "geobrowsers" (i.e., Google Maps and Foursquare) and of a data-mining technique to quantify the sentiment of online reviews about parks. The underlying interest for this study comes from the growing literature suggesting that living near parks or other open spaces contributes to higher levels of physical activity and to lower levels of stress and fewer mental health problems. Mecklenburg County (North Carolina), which encompasses the City of Charlotte, is used as a case study. In a comparison among 97 cities in the USA, The Trust for Public Land ranks Charlotte's park system at the very bottom and reports their spending per resident on their park system among the lowest 20% of these cities. Considering their lower spending, the city government may be particularly interested to leverage publicly available data from social media to complement the assessments they already perform about their park system, such as satisfaction surveys or quality assessments. Nevertheless, Charlotte's low ranking – although unfortunate – indicates an opportunity for the city to improve its park system, which in turn could engage residents in more physical activity and, in doing so, create positive community health outcomes.

## 1 Introduction

There is some evidence in the literature showing that living near public open spaces contributes to higher levels of physical activity and to lower levels of stress and fewer mental health problems (Bedimo-Rung et al. 2005; Lopez and Hynes 2006). Given both physical and mental health benefits, improving access to public open spaces such as public parks, can become a prevention strategy to reduce heart disease. The US Department of Health and Human Services (2008) started reporting the need to improve access to facilities supporting physical activity and to the built

C. C. Dony (✉) · E. Fekete
American Association of Geographers, Washington, DC, USA
e-mail: cdony@aag.org



Y. Lu, E. Delmelle (eds.), *Geospatial Technologies for Urban Health*,
Global Perspectives on Health Geography, https://doi.org/10.1007/978-3-030-19573-1_9

environment – such as sidewalks, bike lanes, trails, and parks – as it recognizes the positive effect on physical activity.

In many cities, the majority of parks and recreational areas are managed by the city government (e.g., planning department, department of parks and recreation). Part of their mission is to monitor the needs of their residents and the use of their parks and recreational areas. These are often accomplished through public satisfaction surveys, which are costly and time consuming. Due to population growth and increased migration to cities (and in between cities), survey results can quickly become outdated because, when a population changes, its needs and satisfaction change too. Therefore, we argue that city governments should leverage social media platforms as an additional source of quantitative and qualitative data, which is inexpensive, and is generated by their residents continuously and captures their suggestions and satisfaction of public spaces. In this chapter, we showcase the use of data available on geobrowsers Google Maps and Foursquare and the application of sentiment analysis to these data to demonstrate a supplementary approach to monitor the perception of public parks in Charlotte, North Carolina.

## 2 Literature Review

In this literature review, three major topics are addressed. The first section provides the literature on place-based approaches toward health prevention. Then, in the second section, the limited research evidence on the effectiveness of place-based approaches is summarized. Finally, the last section of this literature review provides an overview of social media use and the potential of social media for garnering public opinion about urban development.

### *2.1 Place-Based Strategies to Increase Physical Activity*

There has been a worldwide decrease in physical activity which has been associated with a global increase in noncommunicable diseases such as heart disease and chronic diseases (Bauman and Craig 2005). Finding effective strategies that engage more people in regular physical activity, however, has proven challenging. Today, fewer jobs require physical labor and spare time is spent on more sedentary activities such as watching television (Hill et al. 2003). Major changes in urban planning, such as car-centric design, escalators, and automatic doors, have shifted the way we interact with our environment toward less physical activity. These changes make our lives easier and more accessible to anyone, but now also require us to supplement our days with artificial physical exercise in order to maintain a healthy physiology.

Strategies to improve population health generally focus on "individual-based" approaches (Koohsari et al. 2013) and formal healthcare settings (Moon and Gillespie 1995). About 25 years ago, McGinnis and Foege (1993) estimated that only 10–15% of mortality in the USA could be prevented by improving healthcare availability and treatments, while 40% could be prevented by behavioral changes. Yet, in a later study, McGinnis, Williams-Russo, and Knickman (2002) estimated that, at the time, 95% of the national healthcare budget was still invested in medical treatments.

To shift the focus of healthcare prevention from individual-based approaches to strategies that would impact changes in people's behavior, CDC director Thomas Frieden (2010) developed the health impact pyramid (see Fig. 1). This pyramid, based on the "socio-ecological model" popular in public health and prevention research, is designed to show that diseases are caused by a range of factors that play at different levels, from molecules, to the individual, and all the way to their communities, cultures and societies. With his pyramid, Frieden (2010) wants to make it clear that most action should happen at the base of the pyramid (i.e., socioeconomic factors), if we want to have a more effective impact on health outcomes.

The second layer of the pyramid represents efforts to change the context to make individual's default decisions healthier. Monitoring the quality of parks and recreational areas as an indicator for the health of a city and its residents–which is the focus of this chapter, would fit at this stage of the pyramid. These strategies are also referred to as "place-based" approaches toward health prevention and make use of the community and its environment to impact our behavior. The hypothesis here is that making changes at the community level can benefit more individuals without direct intervention of health professionals, which can be more cost-effective and reach more socioeconomic groups at once.

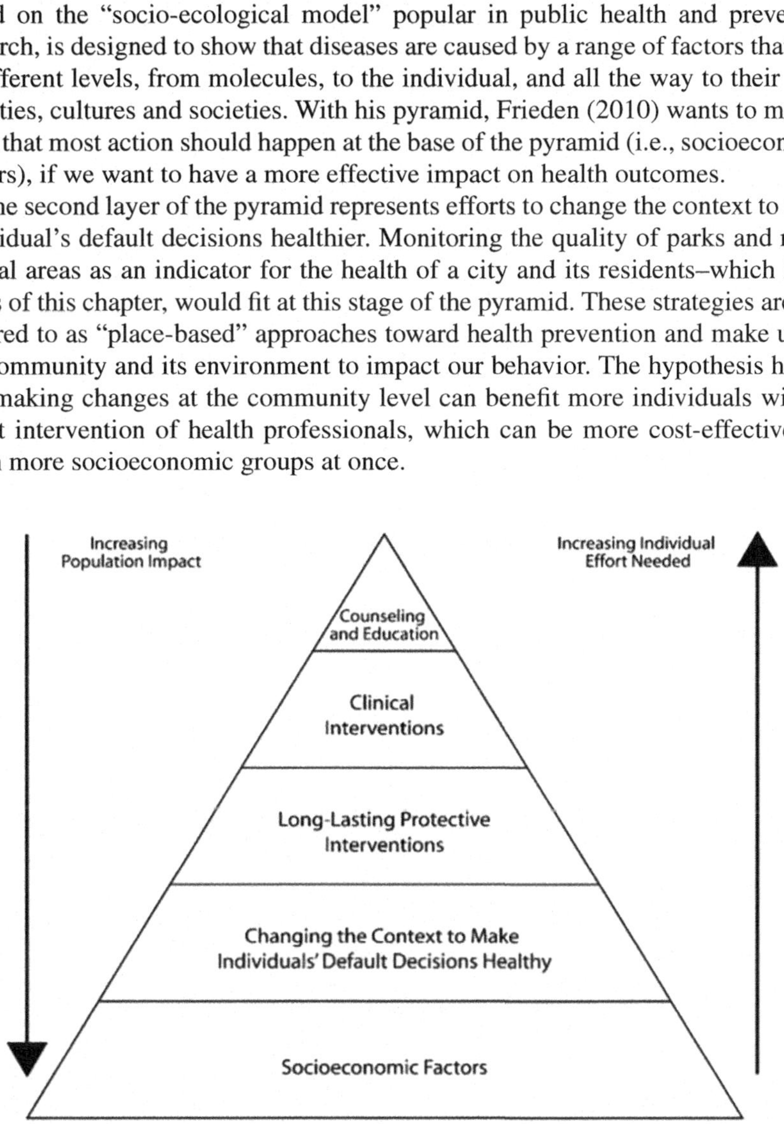


**Fig. 1** The health impact pyramid (Frieden 2010)

## 2.2 Studies on Parks on Community Health

Measuring the direct impact that parks can have on community health remains challenging and as a consequence, there is a lack of research that measures and tests these direct impacts. Rosenberger et al. (2005) used spatial regression (spatial lag models) to understand the link between hospital expenditures, physical inactivity, and recreation availability in West Virginia, controlling for differences in healthcare availability and socioeconomic status between its counties. Their study finds that counties with more active residents were associated with higher availability for recreation and with lower hospital expenditures. Also, Rosenberger et al. (2005) find that more recreation opportunities were associated with less health expenditures per county, which they use as a supporting argument to invest in the supply of recreational opportunities. In a recent study, Mueller et al. (2016) show that 20% of preventable, natural, all-cause deaths in Barcelona (Spain) are attributed to a combination of (1) physical inactivity among residents; (2) their exposure to higher than recommended levels of air pollution, noise and heat; and (3) their access to greenspace. The definition for "access to greenspace" in their study is the one recommended by the European Commission in 2001 and the WHO in 2016, which is defined as a 300-meter linear distance to a green space that is greater than or equal to 0.5-hectare (which is comparable to the size of half a soccer field). Although this definition seems questionable, particularly in a city like Barcelona, their research findings are noteworthy of a mention as a showcase of the kind of literature on this topic and associated research challenges.

With similar motivations, the Trust for Public Land (TPL 2010), a non-profit with a focus on land conservation, started calculating a *ParkScore®* for cities across the USA since 2012. Their intent with *ParkScore®* is to value and compare urban park systems and, in doing so, encourage cities to improve their score. The score is based on criteria such as median park size, percent parkland within city limits, and percent of the population living within a ten-minute walk of a public park. One major weakness of their method, is that it relies on the availability of each city's Open Data, which does not have a consistent quality and mostly provides quantitative data about urban park systems (e.g., total number of parks and their sizes). In order to make fair comparisons between urban systems, however, the local context and qualitative park measures are as important as quantitative measures (Dony et al. 2015). For example, the City of Charlotte is about 8.5 times the size of Washington DC, and its population density is about 6.5 times lower than that of the District. Thus, it comes as no surprise that only 25% of Charlotteans have a 10-minute walking access to public parks compared to 98% of DC's population, yet this criteria accounts for a quarter of the city's total *ParkScore*. The sparsely populated and suburban nature of Charlotte, however, is a draw for many Americans deciding to live there. In that respect, is it realistic or even desirable to expect that Charlotteans have a public park within a 10-minute walk of their home? Charlotte's *ParkScore* is hugely impacted by that criteria ranking them 97th, while DC is ranked 3rd. In this chapter, we hope to showcase one way to access qualitative data about public parks, which could also find use at the TPL to improve the fairness of their *ParkScore*.

A growing body of literature evaluating whether or not access to public open spaces is equivalent across socioeconomic and ethnic/racial groups and between rural and urban areas suggests that access to public open spaces tends to be unequal (e.g., Parks et al. 2003; Dai 2011; Wolch et al. 2014). By quantifying access to urban green spaces using a Geographic Information System (GIS) and evaluating disparities among racial/ethnic and socioeconomic groups in Atlanta, GA, using linear regression, Dai (2011) confirms significantly poorer access to urban green spaces among neighborhoods with a higher concentration of African Americans. A cross-sectional study by Parks, Housemann, and Brownson (2003) shows that US residents living in lower-income areas report lower levels of physical activity. Furthermore, they found that levels of activity were highest among suburban residents and lowest among urban and rural residents, which they reported to coincide with other national cross-sectional studies (Parks et al. 2003).

Based on these ideas and early studies on this subject, Koohsari et al. (2013) have suggested reintroducing public health as a priority for urban planners, arguing that planners are well positioned to study impacts of the built environment on physical activity at different scales and for different populations. The authors further argue that planners have a better understanding of urban issues and that the "public health planner" would find the most cost-effective program(s) at each level of the health impact pyramid (Fig. 1) to maximize improvements in health outcomes. They suggest that a more top-down approach would result in better health outcomes and provide an overall increase in physical activity not just for particular neighborhoods, but for a city as a whole.

Arguments for a top-down approach by public planners are not without criticism. Curran and Hamilton (2012) suggest a more qualitative, small-scale, community-driven approach to the development of public green spaces. They argue that some current planning practices have a narrow focus on the aesthetics rather than on the functionality of neighborhoods and fail to account for environmental injustices within their city. Working with residents of a Brooklyn neighborhood, they were able to make their community "just green enough" to both remedy health concerns and dissuade the undesired effects of gentrification. This research highlights the importance of the local context and the need for qualitative measures for park quality.

## 2.3 *Leveraging Social Media as a Snapshot of Urban Development*

Social media platforms can be a feasible way to collect qualitative perceptions about parks by the local population and to better understand the local context of each public park. While social media sites are not necessarily always representative of the general population (see Fekete 2017), the growth in their daily usage among increasingly large sections of the US population (69% of US adults use social media) make them a useful tool to obtain quick, cost-effective opinions from the general public. While it is true that younger populations are more likely to be on social media (88% of adults aged 18–29 use these sites), the age gap among users of social media has started to close. As of early 2018, 78% of adults aged 30–49 used social media, 64% of adults aged 50–64 used these sites, and 37% of people over the age of 65 also engaged in social media use (Pew Internet Research 2018). Despite some recent concerns over privacy protection on social media, studies have shown that many people continue to stay on these sites because of improved connections among people and organizations important in their lives as well as convenience in accessing information and news (Rainie 2018).

Social media data has been identified as viable sources for uncovering and addressing various social and geographical problems. A variety of geospatial research now utilizes social media sites as a source of data. For example, hazards and emergency planning investigations have explored the potential to use social media data as a real-time human sensory network to both locate immediate disasters and identify areas in need of humanitarian aid or political action (Crooks et al. 2013; Shelton et al. 2014). Data from social media has also been used to understand local neighborhood development, riots and social protests, and the relationship between mapmaking and the neoliberal state (Shelton et al. 2015; Crampton et al. 2013; Fekete and Warf 2013; Leszczynski 2012).

Urban planners have also looked to social media data sources. In their book, Ciuccarcelli, Lupi, and Simeone (2014), explore social media as a source of knowledge for urban planning and management, arguing that time-based and geo-located social media data should be complemented by traditional data collection methods such as surveys to provide more complete insights into the social life of urban spaces. Garcia Esparza, O'Mahony, and Smyth (2010) make the observation that real-time data from the web is far from structured, but offer an additional and valuable source of data that can improve recommendations for decision-making. As an example, Barry (2014) used photographs shared by online users of Flickr (a photo sharing platform of Google) to better understand public perceptions of livestock grazing in public spaces. Interestingly, this study showed that opinions and concerns shared on Flickr provided a perspective that is seldom expressed at public meetings or in surveys. Afzalan and Muller (2014) conducted a study where social media was tested as a communication tool to improve public participation in the planning of local green spaces, concluding that these web-based communication tools helped significantly to create a dialogue and build consensus among groups who do not normally participate in the planning process.

Web-based citizen data and social media are important avenues to explore in the context of urban planning and decision-making. These data sources offer constant inflow of citizen data which could support a faster pace of decision-making, which could be particularly valuable in rapidly growing urban areas where needs and desires of the population adjust as they experience changes in their urban fabric and environment. From an urban planning perspective, data (whether quantitative or qualitative) that comes with geographic information adds value because it allows for spatial analyses and geovisualization, which can benefit strategic planning.

Some social media platforms are more consistent than others in providing geographic information. Twitter, for example, allows their users to geolocate their tweets, yet geolocated tweets only account for 1% of all tweets (Morstatter et al. 2013). "Geobrowsers" – coined by Peuquet and Kraak (2002) – are browsers that use location as a first-level filter for a search, rather than keywords. Consequently, location (and its positional accuracy) plays an important role on these platforms. Therefore, geobrowsers such as Google Maps and Foursquare can provide more suitable social media data for geospatial research.

# 3 Case Study

Mecklenburg County, North Carolina (which encompasses the City of Charlotte) is used as a case study and to persuade the city government to consider the use of these newer sources of data and methods to monitor the quality of their parks. This metropolitan area was chosen as a case study for several reasons.

First, compared to other cities, Charlotte's ParkScore® has been ranked at the very bottom since 2012 (TPL 2016). Even if there is any likelihood that their score is being underestimated by TPL; their consistent low ranking may indicate that there is an opportunity for the City of Charlotte to improve its park system, which in turn could engage residents in more physical activity and in doing so, create positive community health outcomes.

Second, TPL (2016) reports that Charlotte spends about $47.14 per resident on their park system (including volunteer hours as well as public and non-profit spending), which is in the lowest 20% among the 97 cities in their study. Considering Charlotte's low spending, the city government may be particularly interested to leverage publicly available data from social media to complement the assessments they already perform about their park system.

Lastly, we – the authors – have a familiarity with the area. Rather than choosing an unfamiliar city as our case study, our knowledge of the local context of Charlotte puts us in a better position to meaningfully interpret results, and provide recommendations.

## *3.1 Study Area: Mecklenburg County and Charlotte, North Carolina*

Residents of Mecklenburg County – which encompasses the City of Charlotte – who report that they do not engage in physical activity have fluctuated between 17% and 22% since 2005 (MCDH 2016). These figures are consistently lower compared to North Carolina and nationwide averages (see Fig. 2). The Physical Activity Council indicates that 27.7% of the US population is inactive (Physical Activity Council 2016).

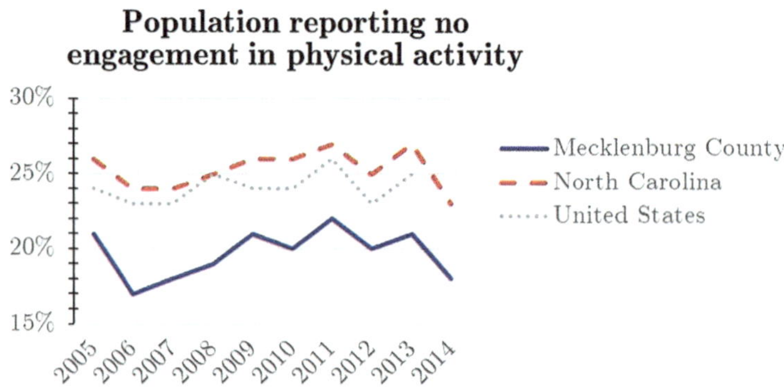


**Fig. 2** Population reporting no engagement in physical activity. (Source: Mecklenburg County Health Department. State of the County Health Report (2007–2015), available at: http://charmeck.org/mecklenburg/county/HealthDepartment/HealthStatistics/Pages/CommunityOverview.aspx)

The Mecklenburg County Department of Park and Recreation (DPR) manages 210 parks and facilities, accounting for over 21,000 acres of parkland (see Fig. 3). They distinguish 5 different public park types, namely neighborhood parks, community parks, regional parks, nature preserves and recently they added one urban park. Neighborhood parks are meant to be proximity parks, they are usually small in size and do not have large parking space. Community parks have more amenities, are larger, often have a recreational center attached to them and provide parking. Regional parks are large and offer several amenities including trails. Various learning activities for youth are organized at regional parks. Some nature preserves are accessible to the public and offer trails. However, many of them are there for conservation purposes or watershed quality assurance. Finally, Romare Bearden park is the first and only urban park so far in Mecklenburg County. This 5th park type is being used since 2015 to designate parks located within the business district of the City of Charlotte, which encompasses 4 out of the 462 neighborhood areas. This label emphasizes the proximity to urban amenities such as walkable shopping areas, museums, etc. For this case study, we are taking into account all park types within Charlotte. Figure 3 also shows "greenway entrances" and "other parks" that are managed by a local government other than the county's DPR. For example, "Park On Wilgrove" (a.k.a. "Mint Hill Municipal Park") which is located in Southeast Mecklenburg County, is managed by the Town of Mint Hill. The locations shown in Fig. 3 were retrieved in 2015 from Mecklenburg County's online Open Mapping resource.

The Charlotte-Mecklenburg Planning Commission is aware of the potential impact of their parks and recreational facilities on their residents' health, which is why, in 2012, they included proximity to parks and recreation as one of the neighborhood quality of life indicators.

The Quality of Life Study for Charlotte and Mecklenburg County – which started in 1993 as the *City Within A City Neighborhood Assessment* – provides neighborhood-level information on social, housing, economic, environmental, and safety conditions. In 1998, The University of North Carolina at Charlotte partnered with the

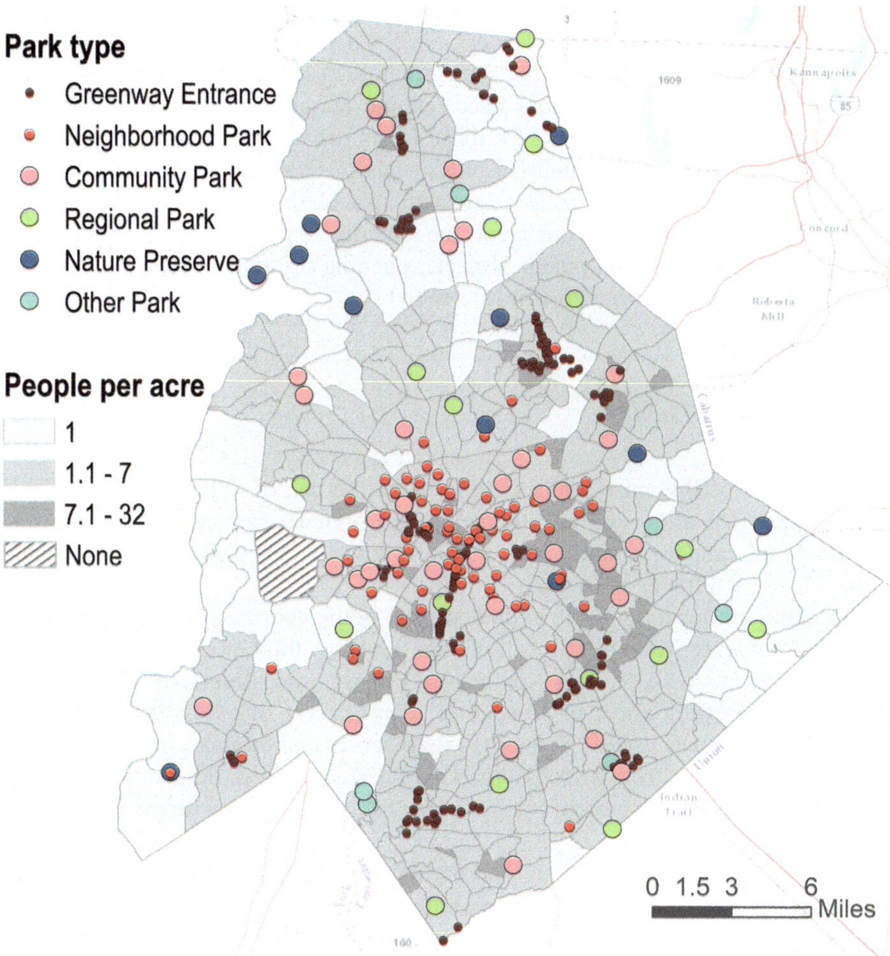


**Fig. 3** Location of public parks in Mecklenburg County (North Carolina)

Charlotte-Mecklenburg Planning Commission to continue and expand this assessment of neighborhoods. Their first report, renamed the *Charlotte Neighborhood Quality of Life Study*, was published in 2000 and has been published every other year since. In 2012, however, instead of a report, the format of the Quality of Life Study was transformed to an interactive dashboard called the Quality of Life Explorer. The underlying information shown in Fig. 3, reflects the population density per Neighborhood Profile Area (NPA), which are geographic units used for the Quality of Life Study.

Charlotte (NC) is an urban area that is rapidly growing in population and that has seen major population and infrastructure changes in the last 5–10 years. Impactful regional changes include the addition of a new highway stretch, a new light rail line, a new ballpark, a new bike share program, and fast real-estate development.

This population growth can lead to a change in needs, lifestyles, and opinions. In this context, it is important that urban leaders develop a clear vision and make decisions in accordance with the evolution of their urban area and its changing population. That includes taking into account needs of the incoming population. In consequence, it is extremely important for those urban centers to acquire tools that can collect data quickly (or continuously) and can automatically interpret incoming data in order to make planning decisions that keep up with the pace at which the region is changing. Data from social media has the potential to provide quick overviews of public opinion about development projects, opening up time and resources for in depth analysis of areas where it is most needed.

## *3.2 Research Design*

The objective of this case study is to explore the use of data available on two geobrowsers that allow digital comments and reviews, namely Foursquare and Google Maps, and to quantify the public opinion about parks in Mecklenburg County using sentiment analysis.

For that, park locations were extracted from both geobrowsers, along with their respective reviews. To extract locations and reviews, the Google and Foursquare APIs were used. An API is an Application Programming Interface which allows registered users to connect to companies' data servers that host their data. The Python language was used to connect to their server and to make specific requests (e.g., extract reviews from a specific park in Charlotte). Once the request is sent in the proper format, the company's server sends back a response with the information requested.

In order to extract locations in (and around) Mecklenburg County, the requests were limited to a bounding box that encompassed Mecklenburg County's boundary delineated by Web Mercator coordinates (34.5498, −81, 4884) and (36.0097, −80.1149). In order to limit the search to locations that qualified as parks, we used each platform's labels. Google created around 90 "place types" which allows users to label a place with one or more of these place types. Among these place types is the type "park." Any location that is digitized in Google Maps is automatically labeled as an "establishment." When a user digitizes a park, they can label it using the "park" type, but they are not required to do so. Foursquare, on the other hand, created around 10 "venue categories" with over 1000 sub-categories. Among the main categories is the "Outdoors & Recreation" category, with "park" being one of its sub-categories. Similarly to Google Maps, any location added to their system is automatically categorized as a "venue," but users can add additional categories that apply.

With over 1000 sub-categories on Foursquare, each venue can be labeled leaving little room for misinterpretation. For example, there are separate sub-categories for "plaza" and "pedestrian plaza" and for "yoga studio" and "Pilates studio." The more limited number of place types in Google Maps, on the other hand, seems to leave more room for interpretation by users. We tested a search using the "park" place type in Google Maps, and the places returned by the API included a significant amount of RV parks, parking lots, cemeteries, and so forth. To filter out most places that do not fit the "park" definition we have in mind for this study, we developed a filter in the Python script that checks all the labels users have assigned to a location and excludes location that are also labeled with the following place types: "parking," "rv park," "cemetery," "place of worship," "church," "gym," "health," "spa," and "zoo." We did not develop a similar filter for Foursquare results.

In a nutshell, using the bounding box around Mecklenburg County and the "park" label on both platforms (with some filtering in Google Maps), locations were extracted together with their associated reviews from online users (see an example from Google reviews in Fig. 4). All requests to the Google and Foursquare APIs were done in May of 2016. The API's servers seem to only send back a subset of all reviews available throughout the entire online history. It is not made clear by the data providers how that selection process works.

Sentiment analysis (a form of data mining) applied to data from reviews left by park users on geobrowsers show new possibilities to monitor park visitor satisfaction in real time. Sentiment analysis has been increasingly used for big data web content

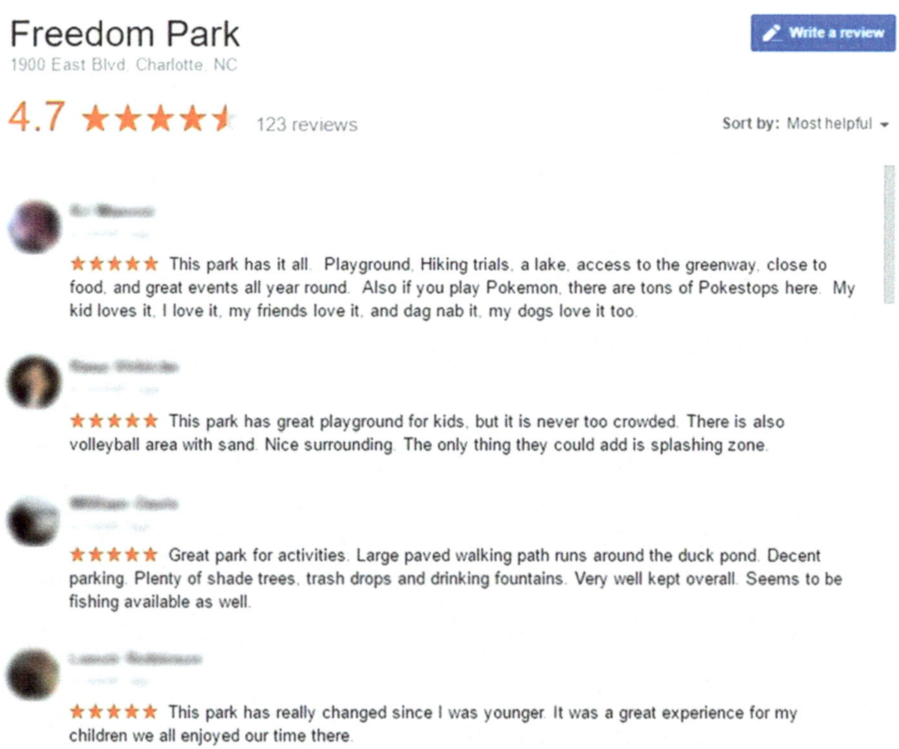


**Fig. 4** Example of reviews left on Google about Freedom Park in Charlotte

(Pang and Lee 2008; Nielsen 2011). It is also increasingly used to measure attitudes toward certain topics on social media, especially from tweets. For example, Paul and Dredze (2011) followed a number of Twitter users and tried to extract messages that were related to disease symptoms. To those tweets, they linked diseases these users could likely be diagnosed with, such as allergies, obesity, or depression and mapped the emergence of certain diseases at the US state level. Twitter messages have also been used as a predictor for stock markets (Bollen et al. 2011) and to better understand the public opinion regarding certain topics, such as vaccination (Salathé and Khandelwal 2011) or the Affordable Care act (Wong et al. 2015).

The opinionfinder algorithm (Wilson et al. 2005), which is a form of data mining, was developed by students and faculty at the University of Pittsburg (PA), which–among other uses, identifies whether certain words included in its preset dictionary are classified to be positive or negative. Based on this sentiment dictionary, it is possible to derive the sentiment of a sentence based on the words it constitutes. The sentiment score $x_t$, of a comment left on online commenting platforms can be analyzed with Eq. 1.

$$x_t = \frac{\text{percent}_t\left(\text{pos.words}\right)}{\text{percent}_t\left(\text{neg.words}\right)}, \tag{1}$$

Where $\text{percent}_t$ (pos. words) is the number of positive words divided by the total number of words constituting the comment and where $\text{percent}_t$ (neg. words) represents the number of negative words divided by the total number of words constituting the comment. The average sentiment scores from reviews posted about one park will determine the overall public sentiment at that park. For this particular method, if this ratio is above 1, the sentiment is considered positive, if it is below 1 it is considered negative and if it is equal to 1 it is considered neutral.

### 3.3 Methodological Limitations

First, online reviews that were extracted from Google Maps and Foursquare do not represent all reviews posted by users throughout the online history. The selection of reviews that is send back by their API server might not be a representative sample of all reviews posted on their respective platforms. Moreover, it is unclear what percentage of reviews is sent back by the server and whether that percentage is constant for each park (e.g., 1% of reviews per park). Second, comments or reviews cannot be extracted from public parks that do not have a "social media presence"; meaning parks that have not yet been digitized on these platforms. Until someone digitizes them, these public spaces will not be listed in any results nor can users leave reviews. Therefore, not all parks shown in Fig. 5 were found on the geobrowsers used in this case study. Third, some parks may be available on either platform, but may not have been labeled as a "park" and therefore would not be returned in the results based on the search terms we used. Moreover, even though we filtered the

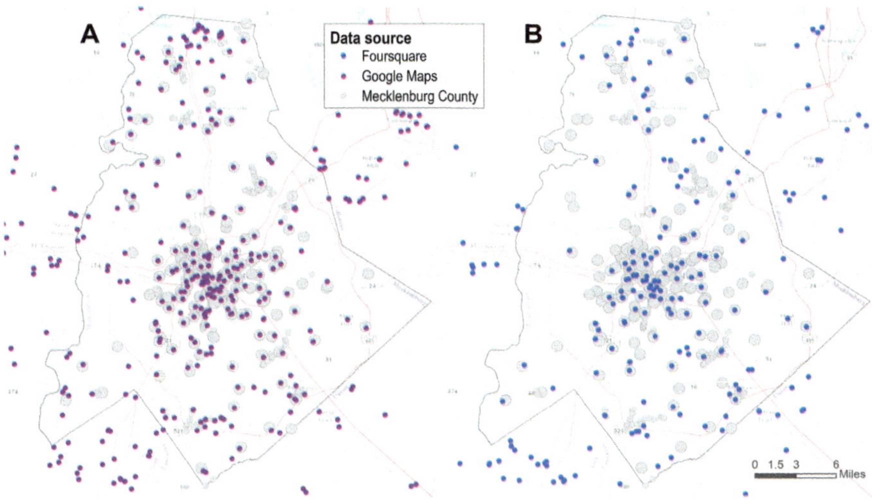


**Fig. 5** Park locations extracted from (**a**) Google Maps and (**b**) Foursquare. The underlying layer of gray dots represents the park locations provided by Mecklenburg County, which are shown in Fig. 3

results to exclude some locations that did not match our definition of a "park" (e.g., RV parks, parking lots, etc.), some locations were still returned in the results that didn't fit our definition. In other words, our filter did not work perfectly. Fourth, residents who never visit parks or do not use social media sites cannot be represented in online reviews. Fifth, the sentiment estimated by the opinionfinder system is based on a dictionary of English words that is not exhaustive. However, the Spanish-speaking community is growing in Mecklenburg County and comments left in Spanish cannot be interpreted using this dictionary. Sixth, the opinionfinder system uses an algorithm to put each word within the context of the entire sentence. Although their algorithm has high accuracy ratings, it cannot be guaranteed that each comments' sentiment is estimated correctly. Reviews written online often contain spelling mistakes and poor sentence structures, which may affect the accuracy rate of the opinionfinder system. Lastly, some parks were only left one comment or none at all, which makes sentiment analysis inadequate for these locations due to the small number problem.

## 3.4 Results

Using the Google Maps API to extract "park" locations, 504 locations were returned, of which 264 (52%) were located within the boundary of Mecklenburg County (see Table 1). Along with these locations, a total of 831 reviews were extracted from Google Maps, ranging from 0 and 12 reviews per location. With the Foursquare

**Table 1** Summary statistics on locations and reviews extracted from the Google Maps and Foursquare APIs

| | Total | *in* Mecklenburg | *outside* Mecklenburg |
|---|---|---|---|
| Extracted from *Google Maps* API | | | |
| *Locations...* | *504* | *264* | *240* |
| ...without any reviews | 253 | 130 | 123 |
| ...with 1 review or more | 251 | 134 | 117 |
| ...with only 1 review | 55 | 25 | 30 |
| *Reviews* | *831* | *388* | *443* |
| Minimim per location | 0 | 0 | 0 |
| Maximum per location | 12 | 12 | 11 |
| Extracted from *Foursquare* API | | | |
| *Locations...* | *436* | *148* | *288* |
| ...without any reviews | 291 | 96 | 195 |
| ...with 1 review or more | 145 | 52 | 93 |
| ...with only 1 review | 62 | 20 | 42 |
| *Reviews* | *357* | *144* | *213* |
| Minimim per location | 0 | 0 | 0 |
| Maximum per location | 17 | 17 | 8 |

API, 436 locations were returned, of which 148 (39%) were located within the boundary of Mecklenburg County. A total of 357 reviews were extracted from these locations, ranging from 0 and 17 reviews per location (see Table 1). Figure 5 shows the locations that were extracted from (a) Google Maps and (b) Foursquare on top of the park locations shown in Fig. 3. This figure shows how much overlap (or agreement) there is between data from the County versus data from the geobrowsers we used for this case study.

Using the opinionfinder system (Wilson et al. 2005) to process all reviews left by online users on Google and Foursquare about parks and recreational facilities, the overall sentiment was identified for each park.

Figure 6a shows all locations extracted from Google Maps. The color of the dot refers to the overall sentiment at that location based on all reviews left by users. A red dot refers to a negative sentiment, a green dot to a positive sentiment, and a yellow dot to a neutral sentiment. The size of each dot represents the number of reviews left by users at that location. Gray dots, however, represent parks at which no reviews were left by users. Figure 6b shows all locations extracted from Foursquare. Here, the same color scheme is used to represent parks with reviews that had positive, negative, or neutral sentiment. Note that in both Fig. 6a and b, parks that only have one review are in a separate category (smallest dot). Since it may not be adequate to measure the sentiment at a park based on only one review, the sentiment at these locations should be taken with a grain of salt.

To summarize the reviews left by Google users at each location, a word cloud (using tagul[1] tools) was generated for 6 parks where the overall sentiment of the

[1] Tagul provides a free and online tool to generate word clouds based on a text: tagul.com

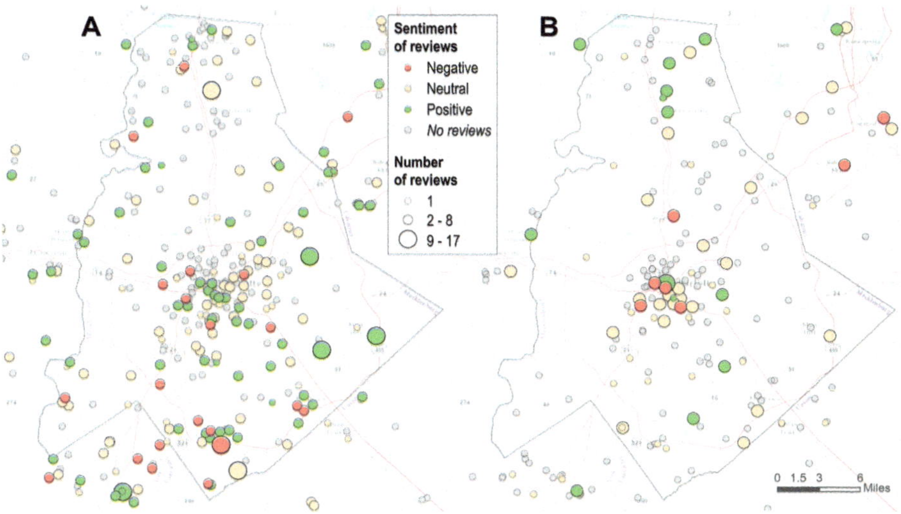


**Fig. 6** Aggregated sentiment from online reviews about parks expressed on (**a**) Google Maps and (**b**) Foursquare

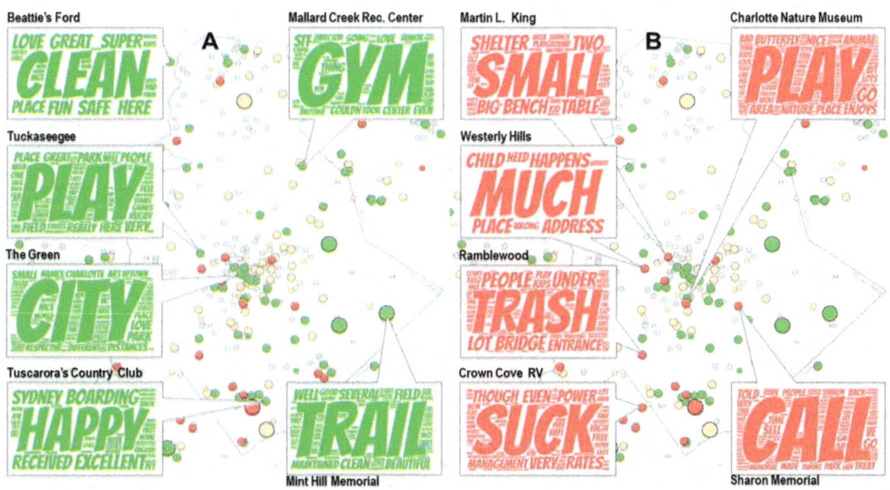


**Fig. 7** Sample of (**a**) positive and (**b**) negative reviews left about parks on Google Maps

reviews was positive (see Fig. 7a). The most common word that appears in reviews about Beattie's Ford Park is "clean," whereas reviews about The Green – which is a plaza near Charlotte's business district – contained the word "city" most frequently. This figure shows that different park locations generate different topics of discussion based on their available amenities. Figure 7b summarizes the reviews left by Google users using a word cloud for 6 parks where the overall sentiment of the reviews was negative. The most common word that appears in reviews about Ramblewood Park is "trash," whereas reviews about Martin Luther King Park contained the word "small" most frequently. Here again, different park locations generate different topics of discussion. To understand why "call" was the most frequent word at Sharon Memorial Park, the comments at that particular park were read fully. From the comments, it became clear that several unique users attempted to call the park before planning their visit and were not pleased because of the rudeness of the people receiving their call. It is important to mention that Sharon Memorial is a cemetery and Crown Cove is a recreational vehicle (RV) park, which shows that the filer we developed excluded some, but not all, locations that didn't fit our definition of a "park."

## 4 Discussion

Urban planners who are responsible for the upkeep of parks must keep in mind the local demographics of an area. While social media use is an easy stepping off point from which to start to understand local attitudes about urban greenspaces, some communities may not be represented through social media reviews. Fortunately, social media use has grown among minority populations in the USA; Hispanics and African Americans now use social media in higher percentages than the white population (Pew Internet Research 2018). However, for check-in and review services such as Foursquare, African Americans are less represented on the platform than whites and Hispanics (Fekete 2015). By extracting online reviews from additional media platforms such as Twitter or Yelp, more content can be collected and analyzed, but it may also improve representation. It is likely, however, that gaps in representation will remain. For example, elderly populations are the segment of the US population that is least likely to use social media (Pew Internet Research 2018).

Reviews extracted from Google Maps and Foursquare do not seem to represent all reviews posted by users throughout the online history. Since it is unclear what percentage of reviews is sent back by the API's server and whether that percentage is constant for each park (e.g., 1% of reviews per park), it is difficult to assess whether any sample is representative and/or comprehensive. Moreover, some parks were only left one comment or none at all, which makes sentiment analysis unfeasible for these locations due to the small number problem. For that reason, word-clouds in Fig. 7, were not made for parks that only have a small number of comments.

Data-mining techniques such as sentiment analysis are freely available and easy to use, which makes this tool feasible for regular monitoring of park reviews. There are, however, limitations that are important to take into account during the decision-making process. For example, due to the current language barriers in the opinion-finder system, non-English reviews should either be singled out or parks located in communities with higher rates of non-English speakers should be assessed manually. This is important because social media use has grown among minority populations in the USA, such as Hispanics (Pew Internet Research 2018) which may express themselves on social media in Spanish rather than English. Testing the performance of the sentiment analysis could be validated by having humans code the sentiment of a sample of reviews and measure the difference between both sentiments obtained for the same reviews.

Finally, it is important to note that when the distribution of sentiment of a location is bi-modal (some very positive and some very negative) or multimodal, the overall sentiment at that location may be shown as having "neutral" sentiment. Neutral may not be an appropriate label for some of these distributions. For example, a bi-modal distribution of sentiment may be better characterized as "polarized" sentiment, rather than "neutral" sentiment. One aspect that may be important to consider is the "seasonality of sentiment." For example, reviews left at parks in the winter may be more negative than those left during the summer. Comments left during a "less ideal" season, may thus negatively impact the overall sentiment about a particular location. These comments are still valuable for planners; however, this is an important consideration to take into account.

# 5 Recommendations

The internet has firmly become entwined with the places and spaces of everyday activity of daily life and social media constitute a part of this web of connections (Kitchin and Dodge 2011). If city parks want to remain viable options for people to visit, thereby improving the overall health of an area, parks should not only pay attention to online reviews, but also have an active presence on social media sites. In an era of smart cities and big data, the actions and searches people are conducting online have a direct effect on the daily leisure activities those same people are performing offline. City governments need to ensure that their amenities are not only physically accessible, but virtually accessible as well, first by making sure all their parks are digitized and labeled as "parks" on different social media platforms. By increasing their online presence, visitation to urban parks could also be encouraged.

Extracting data solely from social media will not provide a good representation of the overall population. Therefore, these data should be collected as a complement to already existing data collection methods such as public surveys, rather than as a substitute. One major benefit city governments should leverage from this suggested complementary data collection is to learn from data provided through social media first, before doing costly surveys. Missing representation in population or topics can be identified through social media and can help target the money, time and effort spent on conducting surveys or interviews by extending it to groups of people that will ensure a more representative assessment.

Data analysts should be aware of the limitations of these techniques within the local context. Having a panel of two or three human judges to manually code a small sample of the reviews would provide both a reasonable estimate of the method's accuracy and invaluable qualitative commentary concerning the sort of urban-planning insights available in social media data. The sentiment estimated by the opinionfinder method is based on a dictionary of English words that is not exhaustive. In Mecklenburg County, the Spanish-speaking community is growing, making non-English reviews more likely. The needs of this community must to be included in the decision-making process; therefore the assessment of parks located in communities with higher rates of non-English speakers should be assessed using a data-mining technique that can handle other languages or through public surveys.

If city governments decide to analyze data from social media, they should share their intentions with their residents clearly and straightforwardly for at least two reasons. First, residents will likely want to know how their information will be used because of growing privacy concerns around the use of such data by companies or governments. Second, someone visiting a park is probably not as likely to leave a review as someone visiting a coffee shop. When we provide an online review of a coffee shop we have a certain confidence that the owner will read it. At a park on the other hand (or any other public amenity), leaving an online review may not feel worthwhile because it of a perception that it will not be read. City governments should encourage visitors to leave reviews online and assure them their comments will be taken into account. This could increase the number of park reviews and thus provide additional data to analyze.

Charlotte is an urban area that is rapidly growing in population and that needs more strategic decision-making tools. Social media is an excellent source of real-time, cost-effective data to monitor the satisfaction of their residents and the health of their urban system. This will require city governments to incorporate such approaches within their existing structure, which will be challenging considering that they are likely trying to keep up with the pace at which their neighborhoods are currently developing. However, it is likely that other departments already leverage data from social media for other purposes. For examples Twitter data are increasingly used by governments for disaster management. Therefore, consolidating a team of geographers and computer scientists that can handle data analytics for different aspects of city governments could be most effective.

## 6 Conclusion

Measuring the health of urban areas will require us to track health indicators at the city level and monitor those over time. Provided with research findings in the literature that show the positive physical and mental health outcomes of people living nearby public and open spaces, the case study and arguments made in this chapter are to encourage cities to monitor resident's satisfaction with public amenities as an indicator for urban health. Public surveys are still the predominant data collection method used by local governments to monitor residents' satisfaction with their services, which are costly and time consuming. Moreover, due to population growth and increased migration to cities (and in between cities), survey results can quickly become outdated. Mecklenburg County, North Carolina (which encompasses the City of Charlotte), was used as a case study to explore the use of data available on geobrowsers – Google and Foursquare – and to showcase the application of sentiment analysis to extract the public's perception about their parks. The use of these and other social media platforms is recommended to city governments as an additional source of quantitative and qualitative data, which is generated by urban residents continuously, in real time, and captures needs, suggestions, and satisfaction of public spaces. Leveraging social media is not only a cost-effective complement to already existing data collection methods, but it also offers cities new ways to engage with their residents. Finally, studies on park satisfaction at the national level – such as the Trust for Public Land (TPL) that has been calculating a *ParkScore®* for cities across the USA since 2012 – may provide interesting comparisons but only rely on quantitative data and may not be appropriate or useful for local decision-making. City governments possess knowledge about the local context and have the means to collect and interpret qualitative data, which are important to analyze and take into account in local decision-making. The consistent low ranking by the TPL, of Charlotte's park system compared to other US cities, however, indicates an opportunity for the City of Charlotte to improve its public park system, which in turn could engage residents in more physical activity and in doing so create positive community health outcomes.

## References

Afzalan, N., & Muller, B. (2014). The role of social media in green infrastructure planning: A case study of neighborhood participation in park siting. *Journal of Urban Technology, 21*(3), 67–83.

Barry, S. J. (2014). Using social media to discover public values, interests, and perceptions about cattle grazing on park lands. *Environmental Management, 53*(2), 454–464.

Bauman, A., & Craig, C. L. (2005). The place of physical activity in the WHO Global Strategy on Diet and Physical Activity. *International Journal of Behavioral Nutrition and Physical Activity, 2*(10), 10. https://doi.org/10.1186/1479-5868-2-10.

Bedimo-Rung, A. L., Mowen, A. J., & Cohen, D. A. (2005). The significance of parks to physical activity and public health: A conceptual model. *American Journal of Preventive Medicine, 28*(2), 159–168. https://doi.org/10.1016/j.ampre.2004.10.024.

Bollen, J., Mao, H., & Zeng, X. (2011). Twitter mood predicts the stock market. *Journal of Computational Science, 2*(1), 1–8.

Ciuccarelli, P., Lupi, G., & Simeone, L. (2014). *Visualizing the data city: Social media as a source of knowledge for urban planning and management*. Cham: Springer Science & Business Media.

Crampton, J., & others. (2013). Beyond the Geotag: Situating 'big data' and leveraging the potential of the GeoWeb. *Cartography and Geographic Information Science, 40*(2), 130–139.

Crooks, A., Croitoru, A., Stefanidis, A., & Radzikowski, J. (2013). Earthquake: Twitter as a distributed sensor system. *Transactions in GIS, 17*(1), 124–147.

Curran, W., & Hamilton, T. (2012). Just green enough: Contesting environmental gentrification in Greenpoint, Brooklyn. *Local Environment, 17*(9), 1027–1042.

Dai, D. (2011). Racial/ethnic and socioeconomic disparities in urban green space accessibility: Where to intervene? *Landscape and Urban Planning, 102*(4), 234–244. https://doi.org/10.1016/j.landurbplan.2011.05.002.

DHHS, United States. Department of Health. (2008). Physical Activity Guidelines for Americans: Be Active, Healthy, and Happy! (Vol. 36). Government Printing Office, Washington, DC.

Dony, C. C., Delmelle, E. M., & Delmelle, E. C. (2015). Re-conceptualizing accessibility to parks in multi-modal cities: A Variable-width Floating Catchment Area (VFCA) method. *Landscape and Urban Planning, 143*, 90–99.
Fekete, E. (2015). Race and (online) sites of consumption. *Geographical Review, 105*(4), 472–491.
Fekete, E. (2017). Foursquare in the city of fountains: Using Kansas City as a case study for combining demographic and social media data. In *Thatcher, J., Eckert J., and A. Shears (Eds.), Thinking big data in geography: New regimes, new research,* (pp. 165–88). University of Nebraska Press, Lincoln, NE, USA.
Fekete, E., & Warf, B. (2013). Information technology and the "Arab Spring". *The Arab World Geographer, 16*(2), 210–227.
Frieden, T. R. (2010). A framework for public health action: The health impact pyramid. *American Journal of Public Health, 100*(4), 590–595. https://doi.org/10.2105/AJPH.2009.185652.
Garcia Esparza, S., O'Mahony, M. P., & Smyth, B. (2010, September). On the real-time web as a source of recommendation knowledge. In *Proceedings of the fourth ACM conference on recommender systems (RecSys '10), Barcelona, Spain (pp.305-308)*. Association for Computing Machinery, New York, NY. https://doi.org/10.1145/1864708.1864773.
Hill, J. O., Wyatt, H. R., Reed, G. W., & Peters, J. C. (2003). Obesity and the environment: Where do we go from here? *Science, 299*(5608), 853–855. https://doi.org/10.1126/science.1079857.
Kitchin, R., & Dodge, M. (2011). *Code/space: Software and everyday life*. Boston, MA: MIT Press.
Koohsari, M. J., Kaczynski, A. T., Giles-Corti, B., & Karakiewicz, J. A. (2013). Effects of access to public open spaces on walking: Is proximity enough? *Landscape and Urban Planning, 117*, 92–99. https://doi.org/10.1016/j.landurbplan.2013.04.020.
Leszczynski, A. (2012). Situating the GeoWeb in Political Economy. *Progress in Human Geography, 36*(1), 7289.
Lopez, R. P., & Hynes, H. P. (2006). Obesity, physical activity, and the urban environment: Public health research needs. *Environmental Health, 5*(25). https://doi.org/10.1186/1476-069X-5-25.
MCDH, Mecklenburg County Department of Health. (2016) 2015 Mecklenburg County State of the Country Health Report, Mecklenburg County Department of Health, Health Statistics and Epidemiology. Available at: charmeck.org/mecklenburg/county/HealthDepartment/HealthStatistics
McGinnis, J. M., & Foege, W. H. (1993). Actual causes of death in the United States. *JAMA, 270*(18), 2207–2212. https://doi.org/10.1001/jama.1993.03510180077038.
McGinnis, J. M., Williams-Russo, P., & Knickman, J. R. (2002). The case for more active policy attention to health promotion. *Health Affairs, 21*(2), 78–93. https://doi.org/10.1377/hlthaff.21.2.78.
Moon, G., & Gillespie, R. (1995). *Society and health: An introduction to social science for health professionals*. Routledge, London, UK. ISBN-13: 978-0415110228.
Morstatter, F., Pfeffer, J., Liu, H., & Carley, K. M. (2013, July). Is the sample good enough? comparing data from twitter's streaming API with Twitter's Firehose. In *Seventh international AAAI conference on weblogs and social media, Cambridge, MA, USA (pp. 400-408)*. Association for the Advancement of Artificial Intelligence, Palo Alto, CA, USA. https://www.aaai.org/ocs/index.php/ICWSM/ICWSM13/paper/view/6071/6379.
Mueller, N., Rojas-Rueda, D., Basagaña, X., Cirach, M., Cole-Hunter, T., Dadvand, P., et al. (2016). Urban and transport planning related exposures and mortality: A health impact assessment for cities. *Environmental Health Perspectives, 125*, 89–96. https://doi.org/10.1289/EHP220.
Nielsen, F. A. (2011). A new ANEW: Evaluation of a word list for sentiment analysis in microblogs. In *Proceedings of the EWSC2011 workshop on 'Making Sense of Microposts': Big things come in small packages, Heraklion, Greece*. (pp. 93–98). https://arxiv.org/abs/1103.2903.
Pang, B., & Lee, L. (2008). Opinion mining and sentiment analysis. *Foundations and Trends in Information Retrieval, 2*(1–2), 1–135.
Parks, S. E., Housemann, R. A., & Brownson, R. C. (2003). Differential correlates of physical activity in urban and rural adults of various socioeconomic backgrounds in the United States. *Journal of Epidemiology and Community Health, 57*, 29–35. https://doi.org/10.1136/jech.57.1.29.

Paul, M. J., & Dredze, M. (2011, July). You are what you Tweet: Analyzing Twitter for public health. In *The Fifth International AAAI Conference on Weblogs and Social Media (ICWSM-11), Barcelona, Spain,* (pp. 265–272). Association for the Advancement of Artificial Intelligence, Palo Alto, CA, USA. https://www.aaai.org/ocs/index.php/ICWSM/ICWSM11/paper/view/2880/3264.

Peuquet, D. J., & Kraak, M. J. (2002). Geobrowsing: Creative thinking and knowledge discovery using geographic visualization. *Information Visualization, 1*(1), 80–91.

Pew Internet Research. (2018). *Social media fact sheet.* Available at: http://www.pewinternet.org/fact-sheet/social-media/. Accessed 31 July 2018.

Physical Activity Council. (2016). *2016 participation report.* Available at: physicalactivitycouncil.com. Accessed 20 May 2016.

Rainie, L. (2018). Americans' complicated feelings about social media in an era of privacy concerns. *Pew Internet Research.* Available at: http://www.pewresearch.org/fact-tank/2018/03/27/americans-complicated-feelings-about-social-media-in-an-era-of-privacy-concerns/. Accessed 31 July 2018.

Rosenberger, R. S., Sneh, Y., Phipps, T. T., & Gurvitch, R. (2005). A spatial analysis of linkages between health care expenditures, physical inactivity, obesity and recreation supply. *Journal of Leisure Research, 37*(2), 216.

Salathé, M., & Khandelwal, S. (2011). Assessing vaccination sentiments with online social media: Implications for infectious disease dynamics and control. *PLoS Computer Biology, 7*(10), e1002199.

Shelton, T., Poorthuis, A., Graham, M., & Zook, M. (2014). Mapping the data shadows of Hurricane Sandy: Uncovering the sociospatial dimensions of big data. *Geoforum, 52*, 167–179.

Shelton, T., Poorthuis, A., & Zook, M. (2015). Social media and the city: Rethinking urban socio-spatial inequality using user-generated geographic information. *Landscape and Urban Planning, 142*, 198–211.

TPL, Trust for Public Land. (2010). *The economic benefits of the park and recreation system of Mecklenburg County, North Carolina.* Available at: tpl.org/charlottemecklenburg-county-park-value-report. Accessed 15 Mar 2014.

TPL, Trust for Public Land. (2016). *ParkScore index.* Available at: parkscore.tpl.org. Accessed 16 May 2016.

Wilson, T., Wiebe, J., & Hoffmann, P. (2005, October). Recognizing contextual polarity in phrase-level sentiment analysis. In *Proceedings of the conference on human language technology and empirical methods in natural language processing (HLT'05), Vancouver, Canada* (pp. 347–354). Association for Computational Linguistics, Stroudsburg, PA, USA. https://doi.org/10.3115/1220575.1220619.

Wolch, J. R., Byrne, J., & Newell, J. P. (2014). Urban green space, public health, and environmental justice: The challenge of making cities 'just green enough'. *Landscape and Urban Planning, 125*, 234–244. https://doi.org/10.1016/j.landurbplan.2014.01.017.

Wong, C. A., Sap, M., Schwartz, A., Town, R., Baker, T., Ungar, L., & Merchant, R. M. (2015). Twitter sentiment predicts affordable care act marketplace enrollment. *Journal of Medical Internet Research, 17*(2), e51.

**Coline C. Dony** is a Senior Geography Researcher at the American Association of Geographers (AAG). She holds a Ph.D. from the University of North Carolina at Charlotte in Geography and Urban Regional Analysis, where her research focused on understanding access to parks and recreation as a way to incite Charlotteans' health. Her research focuses on health geography and on how to leverage new sources of data, such as those made accessible by geospatial data providers and social media platforms. At the AAG, her research focuses on ways to modernize geography education and make it more inclusive.

**Emily Fekete** is the Social Media and Engagement Coordinator at the American Association of Geographers. Her research focuses on the geographies of media and communication, particularly social media, cyberterrorism, and online spaces of retail. She holds a PhD in geography from the University of Kansas.

# Part IV
# Health Policies and Urban Health Management

# Spatiotemporal Analysis and Data Mining of the 2014–2016 Ebola Virus Disease Outbreak in West Africa

**Qinjin Fan, Xiaobai A. Yao, and Anrong Dang**

**Abstract** This study investigates the spatiotemporal pattern of the 2014 Ebola virus disease (EVD) epidemic in the most heavily affected countries in West Africa and also mines the spatial associations between such pattern and other geographically distributed factors. Utilizing the publicly available open-source data, this study demonstrates a research design that integrates various geospatial data processing, analysis, and data-mining techniques to achieve the research objectives. For the 2014 EVD epidemic, spatiotemporal patterns were analyzed and visualized. Fine-grained population data were obtained through a population interpolation method to conduct healthcare accessibility analysis. Finally, associations between the spatiotemporal patterns of the incidences and healthcare accessibility as well as other factors were examined. The results suggest that (1) poor accessibility to healthcare facilities and EVD clusters are identified in many urban areas as well as some remote areas; and (2) EVD cases were more likely to be found in border areas of these countries where accessibility to healthcare facilities is poorer.

## 1 Introduction

Ebola virus disease (EVD), also called Ebola hemorrhagic fever, is a severe and deadly disease in humans and other primates. The virus can be transmitted through contact with a contagious person's bodily fluids such as blood, feces, and vomit which are most infectious (Green 2014). Symptoms of EVD typically start from 2 to 21 days after contacting the Ebola virus (Ganguly 2014). There had been more than 30 known EVD outbreaks since the first discovered in 1976 until the outbreak

Q. Fan (✉) · X. A. Yao
Department of Geography, University of Georgia, Athens, GA, USA
e-mail: qjfan@uga.edu; xyao@uga.edu

A. Dang
School of Architecture, Tsinghua University, Beijing, China
e-mail: danrong@mail.tsinghua.edu.cn

 
Y. Lu, E. Delmelle (eds.), *Geospatial Technologies for Urban Health*, Global Perspectives on Health Geography, https://doi.org/10.1007/978-3-030-19573-1_10

in 2014. The fatality rate of EVD cases varied between 25 percent and 90 percent in past outbreaks and the average was about 50 percent (Singh and Ruzek 2013). The rate during the 2014 outbreak was reported at slightly above 70%. The 2014 EVD in West Africa is by far the largest in history (CDC 2016). It started in December 2013 and was declared over in June 2016 by the World Health Organization (WHO). The peak time was in late 2014. Guinea, Sierra Leone, and Liberia were the most heavily affected countries with widespread and intense transmission. Figure 1 shows the general distribution of EVD cases at the district level in these most heavily affected three countries during this period. The map in Fig. 1 shows the total numbers of EVD cases in each district, providing a general first look without detailed consideration of context information such as population distribution and health facilities. The rate of spread also varied temporally in these regions. Figure 2 shows the weekly changes in the number of new cases in each of the three countries. The figure reveals that the EVD outbreak started at roughly the same time period in the three countries, while the intensity of the new cases varied a lot along the timeline.

Researchers have been making great efforts to understand the spreading process during the outbreak. Since the beginning of the epidemic, a good number of studies have been conducted by researchers in various disciplines. Much of the attention was paid on the biological or ecological perspectives of the Ebola virus itself and the transmission of EVD (e.g., Gatherer 2014; Baize et al. 2014; Carroll et al. 2015), as well as the surveillance of the disease (e.g., WHO Ebola Response Team 2014;

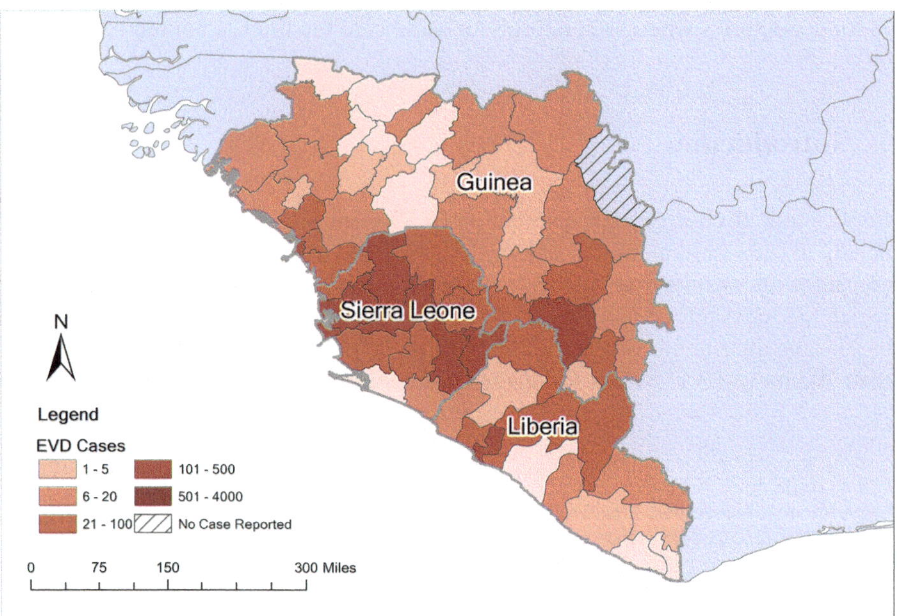


**Fig. 1** Cumulative EVD cases in 2014–2016 for Guinea, Sierra Leone, and Liberia

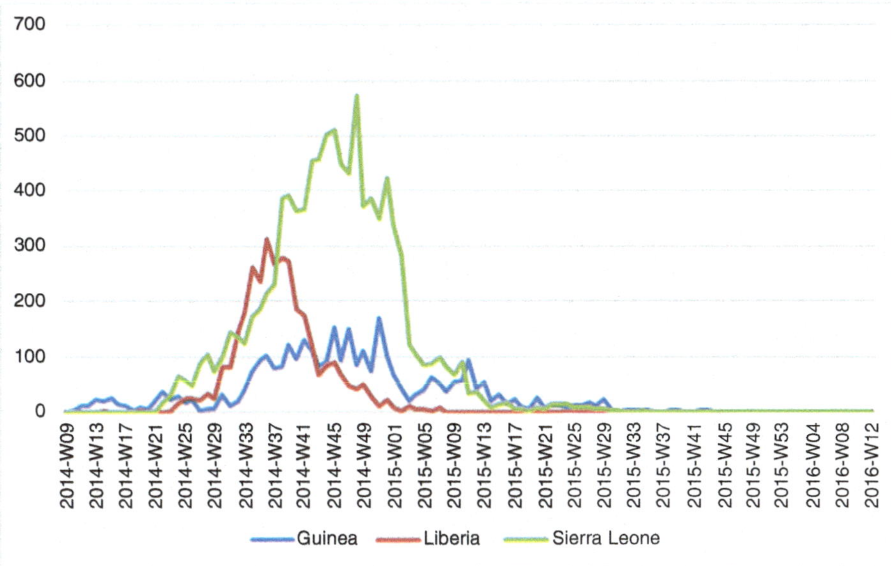


**Fig. 2** Weekly new cases in each country, February 2014–March 2016 (based on WHO weekly reports)

Bawo et al. 2015). A few recent studies examined the geographical pattern of the transmission and the epidemic path of the 2014 EVD outbreak (e.g., Kramer et al. 2016; D'Silva and Eisenberg 2017; Yang et al. 2015; Chowell and Nishiura 2015). While these studies proved that geographical features have significant impact on the transmission of the Ebola virus, prior studies provide little information about the observed geographical patterns. Likewise, to the best of our knowledge, no previous studies examined whether and how socioeconomic, health infrastructural, or geopolitical factors may attribute to the spatiotemporal patterns of the epidemic. However, identifying the spatiotemporal pattern of the epidemic and developing an understanding of the potential contributing factors are critical to effect planning of efforts to combat this or other outbreaks. The limited availability of necessary data for such analysis can be a major obstacle for timely investigation. A great data source is the World Health Organization (WHO) which provides weekly statistics of new EVD cases at the district level. However, very limited types of georeferenced data can be found for West Africa about the demographic, socioeconomic, or health infrastructural situation. Even if some data can be identified, the data type or spatial granularity may not suit the need for GIS analysis and spatial data mining. Therefore, the study has twofold objectives. Firstly, the study develops a methodological framework that can integrate those publicly available open-source data and, by taking advantage of various geospatial data analyses and data-mining techniques, make them sufficient and suitable for the geographical investigation for the second objective. The framework can be generally applicable for future studies of similar epidemic process, so that timely analysis can be performed based on readily available open data.

Secondly, the study aims to investigate the spatiotemporal pattern of the 2014 EVD outbreak and to examine possible associations between the pattern and other factors in the geographical context. In addition to the above two figures of general patterns, there are more nuanced variations at finer geographical scales. The study area includes the three most heavily affected countries in the outbreak, namely Guinea, Sierra Leone and Liberia. Such investigation will shed lights on our understanding of the spreading process. This may consequently help decision-makers and practitioners to make better planning for future combats, which is important for future practices to minimize the morbidity and mortality of EVD and other epidemics.

The paper is organized as follows. The following section reviews prior studies in the literature on epidemic diseases and the geospatial techniques to investigate patterns and relationships. Section 3 introduces the data and study area. The research design and methods of analysis are described in Sect. 4. Results and interpretations of them are presented in Sect. 5. The paper concludes with discussions of the findings and future research directions.

## 2 Prior Studies

A better understanding of the spreading process of an epidemic is essential for making control strategies and for stopping the outbreak. One of the core interventions to control the outbreak of EVD in West Africa was improving disease surveillance and outbreak detection (Bawo et al. 2015). However, research on the current West African EVD outbreak is so far insufficient and incomplete.

Several early studies were carried out within several months since the first EVD case was reported. The WHO Ebola response team conducted a surveillance for the first 9 months of the epidemic and estimated the fatality rate, incubation time and reproduction number (WHO Ebola Response Team 2014). Most of these early studies focused on the biological characteristics and examined the ecology of Ebola virus. Early investigations based on phylogenetic and epidemiologic analysis found the origin of EVD in Guinea (Gatherer 2014; Baize et al. 2014), tried to track early spread and to detect the dynamics of EVD based on models in epidemiology (Kiskowski 2014) followed by studies that aimed to gain insight into the evolution of the virus (Carroll et al. 2015). However, their results are specific to, and limited by, the models and observational data in use. For instance, their findings suggested that the EVD would slowly grow in West Africa and Liberia should have lower growth rates than those in Guinea and Sierra Leone (Shaman et al. 2014). In fact, Fig. 1 suggested otherwise. The rate of growth was higher in Liberia in the middle of 2014 and the country reached its peak earlier than the other two countries.

More recently, the geographic transmission pattern and epidemic path of Ebola virus were investigated with the use of different gravity models at a range of spatial scales (Kramer et al. 2016; D'Silva and Eisenberg 2017; Yang et al. 2015; Chowell and Nishiura 2015). D'Silva and Eisenberg (2017) established a spatial transmission model in a gravity-model framework to explain spatial temporal dynamics of

EVD in West Africa on both national and district scales. Yang et al. (2015) developed a spatial temporal inference method to investigate the spatial temporal progression of the infectious disease. Their findings proved that geographical characteristics have a strong impact on the transmission of Ebola virus. However, the previous studies provide little information about the observed patterns of EVD and did not investigate the specific impacts. Our study will fill this gap by investigating the spatial and temporal patterns of the outbreak and by learning their associations with the geographical contextual characteristics.

Understanding where space-time clusters of EVB occurred is important for public health intervention planning (Meliker and Sloan 2011). Such studies of spatiotemporal analysis can help researchers and investigators to simultaneously study the spread of disease over time and to illuminate the dynamics and unusual patterns of vector-borne diseases (Eisen and Lozano-Fuentes 2009). Scan statistics are widely used in disease surveillance to detect epidemic clusters, particularly in the context of an outbreak (Robertson et al. 2010). This approach was developed originally for temporal clustering by statistically testing whether the number of disease cases in a temporally defined subset exceeds the expectation given a null hypothesis of no outbreak (Robertson et al. 2010). It was first extended to the spatial dimension for spatial cluster detection by the Geographical Analysis Machine (Openshaw et al. 1987). Kulldorff further extended the scan statistics to space-time (Kulldorff 1997). A three-dimensional cylindrical search window is used in this approach, where the spatial search area is defined by the base of the cylinder and the temporal search area is defined by the cylinder height. The Kulldorff's space-time scan statistic has been used as a major analytical method for outbreak cluster detection (Tango et al. 2011). Studies used space-time scan statistics to detect the clusters of dengue fever (de Melo et al. 2012; Banu et al. 2012; Desjardins et al. 2018), West Nile (Lian et al. 2007), influenza (Ahmed et al. 2010; Mulatti et al. 2010), malaria (Gaudart et al. 2006) and other infectious disease.

Visualizing the space-time patterns of EVD clusters can display meaningful information about regions with the greatest burden of disease and help decision-makers to allocate health recourses. In previous studies, space-time clusters are generally visualized in two dimensions using small multiples. A few studies employed the 3D visualization techniques, such as space-time cube, to show the space-time patterns with 3D maps (Cheng and Williams 2012; Cheng and Wicks 2014). With this visualization approach, time becomes the third dimension reflecting temporal dynamics of disease transmission (Desjardins et al. 2018) or of crime concentrations (Nakaya and Yano 2010). Compared with the traditional visualization method of small multiples which display a series of maps with arbitrary time intervals, the 3D visualization provides better understanding of disease clusters and is particularly useful in identifying the geographical diffusion as well as the movement of clusters.

Accessibility to health facilities is critical for disease prevention and treatment. For infectious diseases, such as Ebola, health facilities can provide appropriate clinical treatments to avert patients' severe outcomes and isolation wards to reduce the chance of subsequent transmission. Access to healthcare is affected by where health services are located (supply) and where people reside (demand), yet neither health services nor population is uniformly distributed (Luo and Wang 2003). In previous research, a number of approaches have been used to estimate spatial access to healthcare facilities, including the distance or travel time-related measures (Hadley and Cunningham 2004; O'Neill 2003; Casas et al. 2017), Kernel density methods (Guagliardo 2004; Leibovici et al. 2007), and gravity-based methods (Joseph and Bantock 1982; Talen and Anselin 1998). Particularly, the two-step floating catchment area (2SFCA) method (Radke and Mu 2000; Luo and Wang 2003) has been used as a primary method to estimate spatial accessibility to health facilities. The fundamental idea of 2SFCA method is to define a service area (catchment area) of health facilities by a threshold travel time (or other types of travel cost) while accounting for the ratio between capacity of each facility and the potential demand for it. The traditional 2SFCA method is limited by the utilization of only a single catchment size within a small geographic area. The method has been modified, enhanced, or customized to suit the special situations of specific research problems (McGrail and Humphreys 2014; Chu et al. 2016). For example, Chu et al. (2016) revised the 2SFCA to account for the supply to minimize variability in spatial accessibility. In this paper, we used an enhanced 2SFCA method with multiple sets of catchment sizes and with consideration of urban/rule differences.

## 3 Data and Study Area

This study focuses on the three most affected countries in West Africa, namely, the three neighboring countries Guinea, Sierra Leone, and Liberia. Data used in this study include the EVD outbreak incidences and statistics, population distribution, road networks, land use and land cover data, as well as the geographical distribution of healthcare facilities.

Data about the EVD outbreak were obtained from the WHO report and a patient database it used. The WHO provides an epidemiological situation report which recorded the EVD cases from the first week of 2014 and regularly updated the situation of outbreak in West Africa. This report provides daily information on the numbers of suspected, probable or confirmed EVD cases at the district level. However, the report does not have more details about incidences or patients. Moreover, there is also the problem of underreporting with this data source. The non-hospitalized cases are not included in the database. The database covers 62 of the 63 administrative districts in the 3 countries, in which Guinea has 33 districts (data for the Mandiana district is not available) and Liberia and Sierra Leone have 15 and 14 districts, respectively. The data for each district includes weekly updates on the number of probable and confirmed EVD cases from January 1, 2014 to March 30, 2016.

Open GIS data of population distribution, health facilities, road networks, and land cover remote sensing images of the study area were collected and pre-processed for further analysis. The demographic data of the three countries were collected from the GeoHive website (www.geohive.com). We obtained the 2014 population data for the three countries at the district level. Health facility data were compiled by the Standby Task Force (www.standbytaskforce.org) from various sources. The dataset includes lists of the health facilities with their names, status, type, and location information. The types of the health facilities include public clinics, hospitals, health centers, Ebola treatment centers, as well as individual practitioner and health posts. There are a total of 1197 health facilities located in Guinea, 789 in Liberia and 1730 in Sierra Leone. Such data are georeferenced and pre-processed in ArcGIS to be ready for further spatial analysis. Figure 3 shows the locations of health facilities,

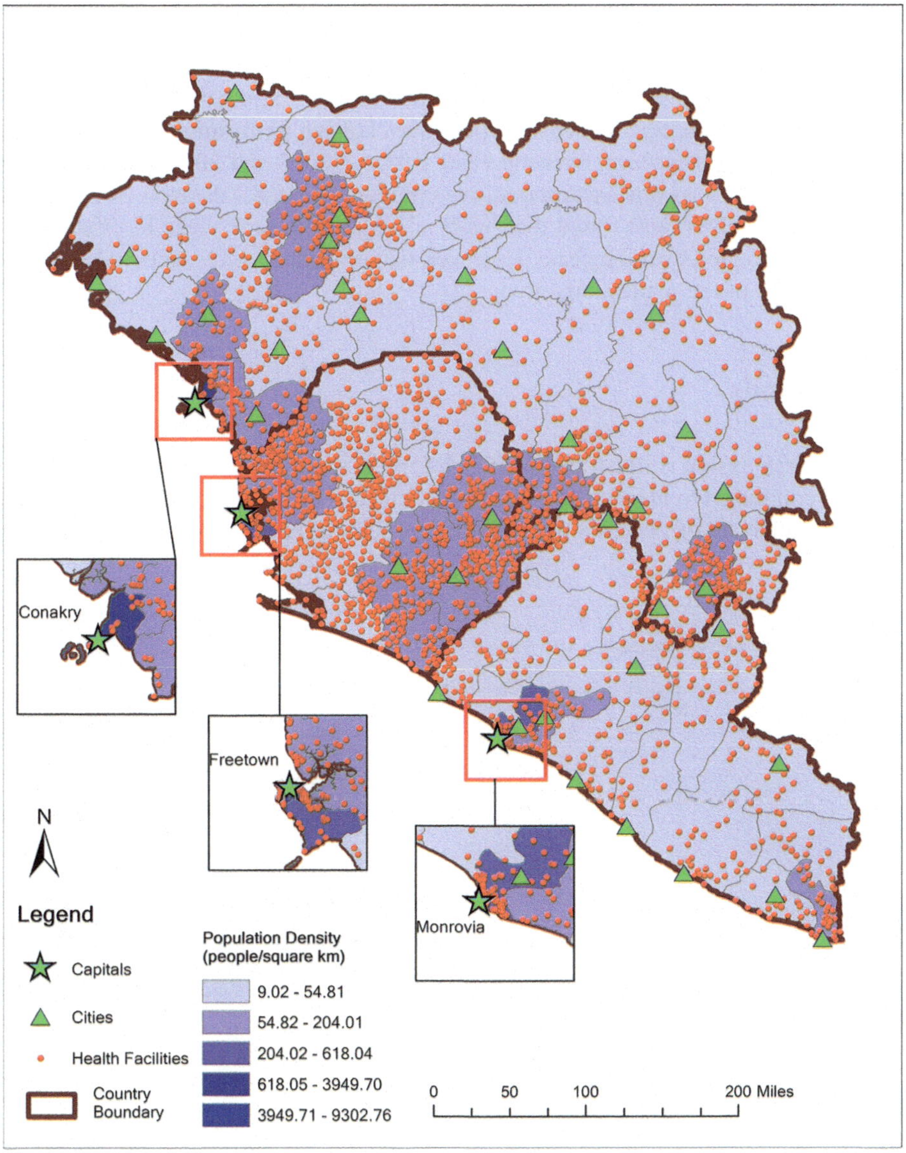


**Fig. 3** Locations of health facilities and cities in Guinea, Sierra Leone, and Liberia

cities, and general population distribution in the study area. Road network data were downloaded from the DIVA-GIS website. The dataset includes all primary and secondary roads in the three countries. Primary roads are those whose speed limit is set as 60 km/hour or above and the secondary road are those whose speed limit is set between 40 km and 60 km/hour. Satellite images of the study area are obtained from the Global Land Cover Facility. We selected the Global Land Cover 2000 product with the spatial resolution of 1 km. We realize that the data of year 2000 are outdated for the study period. However, these are the latest available data for this type of data suitable for population interpolation method. Other more updated land cover datasets, for instance, those from MODIS (Moderate Resolution Imaging Spectroradiometer), are available but have classification schemes that are unsuitable for the purpose of this study. Thus, we decided to use the outdated land cover data with the assumption that land cover types have changed uniformly in the study area at the clan and district level over the years between 2000 and 2014. To understand the degree of bias of this assumption, we compared MODIS datasets of 2000 and 2014 respectively. It was found that major changes only took place in urban areas that have expanded over the years. However, the urban areas only account for a very small portion of the while study area.

## 4 Methods

The research design is illustrated in Fig. 4. First, a spatiotemporal clustering analysis was conducted to identify spatiotemporal clusters of EVD incidence rates. Then a spatial analysis of people's accessibility to healthcare facilities was performed.

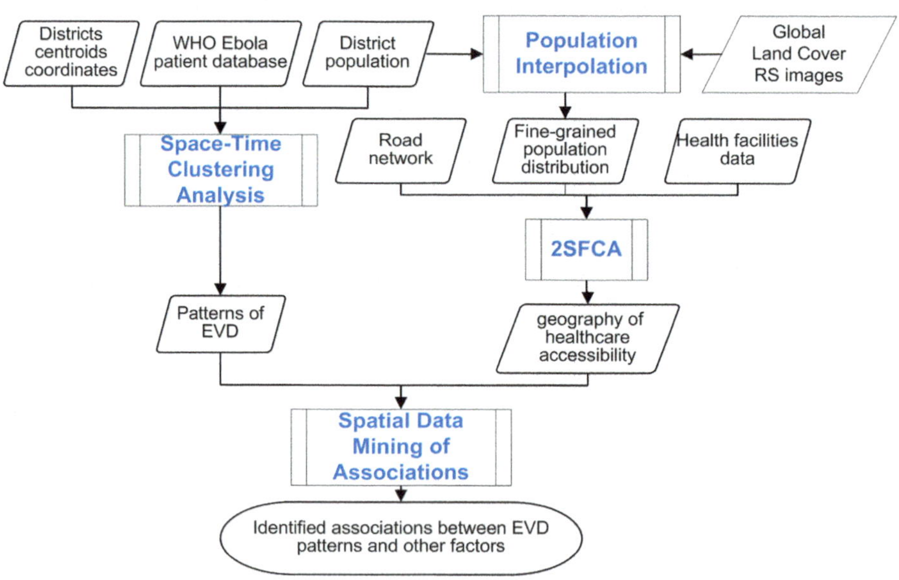


**Fig. 4** Workflow of the Research Method

Apparently, for the accessibility analysis, it is desirable to have the population data of a spatial granularity comparable to that of the EVD incidence data. However, the population data from open GIS data sources were only available at coarser spatial levels. To solve the problem, in the second step, the study applied a population interpolation technique to estimate the population distribution with the use of ancillary land use-land cover information of the geographical region. In the third step, a revised two-step floating catchment area (2SFCA) method was employed to investigate the geography of spatial accessibility to healthcare facilities. Finally, the study examined the association between the spatial pattern of the outbreak and that of the healthcare accessibility as well as the geography of other factors.

## 4.1 Investigating the Spatial Pattern of the Outbreak: Spatiotemporal Pattern Analysis

To examine the spatial pattern of the outbreak, a space-time scan statistic was applied on the geocoded EVD incidence data. This study applied the SatScan statistical method and the associated program to conduct space-time clustering analysis. This statistical approach examines the change of incidence rates in both spatial and temporal dimensions. A cylindrical scanning window in the space-time dimensions was used to scan over the map of disease cases. The base of the cylinder corresponds to the spatial area and the height of the cylinder corresponds to the temporal range of each search (Kulldorff et al. 2004). The size of the cylinder corresponds to the smallest spatial and temporal unit for the clustering analysis. Observed and expected numbers of cases inside and outside of cylinder were calculated and compared. Based on the assumption that the expected number of cases follow a Poisson distribution, a statistical hypothesis testing was then performed to identify the presence or absence of a cluster. The null hypothesis was that the risk of a disease is constant over space and time and thus should be the same inside and outside of the cylindrical scanning window. A log-likelihood ratio is defined in Eq. (1):

$$\mathrm{LLR}(Z) = \ln \frac{L(Z)}{L_0} = \left(\frac{c_z}{n_z}\right)^{c_z} \left(\frac{C - c_z}{C - n_z}\right)^{C - c_z} \quad (1)$$

where $L(Z)$ is the likelihood function for cylinder $Z$, $c_z$ is the number of observed EVD cases inside the cylinder, $n_z$ is the number of expected cases in it, and $C$ is the total number of observed cases in the study area for the entire time period. $L_0$ is the likelihood under the null hypothesis, which is a constant for the study area. The area inside the space-time cylinder may have an elevated risk (cluster) if the likelihood ratio is greater than 1. The technique uses different cylinder sizes, and the cylinder with the highest likelihood ratio is the most likely cluster. Following the method as implemented in SaTScan (Kulldorff et al. 2007; Kulldorff 1997), the maximum log-likelihood for EVD within each cylinder is calculated in Eq. (2):

$$T = \max_{z} \text{LLR}(Z) \tag{2}$$

For a cylinder *Z*, the maximum LLR is the most likely cluster and the corresponding *p*-value is used to determine whether it is considered a statistically significant cluster. The *p*-value is calculated through Monto Carlo hypothesis testing with 999 permutations. As the scanning cylinder is constantly and systematically moving its location over space and time, the approach can be used to detect all potential spatio-temporal epidemic clusters and provide early warning of the outbreaks.

## 4.2 Population Interpolation

The population data available from open-sources were aggregated at the district level for the three countries of study. In order to evaluate the spatial accessibility of people in each spatial unit to health services, population distribution at a finer level of spatial granularity is desirable. Thus, a dasymetric mapping algorithm was applied as a spatial interpolation technique (Kim and Yao 2010) to estimate population at the subprefecture (or clan, used interchangeably hereafter) level. The dasymetric mapping method refines the spatial granularity of population count data with the use of ancillary spatial information. The most commonly used ancillary information is land cover data, which are processed from remote sensing images and then stored as raster data. In this study, the land cover types were then reclassified into residential and nonresidential types. To improve accuracy, the study applied a multi-class classification scheme. The residential land cover classes were categorized as high-density, medium-density, and low-density residential areas, as shown in Table 1. With the multi-class dasymetric mapping method, population counts were redistributed to residential cells only, while the sum of the cell-based counts were kept consistent with the original population count data for each district following the method developed by Kim and Yao (2010).

In this study, satellite images with 1 km spatial resolution from the Global Land Cover 2000 product were used to produce the land use-land cover ancillary information. The land cover distribution of the data is shown in Fig. 5. Following similar studies in the literature, in the image classification process, training areas were

**Table 1** Reclassification of land cover classes into population density categories

| Class name | Land cover types |
|---|---|
| High density | Cities |
| Medium density | Mosaic forest/croplands/croplands with open woody vegetation/irrigated croplands |
| Low density | Closed evergreen lowland forest/degraded evergreen lowland forest/submontane forest (900–1500 m)/mangrove/deciduous shrub land with sparse trees/closed grassland/deciduous woodland/swamp bush land and grassland |
| No density | Waterbodies |

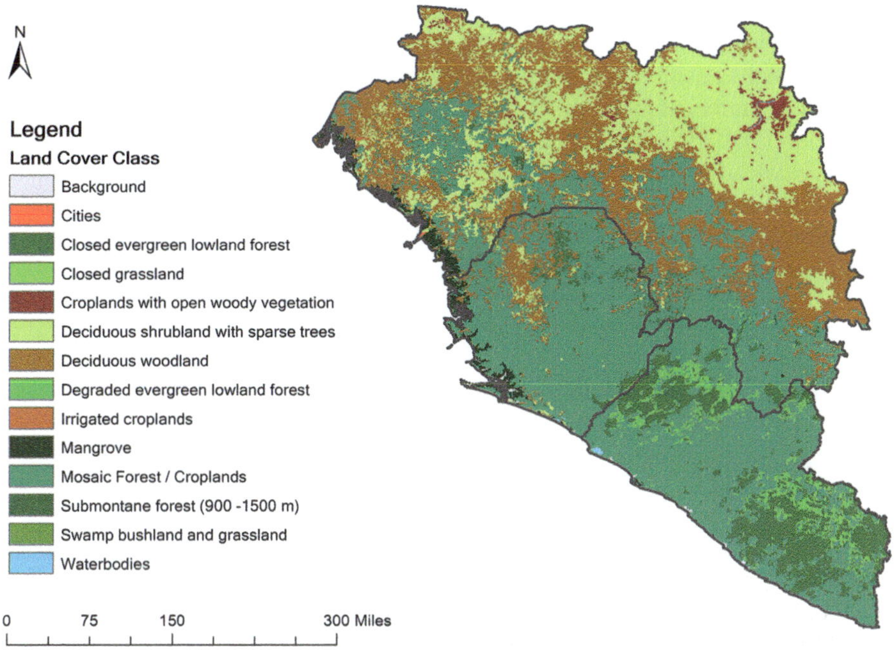


**Fig. 5** Land cover types in Guinea, Sierra Leone and Liberia based on GLC2000

identified where one of the density classes was dominant (Zandbergen and Ignizio 2010). Estimated population densities of particular land cover classes were derived from these training areas. The derived density estimates were then used in the redistribution of population from source areas to dasymetric zones. Each cell in the raster GIS data received an estimated population count according to its density class. Adjustment took place in the process, following the governing criterion which is to preserve the total population count in each district (the source data) in the process.

## *4.3 Measuring Spatial Accessibility to Health Facilities*

The two-step floating catchment area (2SFCA) method was developed in early 2000s (Radke and Mu 2000; Luo and Wang 2003) and has been used as a primary method to estimate spatial accessibility to health facilities. The main idea was to define a service area (so-called catchment area) of health facilities by a threshold travel time while accounting for the ratio between capacity of each facility and the potential demand for it. The capacity of service at a health facility is represented by the type and size of the facility. The potential demand is surrogated by the population in the service area. The method has been modified, enhanced, or customized to suit the special situations of specific research problems (e.g., McGrail and Humphreys 2014;

Luo and Qi 2009). The process in this study was mostly based on the modified version by McGrail and Humphreys (2014). The method was implemented in the following two steps with consideration of specific transportation situation in West Africa:

Step 1. Identify catchment areas of health facilities and calculate the provider-to-population (PtP) ratio in each catchment area. In the computing environment of a geographical information system (GIS), for each health facility location $j$, find all population locations ($i$) that are within an initial travel time ($d_{\text{init}}$) and a maximum travel time ($d_{\max}$). An impedance function $f(d_{ij})$ is added to reflect the fact that access is not uniform within the catchment area, where

$$
\begin{aligned}
f(d_{ij}) &= 1 && \text{for all } 0 < d_{ij} < d_{\text{init}} \\
f(d_{ij}) &= \frac{(d_{\max} - d_{kj})}{(d_{\max} - d_{\text{init}})^{\beta}} && \text{for all } d_{\text{init}} < d_{ij} < d_{\max} \\
f(d_{ij}) &= 0 && \text{for all } d_{ij} > d_{\max}
\end{aligned}
\tag{3}
$$

The provider-to-population ratio of facility $j$, denoted as $R_j$, is calculated for the catchment area by Eq. (4).

$$
R_j = \frac{S_j}{\sum_{i \in \{d_{ij} \le d_0\}} f(d_{ij}) P_i} \tag{4}
$$

where $P_i$ is the population at location $i$ whose centroid falls within catchment $j$ ($d_{ij} \le d_0$), $S_j$ is the number of health facilities at location $j$, and $d_{ij}$ is the travel time between $i$ and $j$. The step assigns an initial PtP ratio to each catchment area centered at all health facility locations.

Step 2. Identify the catchment areas of population locations and calculate the healthcare accessibility at each location. For each population location $i$, find all health facility locations $j$ that are within the maximum travel time ($d_{\max}$). If the number of facilities $n$ is greater than a threshold, the nearest $N$ facilities are considered. Then the impedance function is used to assign weights to each health facilities. The PtP ratios at these locations are summed up to calculate the accessibility from population centroid $i$ to heath facilities, following Eq. (5).

$$
\begin{aligned}
A_i^F &= \sum_{j \in \{d_{ij} < d_{\max}\}} f(d_{ij}) R_j && \text{if } n \le N \\
A_i^F &= \sum_{j \in \{d_{ij} < d_{iN}\}} f(d_{ij}) R_j && \text{if } n > N
\end{aligned}
\tag{5}
$$

where $A_i$ represents the accessibility at population centroid $i$ to all health facilities within the catchment area, $d_{ij}$ is the travel time between $i$ and $j$. For each population location $i$, a larger value of $A_i$ indicates better access to health facilities in the catchment area. The calculation of travel time is performed in TransCAD, a GIS-Transportation program. In the program, all population centroids and facility sites can be automatically connected to the road network with program-generated short connections between each of the centroids/sites and the closest node in the road network. These generated short connections can be interpreted as pseudo roads that implicitly account for other types of local roads or other modes of transportation to bring people to the closest primary or secondary roads.

The threshold values were chosen as follows: The initial threshold travel time $d_{init}$ was set to 30 minutes for cities. When a facility is within 30 minutes of travel to a population demand point, no impedance of distance will need to be considered for the accessibility between the two locations. The maximum time $d_{max}$ was set to 60 minutes for cities. For the area between the initial travel time and the maximum travel time, a distance decay function was used to calculate the impedance of distance. Facilities outside of the maximal travel time were not considered accessible. For rural areas, the initial threshold value was 60 minutes and the maximum was set to 120 minutes. The choices were made based on the consideration of reduced transportation services in the countries and the urban-rural differences. Special considerations for rural areas were often taken in studies of spatial access to healthcare facilities and similar time windows were used (Mcgrail et al. 2015).

### *4.4 Examine Spatial Associations*

In order to find possible associations between the spatial access to healthcare facilities and the clusters of EVD cases, the study employed a technique of spatial association rule mining to investigate it. A spatial association rule (SAR) describes the possible implication of one set of features (or characteristics) by another set of features (or characteristics). It can be expressed in the form of "$X \rightarrow Y$," where $X$ and $Y$ are sets of spatial predicates (Koperski and Han 1995). For example, "Ebola clusters are often located in areas of low accessibility to healthcare" is an SAR. An SAR is not a deterministic rule, but instead is tested statistically. Two important statistical concepts are defined by Koperski and Han (Koperski and Han 1995) to measure the statistical significance of an SAR. They are called *support* and *confidence* respectively. The support of a rule $X \rightarrow Y$ in a set of spatial objects $S$ is the probability that a member of $S$ satisfies pattern $X$. The confidence of $X \rightarrow Y$ is the probability that pattern $Y$ also occurs if pattern $X$ is found.

## 5 Results and Discussion

### *5.1 The Spatiotemporal Pattern of EVD Outbreak*

The retrospective space-time module in the SaTScan software was used to detect the geographic pattern of EVD during our study period from January 1, 2014, to March 30, 2016. The analysis was run for each week at the district level. The incubation period of EVD ranged from 2 to 21 days. This study chose the longest incubation period, which is 21 days, as the maximum temporal cluster size for the space-time clustering analysis. The maximum size of a spatial cluster was set at 50% of the population at-risk. A total of 15 clusters were detected during the study period at the statistical significance level of 0.01, which are listed in Table 2.

A graphical representation of the clusters is given in Fig. 6. Each circle represents an EVD cluster. The size of the cluster corresponds to the districts included in it. Among the 15 clusters, Cluster #1 is the most likely cluster, and it centers in Moyamba (Sierra Leone) from Week 39 to Week 41 in 2014. Cluster #2 is detected between weeks 45 and 47 in 2014 and it centers in Grand Cape Mount. Clusters #5 and #7 are centered in Grand Kru (Liberia) and Boffa (Guinea) respectively. Clusters #3 and #6 overlap with each other in space, and so do Clusters #9 and #10. These clusters vary in size. The largest cluster (Cluster #9) has a radius of 91 kilometers covering three districts. Clusters #4, #8, #11, #12, #13, #14 and #15 are all concentrated within one district respectively and do not have a radius.

To add the temporal dimension to the visual exploration of results, Fig. 7 uses a 3D visual model to show the clusters in space and time. The planar base represents the geographical space with a base map showing the three affected countries, while the vertical axis is the timeline. The 3D maps visualize the dynamics of the EVD clusters including their size, duration, and changes over time. It can be found that space-time clusters occurred during weeks between 38 and 49 in 2014. Along the time line, two major time periods saw the majority of EVD clusters. The time period of weeks 38–40 in 2014 has Clusters #3, #4, and #5 in it. The time period between weeks 45 and 47 in 2014 has 8 clusters in it including #2, #6, #9, #7, #11, #12, #14 and #15.

Figure 8 shows the spatial distribution of relative risks in districts that are located within space-time EVD clusters. A relative risk that is great than 1 means the number of observed EVD cases is higher than the expected cases in the respective district. It can be found that 13 out of 63 districts have a relative risk greater than 1 which suggests increased risks of EVD outcome in these areas. Bonthe in Sierra Leone has the highest relative risk (RR = 13.48) and is located in Cluster #1. For Cluster #2, Pujehun reported the highest relative risk of 13.09.

## 5.2 The Healthcare Service Shortage Areas

To examine people's accessibility to healthcare facilities, population interpolation was first conducted using the dasymetric mapping method as described in the Methods section. Figure 9 shows the population distribution in the three countries before (Fig. 9a) and after (Fig. 9b) interpolation. It is obvious that more detailed spatial variations can be captured after using the additional land use land cover information.

The 2SFCA analysis was performed in ArcGIS to estimate the shortage areas of health facilities. In the first step, a provider-to-population (PtP) ratio was generated for each health facility. The second step sums up PtP ratios within the proximity area

**Table 2** Spatiotemporal clusters based on patient data ($p = 0.001$)

| Cluster ID | Outbreak period | Outbreak districts | Observed EVD cases | Expected EVD cases | Relative risk |
|---|---|---|---|---|---|
| 1 | 2014 Week 39–41 (9/22/2014–10/12/2014) | Moyamba, Bonthe, Western Rural, Bo, Portloko, Tonkolili | 846 | 29.16 | 31.25 |
| 2 | 2014 Week 45–47 (10/27/2014–11/23/2014) | Grand Cape Mount, Bomi, Pujehun | 273 | 6.97 | 40.11 |
| 3 | 2014 Week 38–40 (9/15/2014–10/05/2014) | Lofa, Macenta, Gbarpolu | 230 | 10.44 | 22.47 |
| 4 | 2014 Week 38–40 (9/15/2014–10/05/2014) | Mali | 170 | 4.6 | 37.48 |
| 5 | 2014 Week 38–40 (9/15/2014–10/05/2014) | Grandkru, Maryland | 137 | 3.07 | 45.12 |
| 6 | 2014 Week 45–47 (10/27/2014–11/23/2014) | Yomou, N'zerekore, Bong | 109 | 14.36 | 7.65 |
| 7 | 2014 Week 45–46 (10/27/2014–11/16/2014) | Boffa, Fria, Dubreka, Boke | 96 | 11.47 | 8.43 |
| 8 | 2014 Week 47–49 (11/23/2014–12/07/2014) | Grandgedeh | 41 | 1 | 41.18 |
| 9 | 2014 Week 45–47 (10/27/2014–11/23/2014) | Koinadugu, Faranah, Bombali | 83 | 15.13 | 5.52 |
| 10 | 2014 Week 45–46 (10/27/2014–11/16/2014) | Mamou, Dalaba, Dabola | 48 | 6.74 | 7.15 |
| 11 | 2014 Week 45–46 (10/27/2014–11/16/2014) | Gaoual | 24 | 2.05 | 11.72 |
| 12 | 2014 Week 45–46 (10/27/2014–11/16/2014) | Dinguiraye | 24 | 2.07 | 11.63 |
| 13 | 2014 Week 46–48 (11/16/2014–11/30/2014) | Kankan | 16 | 0.99 | 16.13 |
| 14 | 2014 Week 45–46 (10/27/2014–11/16/2014) | Gueckedou | 24 | 3.08 | 7.80 |
| 15 | 2014 Week 45–46 (10/27/2014–11/16/2014) | Beyla | 24 | 3.44 | 6.99 |

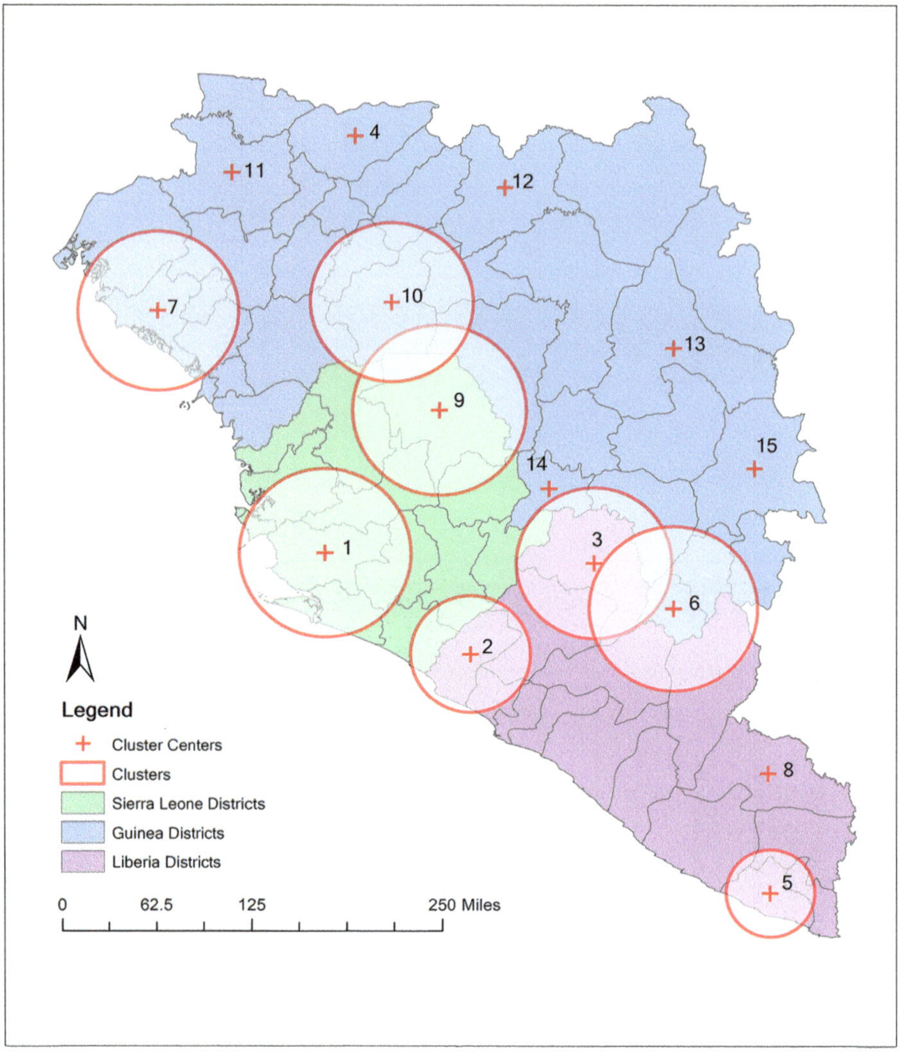


**Fig. 6** Spatiotemporal EVD outbreak clusters in Guinea, Sierra Leone, and Liberia 2014–2016

(catchment area) of each population location. This is to assess the spatial accessibility of that population location to healthcare. The analysis was performed separately for each country because the facilities are not accessible across borders during the outbreak. The health accessibility maps of the three countries at the clan level were then consolidated into one map in Fig. 10a. A darker symbol color indicates more severe shortage of healthcare services in the associate spatial unit. For spatial data mining in the next step, the spatial pattern of accessibility was compared with

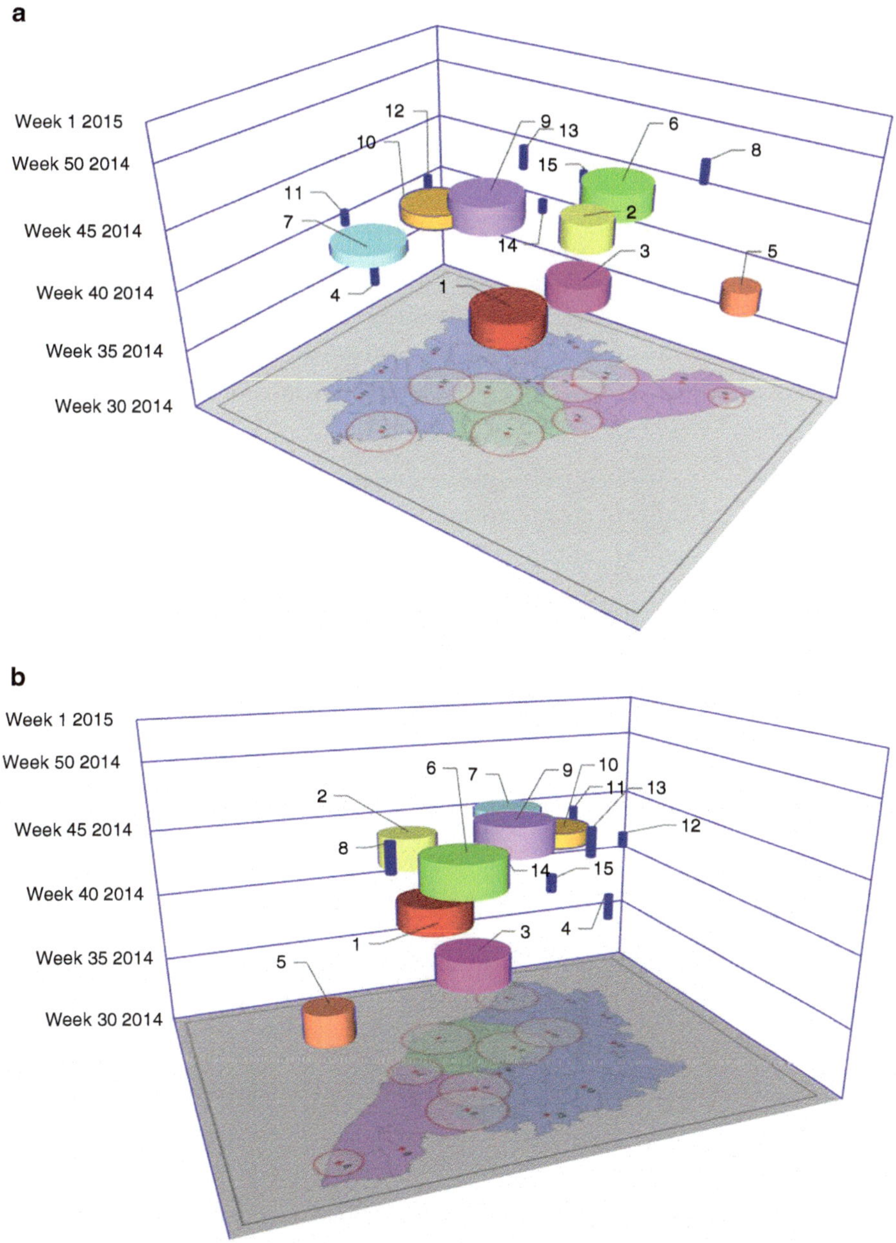


**Fig. 7** Three-dimensional visualization of the spatiotemporal clusters of EVD from two perspectives (**a**) viewing from the ocean (west) and (**b**) viewing from the continent (East)

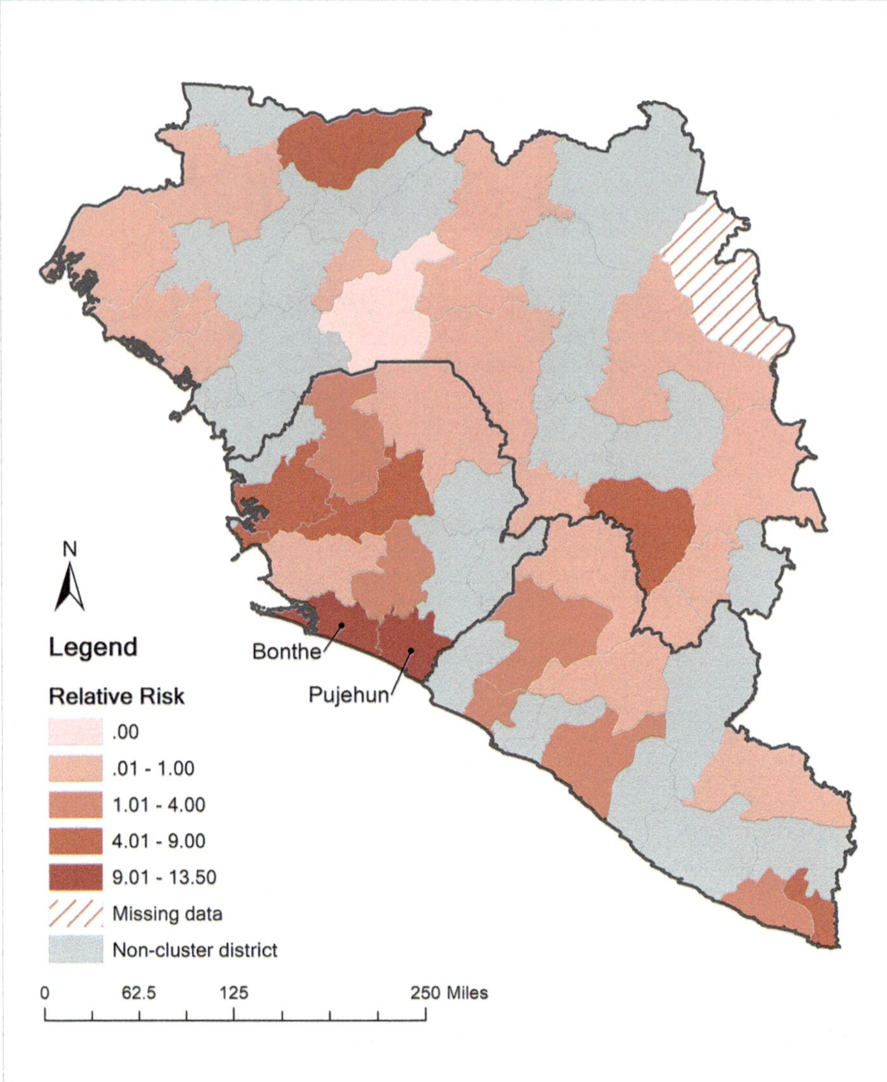


**Fig. 8** Statistically significant relative risk per district for Guinea, Sierra Leone and Liberia

patterns of other variables which are only available at the district level; thus, the clan-level accessibility measures were also aggregated to the district. The *PtP* ratios were aggregated using population weighted average. Using the natural breaks classification method, the resulting accessibility values for the districts were classified into five categories in order to show the varying levels of accessibility in space. Figure 10b shows the spatial accessibility to healthcare service at the district level vis-à-vis the EVD clusters.

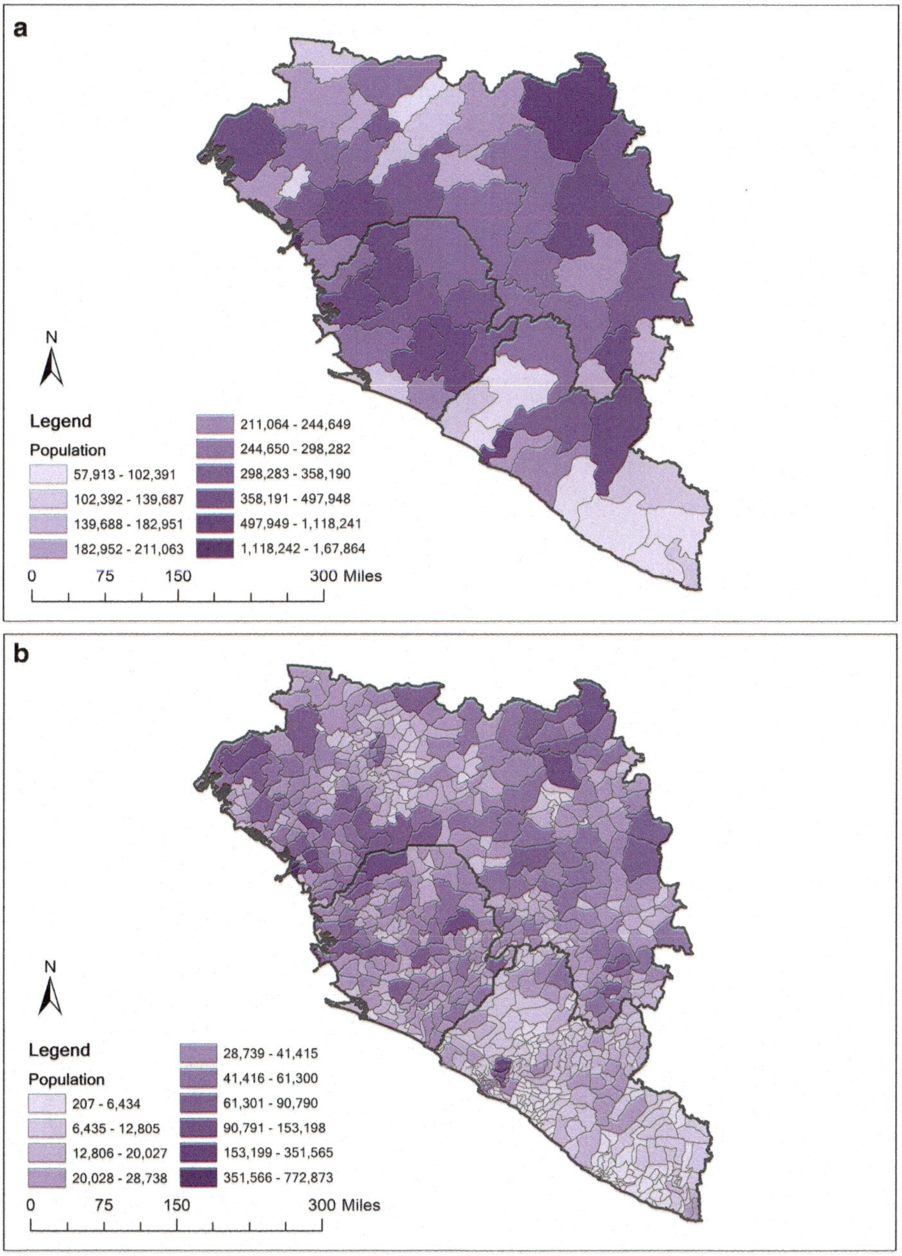


**Fig. 9** Population distribution at (**a**) district level and (**b**) subprefecture (clan) level

## 5.3 Spatial Data Mining

The study selected three sets of spatial characteristics of places and examined the possible associations among them. The first characteristic is whether a place is an EVD cluster. The second characteristic is about the geographic context of the

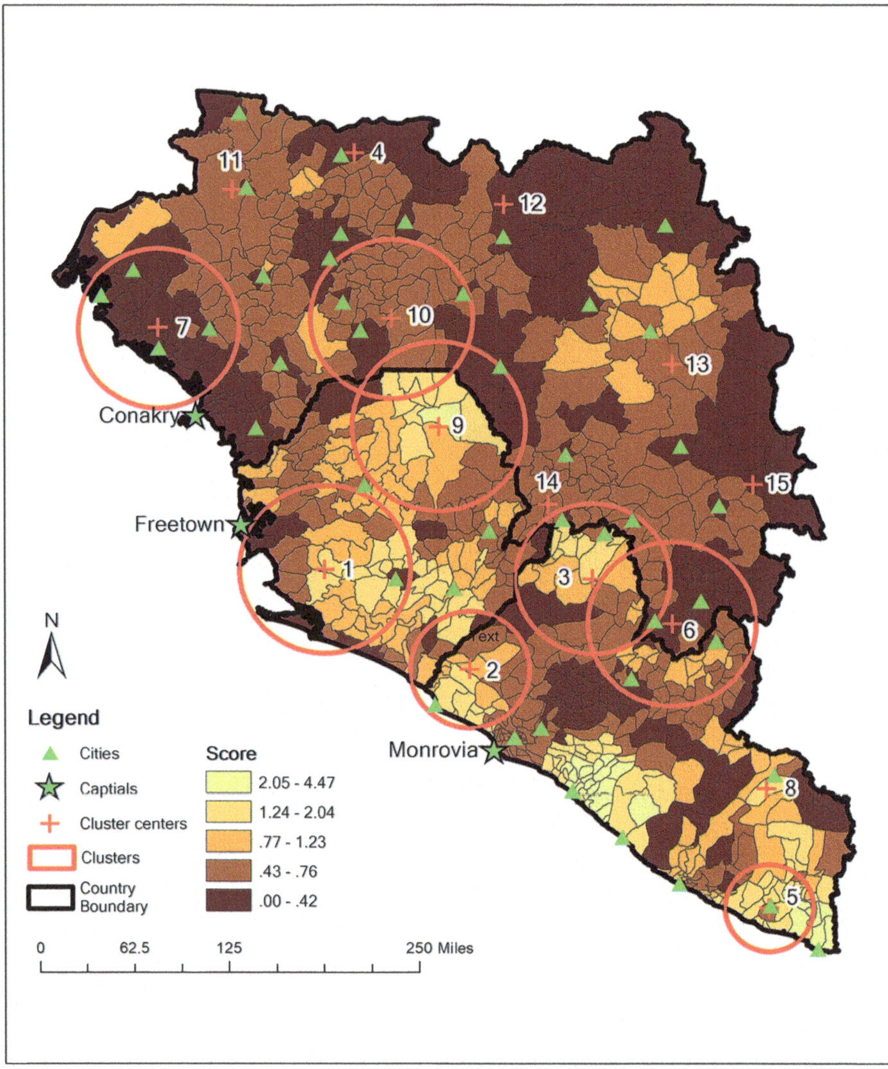


**Fig. 10** Spatial accessibility to health facilities in Guinea, Sierra Leone, and Liberia vis-à-vis EVD clusters that were identified using population data at (**a**) subprefecture (clan) level and (**b**) district level

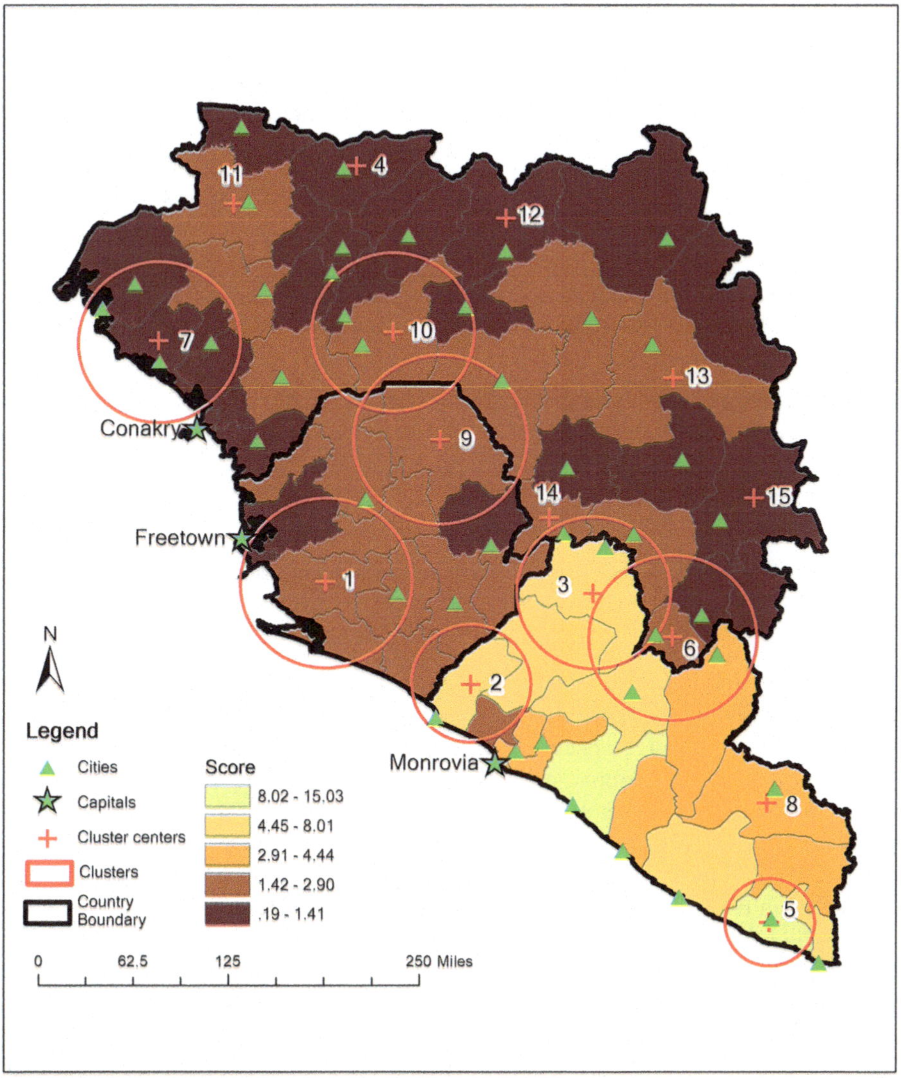


**Fig. 10** (continued)

place, namely whether it an urban area or a rural area, and whether it is located in a border area. The third characteristic is the level of accessibility to healthcare services. In consideration of the modifiable area unit problem (MAUP) and the sensitivity of findings to the spatial unit of data, the association analysis was performed at two levels of spatial units, respectively, namely the clan level and the district level. The clans, or subprefectures, are the finer spatial units nested within districts. It is the third level in the hierarchy of administrative divisions in the three

**Table 3** Identified association rules at the clan level (797 clans in total)

| Rule ID | If A | Then B (is likely true) | Conf. |
|---|---|---|---|
| ***Urban clans*** *(minimum support:0.6 minimum confidence: 0.7, N = 47)* | | | |
| 1 | If the clan has the lowest healthcare accessibility (Level 1) (32) | It is an interior clan (28) | 0.88 |
| 2 | If it is an interior clan (34) | It is more likely to have the lowest accessibility to healthcare service (28) | 0.82 |
| ***Rural clans*** *(minimum support:0.5 minimum confidence: 0.7, N = 750)* | | | |
| 2 | If not located in an EVD cluster (394) | It is an interior clan (298) | 0.78 |

Note: (1) The number in a parenthesis is the corresponding number of clans satisfying the condition; (2) Accessibility levels are in the classification scale of 1 (lowest) to 5 (highest)

**Table 4** Identified association rules at the district level

| Rule ID | If A | Then B (is likely true) | Conf. |
|---|---|---|---|
| *Minimum support: 0.45, minimum confidence: 0.7. N = 64* | | | |
| 1 | If it is in an EVD cluster (30) | It is a border district (25) | 0.83 |
| 2 | If the accessibility level is low (Level ≤ 2) (48) | It is a border district (35) | 0.73 |

Note: (1) The number in a parenthesis is the corresponding number of districts satisfying the condition; (2) Accessibility levels are in classified into five groups, ranging from 1 (lowest) to 5 (highest)

West Africa countries. The open-source data-mining program Weka 3.6 was used for the association rule mining. Data-mining results at the two levels are summarized in Tables 3 and 4, respectively. For the parameter setting, the minimum support was set to 0.45 or above and the minimum confidence was set to 0.7. This means a rule ($X \rightarrow Y$) will not be identified unless at least 45% of all cases satisfy predicate $X$ and at least 70% of those cases that satisfy predicate $X$ also satisfy predicate $Y$. For instance, among all 750 rural clans, 394 (or 52.8%) of them are not located in an EVD cluster (predicate $X$ for the identified Rule 2 in Table 3). Among these 394 clans, 298 (or 75.6%) of them are not interior clans (predicate Y for Rule 2 in Table 3). In general, the confidence is set high (70%) because we want to find a rule that take place at high probability. The support is set differently in different cases. It is 0.6 for urban clans, 0.5 for rural clans, and 0.45 for districts. The reason that the support parameter is set lower is because it is more about the popularity of cases where the identified rule is applicable, and not the validity of the rule. The popularity is controlled by factors that is not directly related to the validity of the rule. For instance, as shown in Table 4, the support is 0.45 because only 30 districts (about 47%) among all 64 districts happen to involve an EVD cluster in it. If we set the support too high (say, 0.5), we will not be able to learn about them.

At the clan level, a good understanding of the data helps us to better interpret the results. There are 797 clans, of which only 5.8% are in cities. About 29.4% of all clans are in border areas. Close to half of them (46.9%, or 375 clans) are either completely or partially located in EVD clusters. Among the 375 EVD-cluster clans, 18 (or 4.8%) of them are urban areas, 131 (or 34.9%) of them are in border areas, and 97 (or 25.9%) of them have the lowest level of accessibility to healthcare (level = 1). Because of the dominant presence of rural clans, association rules are biased toward rural clans if all clans are analyzed together. Therefore, the study conducted association rule mining on urban and rural clans separately. The results are summarized in Table 3. It suggests that, contrary to the common sense, many interior urban clans have low accessibility to healthcare resources. Our interpretation is that high population densities in urban areas can make healthcare resources relatively scarce constrained the capacity of each facility. For rural clans, clusters are very likely found to be associated with rural border clans. For urban areas, only 30% of them belong to border clans. Because the presence of urban border clans is not strong enough to meet the minimum support criterion (0.45), the identified association rules for urban clans are both associated with interior clans.

At the district level, as there was no rural and urban demarcation, the association rule mining was performed on all districts. The most significant findings at this level include the following: (1) Ebola clusters are more likely to be found in border districts, and (2) areas with low accessibility levels are more likely to be located in border districts.

## 5.4 Discussions

The study finds that counterintuitively, many urban clans are found to have very low accessibility to healthcare services. It is probably because inadequacy of healthcare facilities is severe in many urban areas. Contrary to common expectations, many areas in all three capital cities have low accessibility to health service. For instance, Cluster #1 covers the capital city of Sierra Leone and Cluster #2 is detected in the capital city of Liberia. Three reasons maybe accountable for this situation. First, as explained in the section of accessibility analysis, the travel time parameters were set differently for urban and rural areas, and the standards are much higher for urban areas. Thus, for the same geographical distribution of healthcare distribution, it is much more likely for an urban residential area to be classified into a lower level of accessibility than that for a rural area. Secondly, because the accessibility analysis was performed on the existing healthcare facilities, none of those temporarily established Ebola-specific healthcare services were included in the study. Thirdly, although cities are typically provided with more health services, it can still be inadequate due to high population densities and more severe socioeconomic disparities.

Border areas are found to be most vulnerable to EVD. Not all regions of poor health accessibility turn out to be part of the EVD clusters; however, those of them at the border between countries are most likely to be in an outbreak cluster. In fact, most of the identified clusters are along the border lines. The EVD Clusters #3 and #6 are found in the same general region at different time periods. It is the border region among the three countries, which is the primary region of the 2014 EVD outbreak. Many clans in this region have poor accessibility to healthcare facilities. Having multiple outbreak clusters repeatedly in the region suggests that people in these districts have higher chance to be infected. The reason that border areas are more vulnerable is probably associated with its socioeconomic and geopolitical position. The remote areas can be less economically active. In addition, the national and local administrations may have exerted more resources on other border-related matters and made insufficient efforts on providing healthcare services in the region. At the same time, the border areas may also be more likely to have transient population, which increases the chance of transmission. The findings suggest important implications for health management during a fight with disease epidemics. Targeted healthcare interventions may be particularly important for high density urban areas and remote/border areas.

## 6 Conclusion

This study presents a generally applicable framework to examine the spatiotemporal patterns of the 2014 EVD outbreak and the associations between the pattern and other spatially varying factors. The research design employed spatiotemporal clustering analysis, population interpolation, healthcare accessibility assessment, and spatial data mining to achieve the objectives based on readily available open data. It is also one of the first studies that apply the popular 2SFCA method for healthcare accessibility analysis at the multi-nation level that includes rural areas of coarse spatial granularity. The research design can be particularly useful for timely detection of trends before additional and more detailed data become available. The identified patterns and associations lead to two key findings. First, the border areas are most vulnerable to EVD outbreaks. Second, people in some urban areas, and particularly in big cities, are found to lack sufficient access to healthcare services. These findings provide evidences for the dire need of sufficient resources in identified problem areas so as to improve the efficiency of epidemic control. For instance, the findings strongly suggest that planners and practitioners needed to pay particular attention to the border areas and cities of high population densities. Such analysis can be helpful for the regions' governments and healthcare practitioners to make informed decisions to effectively reduce morbidity and mortality rates in future combats with epidemic diseases.

The present study has several limitations, which point out possible future research avenues. Although the study took careful considerations with the design and different parameter choices for urban and rural areas, the method can be further improved in the future. For instance, the current 2SFCA method produces a dichotomous measure, either access or no-access. Locations outside of the catchment areas are assumed to have no access at all. Also, it does not differentiate between different levels of impedances within the first level of catchments (within the minimum threshold). Moreover, few previous studies have applied the 2SFCA method in rural areas or large-scale regions. Therefore, without guidance from learned experiments reported in the literature, the choices of parameter settings, such as the threshold impedance values used for defining the catchment areas and the cut-off values of summed *PtP* ratios, deserve more careful evaluations. For future studies, a sensitivity analysis can be helpful to find out the most appropriate values for these parameters. Secondly, our association rule mining included only a small set of variables including urban/rural type, border/non-border, the level of healthcare accessibility, and status of EVD cluster. More contributing factors, such as the socioeconomic status, education level, and availability of health insurance, can be considered in future studies. In addition, future research efforts can also be made to improve the tools and techniques needed for the study. For instance, our spatial data mining was constrained by the association rule mining tool which only processes categorical data. Moreover, the scan statistics are popular but also have major limitations. For example, the cylindrical shape of the scan window and of the identified clusters may not reflect the true boundary of the outbreaks. In the future, other shapes of scanning windows can be adopted such as the linear, empty center circular, or ring-shaped scan windows. Other type of space-time clustering techniques, such as space-time K-function, can also be explored. Another future research direction is to explore the border effect in the process of epidemic diffusion. This study did find multiple EVD clusters across borders. At the same time, it is also obvious that different incidence rates are observed on the two sides of border lines (see Fig. 1). While this study only explored the patterns and revealed the phenomena, it is still open for further investigation whether the virus is transmitted across border and how the process works.

## References

Ahmed, S. S. U., et al. (2010). The space--time clustering of highly pathogenic avian influenza (HPAI) H5N1 outbreaks in Bangladesh. *Epidemiology & Infection, 138*(6), 843–852.

Baize, S., et al. (2014). Emergence of Zaire Ebola virus disease in Guinea—Preliminary report. *The New England Journal of Medicine, 371*(15), 1418–1425.

Banu, S., et al. (2012). Space-time clusters of dengue fever in Bangladesh. *Tropical Medicine and International Health, 17*(9), 1086–1091.

Bawo, L., et al. (2015). Elimination of Ebola virus transmission in Liberia—September 3, 2015. *Morbidity and Mortality Weekly Report, 64*, 979–980. Available at: http://www.cdc.gov/mmwr/pdf/wk/mm6435.pdf. Accessed 11 Sept 2015.

Carroll, M.W., et al. (2015). Temporal and spatial analysis of the 2014–2015 Ebola virus outbreak in West Africa. *Nature, 524*(7563), 97.

Casas, I., Delmelle, E., & Delmelle, E. C. (2017). Potential versus revealed access to care during a dengue fever outbreak. *Journal of Transport and Health, 4*, 18–29. https://doi.org/10.1016/j.jth.2016.08.001.

Centers for Disease Control and Prevention. (2016). Outbreaks chronology: Ebola virus disease. Available at: http://www.cdc.gov/vhf/ebola/outbreaks/history/chronology.html.

Cheng, T., & Wicks, T. (2014). Event detection using Twitter: A spatio-temporal approach. *PloS One, 9*(6), e97807.

Cheng, T., & Williams, D. (2012). Space-time analysis of crime patterns in central London. *ISPRS – International Archives of the Photogrammetry, Remote Sensing and Spatial Information Sciences, XXXIX-B2*(September), 47–52.

Chowell, G., & Nishiura, H. (2015). Characterizing the transmission dynamics and control of Ebola virus disease. *PLoS Biology, 13*(1), 1–9.

Chu, H. J., et al. (2016). Minimizing spatial variability of healthcare spatial accessibility—The case of a dengue fever outbreak. *International Journal of Environmental Research and Public Health, 13*(12), 1235.

de Melo, D. P. O., Scherrer, L. R., & Eiras, Á. E. (2012). Dengue fever occurrence and vector detection by larval survey, ovitrap and mosquiTRAP: A space-time clusters analysis. *PLoS One, 7*(7), e42125.

D'Silva, J. P., & Eisenberg, M. C. (2017). Modeling spatial invasion of Ebola in West Africa. *Journal of theoretical biology, 428*, 65–75.

Desjardins, M. R., et al. (2018). Space-time clusters and co-occurrence of chikungunya and dengue fever in Colombia from 2015 to 2016. *Acta Tropica, 185*(April), 77–85. https://doi.org/10.1016/j.actatropica.2018.04.023.

Eisen, L., & Lozano-Fuentes, S. (2009). Use of mapping and spatial and space-time modeling approaches in operational control of Aedes aegypti and dengue. *PLoS Neglected Tropical Diseases, 3*(4), 1–7.

Ganguly, S. (2014). Ebola hemorrhagic fever: A review on global facts, concepts and public health issues. *World Journal of Pharmaceutical Research, 3*(9), 401–404.

Gatherer, D. (2014). The 2014 Ebola virus disease outbreak in West Africa. *Journal of General Virology, 95*(Part 8), 1619–1624.

Gaudart, J., et al. (2006). Space-time clustering of childhood malaria at the household level: A dynamic cohort in a Mali village. *BMC Public Health, 6*(1), 286.

Green, A. (2014). Ebola emergency meeting establishes new control centre. *The Lancet, 384*(9938), 118. Available at: http://linkinghub.elsevier.com/retrieve/pii/S0140673614611478.

Guagliardo, M. F. (2004). Spatial accessibility of primary care: Concepts, methods and challenges. *International Journal of Health Geographics, 3*(1), 3.

Hadley, J., & Cunningham, P. (2004). Availability of safety net providers and access to care of uninsured persons. *Health Services Research, 39*(5), 1527–1546.

Joseph, A. E., & Bantock, P. R. (1982). Measuring potential physical accessibility to general practitioners in rural areas: A method and case study. *Social Science & Medicine, 16*(1), 85–90.

Kim, H., & Yao, X. (2010). Pycnophylactic interpolation revisited: Integration with the dasymetric-mapping method. *International Journal of Remote Sensing, 31*(21), 5657–5671.

Kiskowski, M. (2014). Description of the early growth dynamics of 2014 West Africa Ebola epidemic. *arXiv preprint arXiv:1410.5409*.

Koperski, K., & Han, J. (1995). Discovery of spatial association rules in geographic information databases. *In International Symposium on Spatial Databases* (pp. 47–66). Springer, Berlin, Heidelberg.

Kramer, A. M., et al. (2016). Spatial spread of the West Africa Ebola epidemic. *Dryad Digital Repository, 3*, 160294.

Kulldorff, M. (1997). A spatial scan statistic. *Communications in Statistics – Theory and Methods, 26*(6), 1481–1496.

Kulldorff, M., et al. (2004). Benchmark data and power calculations for evaluating disease outbreak detection methods. *Morbidity and Mortality Weekly Report, 53*, 144–151.

Kulldorff, M., et al. (2007). Multivariate scan statistics for disease surveillance. *Statistics in Medicine, 26*(8), 1824–1833.

Leibovici, D., et al. (2007). Extracting Dynamics of Multiple Indicators for Spatial recognition of Ecoclimatic zones in Circum-Saharan Africa. *GISRUK 2007*, 114.

Lian, M., et al. (2007). Using geographic information systems and spatial and space-time scan statistics for a population-based risk analysis of the 2002 equine West Nile epidemic in six contiguous regions of Texas. *International Journal of Health Geographics, 10*, 1–10.

Luo, W., & Wang, F. (2003). Measures of spatial accessibility to health care in a GIS environment: Synthesis and a case study in the Chicago region. *Environment and Planning B: Planning and Design, 30*(6), 865–884.

Luo, W., & Qi, Y. (2009). An enhanced two-step floating catchment area (E2SFCA) method for measuring spatial accessibility to primary care physicians. *Health & Place, 15*(4), 1100–1107.

McGrail, M. R., & Humphreys, J. S. (2014). Measuring spatial accessibility to primary health care services: Utilising dynamic catchment sizes. *Applied Geography, 54*, 182–188. https://doi.org/10.1016/j.apgeog.2014.08.005.
Mcgrail, M. R., et al. (2015). Spatial access disparities to primary health care in rural and remote Australia. *Geospatial Health, 10*, 358.
Meliker, J. R., & Sloan, C. D. (2011). Spatio-temporal epidemiology: Principles and opportunities. *Spatial and Spatio-temporal Epidemiology, 2*(1), 1–9.
Mulatti, P., et al. (2010). Evaluation of interventions and vaccination strategies for low pathogenicity avian influenza: spatial and space–time analyses and quantification of the spread of infection. *Epidemiology & Infection, 138*(6), 813–824.
Nakaya, T., & Yano, K. (2010). Visualising crime clusters in a space-time cube: An exploratory data-analysis approach using space-time kernel density estimation and scan statistics. *Transactions in GIS, 14*(3), 223–239.
O'Neill, L. (2003). Estimating out-of-hospital mortality due to myocardial infarction. *Health Care Management Science, 6*(3), 147–154.
Openshaw, S., et al. (1987). A mark 1 geographical analysis machine for the automated analysis of point data sets. *International Journal of Geographical Information System, 1*(4), 335–358.
Radke, J., & Mu, L. (2000). Spatial decompositions, modeling and mapping service regions to predict access to social programs. *Geographic Information Sciences, 6*(2), 105–112.
Robertson, C., et al. (2010). Review of methods for space-time disease surveillance. *Spatial and Spatio-temporal Epidemiology, 1*(2–3), 105–116. https://doi.org/10.1016/j.sste.2009.12.001.
Shaman, J., Yang, W., & Kandula, S. (2014). Inference and forecast of the current West African Ebola outbreak in Guinea, Sierra Leone and Liberia. *PLoS Currents, 6*. https://doi.org/10.1371/currents.outbreaks.3408774290b1a0f2dd7cae877c8b8ff6.
Singh, S. K., & Ruzek, D. (2013). *Viral hemorrhagic fevers*. London: CRC Press. Available at: https://books.google.com/books?id=WzzOBQAAQBAJ.
Talen, E., & Anselin, L. (1998). Assessing spatial equity: An evaluation of measures of accessibility to public playgrounds. *Environment and Planning A, 30*(4), 595–613.
Tango, T., Takahashi, K., & Kohriyama, K. (2011). A Space-Time Scan Statistic for Detecting Emerging Outbreaks. *Biometrics, 67*(1), 106–115.
WHO Ebola Response Team. (2014). Ebola virus disease in West Africa—The first 9 months of the epidemic and forward projections. *New England Journal of Medicine, 371*(16), 1481–1495. Available at: http://www.nejm.org/doi/abs/10.1056/NEJMoa1411100. Accessed 25 Sept 2016.
Yang, W., et al. (2015). Transmission network of the 2014-2015 Ebola epidemic in Sierra Leone. *Journal of the Royal Society, Interface/the Royal Society, 12*(112), 204–211. Available at: http://www.ncbi.nlm.nih.gov/pubmed/26559683, http://www.pubmedcentral.nih.gov/articlerender.fcgi?artid=PMC4685836.
Zandbergen, P. A., & Ignizio, D. A. (2010). Comparison of dasymetric mapping techniques for small-area population estimates. *Cartography and Geographic Information Science, 37*(3), 199–214.

**Qinjin Fan** is a fifth-year PhD candidate in the Geography Department at the University of Georgia. Prior to arriving at UGA, she earned a master's degree in geography at the State University of New York at Buffalo with a focus on Geographic Information Science. Her doctoral research mainly focuses on the spatial and temporal distribution of female breast cancer in the United States in the past 15 years. She is interested in the relationships between female breast cancer survival and the changes of socioeconomic, environmental and health policy factors. Her recent research involves the Ebola virus disease outbreak and spatial accessibility to healthcare facilities.

**Dr. Xiaobai Angela Yao** is Professor of Geography at the University of Georgia (UGA). Her research interests include geospatial data analytics, network science, location-based big data, and particularly the applications of them to study urban dynamics, human activities, and public health.

She obtained her Ph.D in Geography from the State University of New York at Buffalo, M.S. in GIS for urban applications from the International Institute of Aerospace and Earth Science (ITC) in the Netherlands, and her B.S. degree in GIS for Urban Planning and Management from Wuhan University (formerly WTUSM) in China. Dr. Yao is currently chair of the International Cartographic Association commission on Geospatial Analysis Modeling.

**Dr. Anrong Dang** is a professor of urban planning at the School of Architecture, Tsinghua University. He was first trained as a geographer and now a specialist in GIS applications in urban planning. He obtained his Ph.D. in Cartography and GIS from Chinese Academy of Science in 1997. He has been a professor in urban and rural planning at Tsinghua University since 2006. He published more than 150 papers and five textbooks. In recent years, his research interests focus on smart city and health city using information technology and big data.

# Extending Volunteered Geographic Information (VGI) with Geospatial Software as a Service: Participatory Asset Mapping Infrastructures for Urban Health

**Marynia Kolak, Michael Steptoe, Holly Manprisio, Lisa Azu-Popow, Megan Hinchy, Geraldine Malana, and Ross Maciejewski**

**Abstract** Community asset mapping is an essential step in public health practice for identifying community strengths, needs, and urban health intervention strategies. Community-based Volunteered Geographic Information (VGI) could facilitate customized asset mapping to link free and accessible technologies with community needs in a mutually shared, knowledge-producing process. To address this issue, we demonstrate a participatory asset mapping infrastructure developed with a Chicago community using VGI concepts, participatory design principles, and geospatial Software as a Service (SaaS) using a suite of free and/or open tools. Participatory mapping infrastructures using decentralized system architecture can link data and mapping services, transforming siloed datasets to integrated systems managed and shared across multiple organizations. The final asset mapping infrastructure includes a flexible and cloud-based data management system, an interactive web map, and community asset data stream. By allowing for a dynamic, reproducible, adaptive, and participatory asset mapping system, health systems infrastructures can further support community health improvement frameworks by facilitating shared data and

M. Kolak (✉)
Center for Spatial Data Science, University of Chicago, Chicago, IL, USA
e-mail: mkolak@uchicago.edu

M. Steptoe · R. Maciejewski
School of Computing, Informatics & Decision Systems Engineering, Arizona State University, Tempe, AZ, USA

H. Manprisio · L. Azu-Popow
Community Services/External Affairs, Northwestern Memorial HealthCare, Chicago, IL, USA

M. Hinchy
Consortium to Lower Obesity in Chicago's Children, Ann and Robert H. Lurie Children's Hospital, Chicago, IL, USA

G. Malana
Erie Humboldt Park Health Center, Chicago, IL, USA

 
Y. Lu, E. Delmelle (eds.), *Geospatial Technologies for Urban Health*, Global Perspectives on Health Geography, https://doi.org/10.1007/978-3-030-19573-1_11

decision support implementations across health partners. Such “community-engaged VGI” is essential in integrating previously siloed data systems and facilitating means of collaboration with health systems in urban health research and practice.

## 1 Introduction

Community asset mapping is an essential step in public health practice for identifying community strengths, needs, and ultimately health intervention strategies. The dominant method of community asset collection today incorporates siloed data systems, where each group constructs and maintains their data; within health systems, siloed data likewise challenges collaboration (Groves et al. 2013). While proprietary datasets encoding standardized methodology and data at the neighborhood level are on the rise, not all groups may benefit. While health practitioners increasingly develop interventions geared toward improved outcomes within an eco-social perspective, existing frameworks of community health remain siloed, rather than a desired state of shared ownership and collaboration (CDC 2015). Siloed approaches result in overlapping and redundant work, lack of communication and/or increased competition between groups, and both fragmented and incomplete datasets for all groups.

At the core of this challenge remains a mismatch of domain knowledge and technological expertise. Community organizations retain a deep view of their groups and topics but may not have the budget or programming expertise to abstract this content into data and maps. Tech-savvy groups hired or employed by clinical systems may have the ability to develop databases, maps, and analysis but can be limited in their depth of neighborhood knowledge. While multiple technologies exist for streamlined data management and use, new systems are needed to extend existing Volunteered Geographic Information (VGI) concepts to bridge community groups and health systems in collaboration. Community-based or “community engaged VGI” could facilitate customized asset mapping to link free and accessible technologies with community needs in a mutually shared, knowledge-producing process.

To address this issue, we demonstrate a participatory asset mapping infrastructure developed with a Chicago community using VGI concepts, participatory design principles, and geospatial Software as a Service (SaaS) using a suite of free and/or open tools. Participatory mapping infrastructures using decentralized system architecture can link data and mapping services, transforming siloed datasets to integrated systems managed and shared across multiple organizations. A community-engaged approach defines the infrastructure direction and fuses technological expertise with localized domain knowledge of community assets. Our approach focuses on community-based construction of the VGI process by co-developing a system that works for a specific community, rather than forcing the community to adapt to an existing system. First, we provide a background of participatory asset mapping and the modern data infrastructures available to support improved processes. We delve into the methods and results of the Chicago case study, and conclude with a discussion on the next generation of participatory asset mapping.

# 2 Background

## *2.1 Defining Participatory Asset Mapping*

Community asset mapping is performed to establish an updated inventory of resources available to a community, from food pantries to cultural centers. It is a process of engagement that reconceptualizes communities as inherently resourceful and resilient places characterized by assets to be strengthened, rather than highlighting deficits to be remedied (Kramer et al. 2012; Kerka 2003). Resources identified may belong to the entire community or focus on individuals or groups within the community. Groups may implement asset mapping for varying purposes; for example, a local community organization may collect an inventory of nearby resources to share with clients seeking services, or asset mapping may be used by a healthcare system to develop priorities for community engagement. While there are many forms of asset mapping as a method of research, a common approach is the Kretzmann and McKnight's (1993) asset-based community development (ABCD) strategy for community building and community capacity building where a community maps its assets to develop localized interventions. Asset categorization is unique to the community involved, and data collection can be quantitative or qualitative. In social work and community-based participatory research, asset mapping can simultaneously develop the knowledge base and support stakeholders seeking to develop culturally appropriate interventions (Lightfoot et al. 2014). Participatory asset mapping highlights the role of the community as an integral stakeholder and member in the process.

Geographic information systems (GIS) are used in some asset mapping but tend to be considered an additional technological feature of sophistication rather than an accessible core component. We use the term GIS broadly to refer to technological systems or infrastructures that facilitate spatial data processing, manipulation, and/or visualization, including desktop and web-based technologies. In under-resourced communities, the technological benefits of GIS-facilitated asset mapping are often out of reach. Kramer et al. (2012) reviewed six dominant approaches to community asset mapping: because of the lack of technology in under-resourced contexts, GIS systems did *not* meet the criteria of inclusiveness, collaboration, capacity building, responsiveness, or empowerment. High-cost proprietary GIS software systems, lack of technological capacity in under-resourced settings, and the unavailability and/or lack of skilled GIS users may all serve as barriers to accessibility. However, recent advances in computing have the potential to lower technological barriers for end users by moving complex infrastructure architectures to the back-end or "server-side," and likewise making front-end interfaces more accessible and user-friendly. A Participatory Asset Mapping framework constructed from a modern spatial systems configuration may thus allow for greater collaboration and customization, allowing users to take advantage of GIS capabilities.

### 2.1.1 Participatory Mapping for Urban and Community Health

Volunteered Geographic Information (VGI) and Public Participation GIS (PPGIS) serve as examples of processes using geographic information systems for and with communities. VGI is user-generated content on the Web that has a spatial or geographic component (Goodchild 2007). It can be shared as place-based information by direction of the web user, as done in collaborative real-time web mapping of disasters; or it can be volunteered as additional metadata, as in the case of social media geotagging. PPGIS incorporates public engagement more formally as a way to empower and include marginalized populations, involving the public in decision-making using a GIS (Mandarano et al. 2010). This method has been used by local governments to support and steer new urban planning efforts, often soliciting input through interactive maps and online forms. Even with existing challenges, participatory approaches can make research more relevant, improve policies, and facilitate better knowledge production in environmental health applications (English et al. 2018).

With the increasing promise of digital representation across communities and growth of VGI content, the "knowledge politics" of spatial data infrastructures are likewise challenged (Elwood 2006, 2008, 2009). Pairing community-based expert knowledge can give agency to communities seeking solutions. It can also optimize policy or planner solutions by identifying what matters most to the communities being served, as well as facilitate new innovations. PPGIS and participatory VGI methods are increasingly used in built environment in urban health research, from seeking to better understand how place impacts health to characterizing nuanced food environments. Participatory mapping is used to identify and characterize various forms of urban green (i.e., natural environment) and blue (i.e., water environments) place, like the diversity and spatial patterns of place clusters and characteristics of its users, as well as associated benefits like social interaction, psychological benefits, and physical activities (Korpilo et al. 2018; Raymond et al. 2016; Brown et al. 2018). As a planning tool, PPGIS is used in site suitability analysis to incorporate community support, and to extend or replace surveys and activity logs to capture place behaviors. It has also been used to identify regional food assets or food retail site suitability and better understand phenomena related to accessibility and place relationships (Fast and Rinner 2018; Sadler 2016). Participatory GIS enables and engages expert community knowledge to identify community assets and advocate for locally relevant policies.

VGI methods are additionally core to crisis mapping applications, in which volunteers and/or community members develop maps for mitigation efforts in natural disasters, health crises, and other humanitarian needs. For example, YouthMappers (an international organization of students and youth) use these methods to generate maps prioritized by USAID and affiliated organizations like the World Health Organization in areas of extreme poverty and health crises around the world (Solís et al. 2018). This group serves as part of Humanitarian OpenStreetMap Team (HOTOSM) and associated OpenStreetMap technology to update and add maps to an open and global database, also known as open mapping. Crisis mapping incorporates local knowledge, need, or direction; volunteered mapping contributions; and generally open data infrastructures that facilitate the management and distribution of spatial data and its updates. Big data use and approaches have further enhanced digital humanitarianism efforts, though with varying implications as relationships between participants, users, and developers are complex (Burns 2015).

There remain challenges and unexpected consequences in crisis mapping, VGI, and participatory mapping methods. Much inquiry has emerged to establish how VGI may differ or converge with PPGIS, according to how truly engaged or participatory the underlying processes are structured. While participatory GIS can be used to challenge ideas, priorities, and power, it can also reify existing digital divides and develop participation hierarchies (Cochrane and Corbett 2018; Sieber et al. 2016). Much of existing work has been driven or initiated by researchers and planners, rather than communities themselves, thus limiting project goals and scope of work. Examining the networks of participants and practice show that participatory GIS assemblage and decision-making processes are subject to inclusion, exclusion, and marginalization (Bittner et al. 2013). For example, there remains a lack of local mappers in crisis mapping applications, as main mappers remain outside of the communities they are mapping (Meier 2011; Brandusescu and Sieber 2018). Social processes do not change with new technologies but rather transform and impact the mapping process, potentially in new and differing ways (Glasze and Perkins 2015). Critical GIS studies underscore the process-based and inherently sociopolitical approach of mapping and explore these concepts further.

While new Participatory GIS (PGIS) methods in Web 2.0 technologies are promising, they remain complicated and ripe with opportunities for more complete engagement across system components. Fast and Rinner (2018) note that there remains "a growing need for intermediaries who can bridge the gap between experts in the subject matter and experts in digitally enabled participation." When PGIS is viewed as a complex infrastructure of multiple and sometimes competing participants, goals, and needs, examining the underlying spatial data infrastructure linking both technological and knowledge expertise is crucial. Both data and processes may no longer be defined and standardized by single users or institutions but instead refined and composted through multiple user participation – thereby necessitating more dynamic and distributed infrastructures.

## 2.2 *Dynamic Infrastructures for Data Curation*

Rather than refining a single, siloed dataset of community assets, our goal here is to curate a shared dataset across a network of users. A distributed and decentralized network is connected through service-oriented architecture, blending grid, and cloud computing systems, thus facilitating connections and updates over time. Before delving into the case study, we first provide additional background on dynamic and inverted architecture and how asset mapping can be viewed as either a siloed, managed, or shared data management system.

### 2.2.1 Service-Oriented Architecture and SaaS

Moving from siloed systems to distributed networks necessitates new types of architectures. Service-oriented architectures (SOA), grid, and cloud computing architectures (or "cyberinfrastructure") are technology agnostic and have been successfully used to integrate data across distributed, interoperable infrastructures. Cloud providers such as Amazon Web Services (AWS), Google Cloud Platform, and Microsoft Azure are continuing to see growth across a wide range of their services as companies and organizations (e.g., Shutterfly, Comcast, GoDaddy) move and build on these new architectures. SOA is a set of components that can be invoked, generally as communication protocols over a network, and whose interface descriptions can be published and discovered. Consuming data through Application Programming Interface (API) services within a data infrastructure framework serves as an example use of SOA. SOA is increasingly used to access data available as web services, serving as a standard in much web development. However, it is underutilized in multiple fields, including public health and decision-making, specifically when considering the ability for leveraging SOA to consume and integrate multiple different types of data from different sources.

The underlying challenge of sharing data across distributed systems was initially called the "grid problem," with the goal of creating a more "flexible, secure, coordinated resource sharing among dynamic collections of individuals, institutions, and resources" (Foster et al. 2001). Grid architecture was proposed as a possible solution, incorporating protocols, services, APIs, and software development kits (Ananthakrishnan et al. 2015; Foster and Kesselman 1999). Grids have also been used as systems to integrate resources from different organizations for common shared goals (Bote-Lorenzo et al. 2004), though these systems have been increasingly replaced with cloud computing platforms. Cloud computing is similar to grid computing, but with a few notable differentiations. Following an extensive review of literature from the first years of cloud computing, Vaquero et al. (2008) defined clouds as a "large pool of easily usable and accessible virtualized resources (such as hardware, development platforms and/or services) that can be dynamically reconfigured to adjust to a variable load (scale), allowing also for an optimum resource utilization." Cloud computing tends toward isolated and centralized systems to allow for greater security and interoperability, in contrast to shared and decentralized grid systems (Vaquero et al. 2008). However, Cloud computing generally has greater usability and flexibility, allowing for users to choose their own required architecture as long as the required service is supported. By blending the technical capabilities enabled by Cloud computing services with Grid-inspired shared resources, we can implement a flexible and collaborative decision support system to empower users.

Software as a Service (SaaS) serves as an integral component of cloud computing platforms (Hobona et al. 2012; Wang et al. 2008). SaaS is software distributed and delivered through the internet, facilitating on-demand needs and great flexibility. Service-oriented architecture incorporates SaaS, often from multiple sources, to piece together a dynamic framework. For example, within a web application, a JavaScript library may be called to enable interactive visualizations; a Google Maps API called to serve a customized base map; and a jQuery API called to calculate search queries on-the-fly. Geospatial Software as a Service, or GeoSaaS, leverages cloud computing to distribute and process geospatial services. These services could include geoprocessing of data and modelling services, in addition to visualization and search query features familiar in web maps. This can be seen in work by Zhan et al. (2012), where they have implemented a GIS system using SaaS for logistics vehicle monitoring allowing numerous companies to share the same services. Other work includes decision support systems for the modelling of domestic wastewater treatments solutions and civil engineering design (Qazi et al. 2013; Kang and Lee 2014).

### 2.2.2 Decentralized Spatial Infrastructures

At the beginning of the twenty-first century, traditional geographic information systems (GIS) were considered no longer appropriate for modern, distributed, and heterogeneous network environments because of their closed architecture and inflexible infrastructure (Tsou and Buttenfield 2002). However, they are still commonly used to house spatial data in multiple health sectors and disciplines as one of many isolated data system silos. Spatial system infrastructures have begun to move from closed, desktop systems to more transparent, distributed systems that are flexible enough to accommodate dynamic interaction by users.

Recent developments reflect a radical change in infrastructure architecture, moving toward increasingly inverse systems (Coetzee and Wolff-Piggott 2015). These new, emerging inverse infrastructures can exist alongside or in place of traditional systems, tend to develop independently, and are user-driven (Vree 2003; Egyedi and Mehos 2012; Egyedi et al. 2007; Coetzee and Wolff-Piggott 2015). The "virtual organizations" of Foster's Grid increasingly serve as agents that impact the evolution of a data infrastructure, impacted by new types of data getting produced by new types of producers. If an inverse infrastructure is desired to collect and share data across organizations, then user-centric design must be likewise integrated from the start.

In recent years, there has been increasing popularity of distributed infrastructures and framework that are being widely adopted particularly Apache Hadoop and Apache Spark (Zaharia et al. 2016). Specific to spatial analysis and GIS support, there are several Apache Hadoop-based implementations such as SpatialHadoop and Apache Spark systems such as GEOSpark (Yu et al. 2015). Systems such as GEOSpark provide geometrical operation libraries that allow users to develop spatial data processing applications in a distributed environment. Through the utilization of spatial resilient distributed datasets (SRDDs), it is possible for applications to store data and perform operations across a cluster of machines helping organizations to move away from traditional GIS infrastructures (Zaharia et al. 2012).

### 2.2.3 From Siloed to Shared Systems

By enforcing a participatory design methodology, the ultimate objectives of a data infrastructure can be customized to meet the final needs of the end users. By allowing for a more dynamic systems approach within a data infrastructure, as is enabled when using SaaS technologies, the system can change over time to accommodate new needs as they are uncovered. A desired balance achieves an inherently useful but flexible system, where the systems design may adapt to new types of data streams. This framework can be used to improve community asset mapping in health, for example, by translating a centralized, top-down approach to a distributed, collective model.

In Fig. 1, different approaches to gathering community assets are considered as a (a) siloed, (b) managed, or (c) shared infrastructure. The dominant method of community asset collection today is (a) siloed data systems, where each group constructs and maintains their data. Organizations may record data in spreadsheets, word documents, databases, and/or spatial database systems. Some may geocode locations and convert their data to a map and there may be a high cost associated in required staff expertise and/or software required. While these challenges prevent groups from maximizing their use of the data, the knowledge of data maintained within groups tends to be high. A small group of community workers may know all the food pantry locations and their updated information on a monthly basis, for example, but not have the technical budget or infrastructure to digitize that information. One contemporary approach to this problem has been the emergence of proprietary datasets that establish a baseline of technological and knowledge-base standards, here termed a (b) managed infrastructure. Workers are hired to collect data, who may or may not be familiar with the community and are rarely content experts, who in turn update the database. The data is then sold to large organizations (like hospitals) and occasionally made available in nonprofit settings or as limited views on public systems.

A third approach is an inverted or (c) shared infrastructure that builds from a simple, shared spatial database. In this model, each organization contributes their data with service-oriented methods that can accommodate both nonprofit and corporate settings alike. An instance of such an infrastructure is the contribution of distributed geospatial processing by Yang et al. (2008) for Digital Earth, which allows the sharing of Earth data and computing resources across domains. Another infrastructure is one that mines volunteered geographic information from the web and utilizes a distributed geoprocessing workflow to facilitate gazetteer research (Gao et al. 2017).

## 3 West Humboldt Park Case Study

We implement an innovative full-stack asset mapping infrastructure using participatory design principles and spatially explicit service-oriented architecture. First, an ideal data management strategy was defined with community stakeholders using

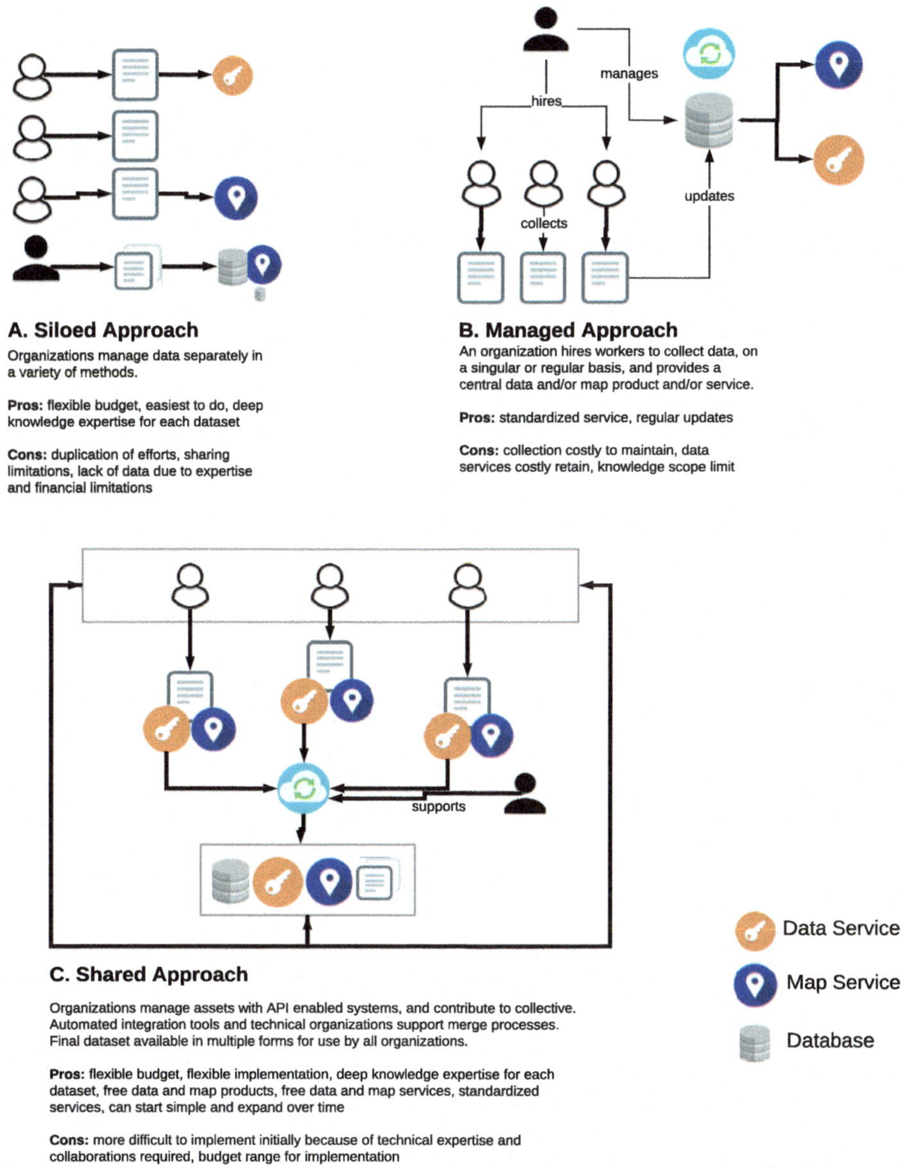


**Fig. 1** Asset management approaches: (**a**) Siloed, (**b**) Managed, and (**c**) Shared

participatory planning and user-centric design. Next, we curated an initial asset dataset using data from multiple organizations, and further updated manually and with web-based tools. Finally, we developed a light, but functional, front-end web application for data sharing, visualization, and exploration.

### 3.1 User-Centric Design

Asset map stakeholders include multiple community organizations and medical provider groups that make up the West Humboldt Park Healthy Community Initiative Coalition, whose mission is to develop "a community with access to resources for wellness, employment, education, and affordable housing in order to help residents achieve a better quality of life in a safe community" (West Humboldt Park Development Council 2013). The coalition is made of multiple organizations that are based in or do work within the West Humboldt community, including community development organizations, nonprofits and charitable groups, small businesses, faith-based organizations, active community members, clinical partners, and community managers from hospitals and insurance groups. Meetings are held regularly from monthly to quarterly time frames, according to the availability and interest of group members. This West Side community represents highly underserved and segregated populations in the City of Chicago. A core component to the West Humboldt Park Healthy Community Initiative Coalition mission is defining and sharing community resources between organizations, as well as with residents of the community. However, there is a lack of sustainable funding available to pay for the software and technical skills required for enterprise-level data management across organizations, as well as funding for the publication of asset maps (online or in print) for the community. Furthermore, data is fragmented; each organization had their own dataset of varying technical sophistication, from post-it note collections to Excel spreadsheets to a geodatabase powered by a student intern license. Finally, data and publications required consistent updates; for example, resources shared for a health fair became out of date for the next year. While proprietary community resource data for the region became available by 2017 (with monthly updates), restrictive data licenses made it costly for some coalition partners to access and/or legally share online in public-facing spaces.

This project originated in 2014 as an extension of West Humboldt Park Healthy Community Initiative Coalition community meetings. A map was requested by academic partners for the community health fair, and conversations to follow posed a web-based map with dynamic updates a more sustainable and effective ideal. The mapping concept, goals, direction, and original data were developed and integrated by Our Lady of Angels, Kelly Hall YMCA, and the West Humboldt Park Development Council (all members of the coalition) in 2014, with additional guidance from the Northwestern Memorial Hospital Department of Community Services and Northwestern University Institute of Public Health and Medicine, with multiple updates in the following years (see Table 1). The mapping interface was updated and reviewed in further coalition meetings, with most updates finalized in 2017. We used a participatory design approach to determine the data and mapping needs of the community coalition. Participatory design, more recently termed cooperative or co-design, is characterized by user input and generally follows three stages: initial exploration of work, discovery process, and prototyping (Spinuzzi 2005). In the first stage, we learned about the various ways organizations managed their own

**Table 1** Iterative data updates for West Humboldt Park Resource Map using Community Health Resource Type categories

| Sequence | Data source | Resource type | Year |
|---|---|---|---|
| Initial data | Our Lady of Angels, Kelly Hall YMCA, West Humboldt Park Development Council | All | 2014 |
| Update 1 | Diabetes Link (Northwestern University) | Healthy living | 2015 |
| Update 2 | La Casa Norte | Food security | 2016 |
| Update 3 | Logan Square Neighborhood Association | Mental health | 2017 |
| Continuous | Multiple: Updates in meetings and from partners | All | Continuous |

Data sources include community organizations from the core coalition participating in the project

community resource data including technologies used and workflow routines. In the next stage, we sought to understand and prioritize goals for an idealized or updated asset data-sharing management strategy. In the third stage, we incorporated regular feedback to iteratively improve multiple prototypes for the final product.

## 3.2 *Asset Data Curation*

Resource data from the West Humboldt Park Pilot was curated and updated using a shared approach, following an inverse infrastructure concept. Through user interviews and community meetings, initial open data sources were first identified to "seed" the asset data collective. Through a defined and iterative web-based process, data was then updated in cycles across participating organizations (see Table 1). The data sources are all community organizations or related members that are part of the West Humboldt Park Healthy Community Initiative Coalition. A complete list of community organizations that contributed as data sources with current web addresses and contact information is available at the code repository (https://github.com/Makosak/HumboldtResources). Resource types include asset data categories that impact or reflect dimensions of community health. For example, food pantry and community meal locations contributed from La Casa Norte in 2014 serve as essential food security resources, and could also proxy areas of nutritional vulnerability. These resources thus reflect dimensions of the social determinants of health, or conditions of the physical or social environment that impact overall health outcomes (Healthy People 2020). The curation of asset data of interest to community groups could additionally aid accessibility analysis, gain insight into the availability and quality of resources, and ultimately advocate for change.

We worked with several community organization representatives from the community health coalition to determine how each group collected and maintained their data, each with varying experience, interest, and institutional capacity for technology. Data management systems were moved to an online, shared web environment that was accessible for community members. Data is thus shared as a service, harvestable through the online format each organization made available.

The final curated asset dataset includes facility name, facility description, source(s), primary and alternate address, geometry, primary and secondary categories, primary and specialty services available, cost schedule (e.g., referral or appointment notes), free service indicator, eligible ages, languages spoken, time schedule, contact name, and phone. We incorporated a basic data model that retained flexibility and included core data essentials that were meaningful to the group (i.e., site name, description, and address source). Only facility name, address, and primary category are required for inclusion in the dataset. This data model was refined with community input, and may still further be updated (Table 2).

To contribute to the data collective, an organization can update an online form, create an online spreadsheet or Fusion Table if they use Google services, or upload their existing spreadsheet to a shared cloud drive. Google Fusion Table is a free albeit proprietary Google data management and visualization service that includes limited spatial data services like geocoding address fields and sharing coordinates

**Table 2** Data Model showing data entity attributes and sample entries

| | Data entities | |
|---|---|---|
| | name | Community Group Fitness Center |
| | description | This organization encourages physical fitness through sports activity. |
| | address | 5445 W. North Ave, Chicago IL |
| | phone | (773) 555-1234 |
| | data source | Diabetes Link |
| | url | http://www.communitygroup.org |
| | services | emergency services, food pantry |
| | AllCategory | Emergency Services |
| | SubCategory | hot meals, emergency services, food pantry |
| | Type | 1 |
| | Prim_Label | orange_blank |
| | cost_schedule | $4/visit open gym, free open swim |
| | has_free_ services | Yes |
| | languages | English, Spanish |
| | eligible_ages | All Ages |
| | time_schedule | Monday-Friday 11:30-12:30 |
| | contact_name | Jim Smith |
| | contact_phone | (773)555-4321 |
| | contact_email | jim.smith@someorg.com |
| | special_services | Prescription Fee Waiver for Adults w/ High BMI and Chronic Disease (ie diabetes) |
| | address2 | 1234 W. Milwaukee Ave, Chicago IL, 60633 |
| Key | postal_code | 60647 |
| Key | ComArea | 23 |
| Key | WardArea | 27 |
| | Comments | last updated on 12/17 by CC |

and nonspatial data features as a web service. Data is then geocoded, merged, and/or updated, according to data standards established. Initially, updates were performed manually to ensure compliance. In another prototype, machine learning techniques were implemented to de-duplicate the structured data using *pgdedupe* (a python package) when new data was shared in bulk. Ultimately, community organizations were interested in a hybrid approach where (Ananthakrishnan et al. 2015) active community organization "super users" would directly update their records in a master Google Fusion Table or (Bittner et al. 2013) keep up their own unique organizational Google Fusion Tables that would automatically update the master; (Bote-Lorenzo et al. 2004) edits to data by other, less active users were done by Google Form (web-based survey administration application) entry, validated and updated into the master Fusion Table using Google Spreadsheet scripts; and (Brandusescu and Sieber 2018) new bulk inserts were incorporated using *pgdedupe*. The updated, shared data stream is then made available on the public website as both product and service. A simplified schematic of this dynamic process for community asset mapping in urban health applications process is shown in Fig. 2.

While proprietary, the ease of use, simplified user interface, and previous examples of open-source integration (as in Eder 2015) made Fusion Tables and connected Google web data service libraries preferred for this application. Other proprietary and open data management and visualization web services by Carto

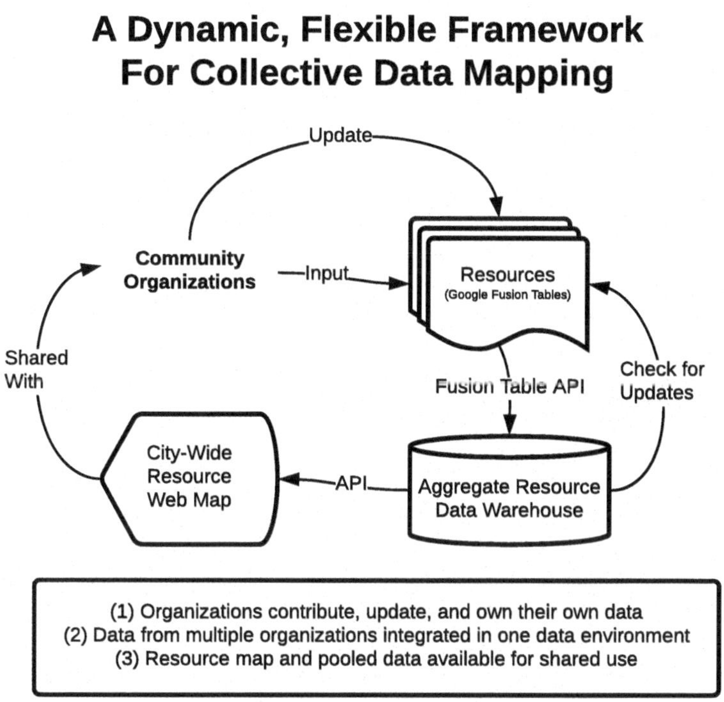


**Fig. 2** A dynamic, flexible framework for collective asset mapping in urban health applications

(https://carto.com/) and Chainbuilder (https://www.chain-builder.net) could be more suitable for future versions and could be easily reconnected into the core web mapping application as an alternate data pipeline. Both proprietary and open-source software and web services come and go in a constantly maturing and ever-changing technoscape; the underlying SaaS architecture must be able to flex with and adapt to these changes.

A comparison of this open, data community collective and proprietary dataset was then conducted to determine completeness and other available metrics of data quality. We used the Purple Binder proprietary community resource dataset, as made available on the Chicago Healthy Atlas by the Smart Chicago Collaborative (Chicago Department of Public Health 2017). This proprietary dataset is only available to be viewed online, and cannot be extracted outside of the web interface without violating data service agreements (without purchase).

## 3.3 Web Mapping Application

An interactive web map was identified as a priority by organizations, and served as an additional focus for development. A light front-end incorporating and building upon existing free and/or open-source spatial software was developed for free, accessible, and long-term use using free and/or open-source software. The "West Humboldt Park Resource Map" serves as a customized version of a Google Fusions searchable web map template (Eder 2015), and utilizes CSS, HTML, and JavaScript. The application leverages Bootstrap and jQuery libraries in addition to Google Maps and Google Fusion Table APIs. It is served and hosted on Github using a gh-pages branch. The Google Maps JavaScript library was used to generate a base map because of extensive, easy-to-read documentation and flexibility of base map configuration within JavaScript. It additionally facilitated easy links between other Google-based libraries like Fusion Tables, Forms, and Charts. The Google Maps interface was additionally familiar to community members as its free-of-use way-finding tool is well established. The Leaflet library was also considered, which uses OpenStreetMap tiles, and may be used in future iterations. The dynamic, iterative nature of the infrastructure allows these changes over time.

The "West Humboldt Park Resource Map" serves as both a data integration and dynamic asset mapping application, with code opened and available at the Kolak and Stepanoe (2016) repository. The web application allows users to query resource data with simple buffer analysis, with immediate results made available for interaction and exploration. In both cases, data can be downloaded on the site in multiple formats. The front-end uses spatially explicit SaaS to add, visualize, and query data as well as supply customized base maps and reports as a service. By design, the final structure retains shared ownership and collaboration across multiple stakeholders, including community groups and health systems.

## 3.4 Results

Following regular discussions at coalition meetings and feedback from early prototypes, it was determined that the group needed an asset data-sharing management strategy that facilitated the following: (Ananthakrishnan et al. 2015) low technology cost, (Bittner et al. 2013) minimal upkeep needs for any staff member or volunteer, and ability for organizations to easily (Bote-Lorenzo et al. 2004) update, (Brandusescu and Sieber 2018) map, and (Brown et al. 2018) explore their collective data. Furthermore, each of these criteria applies to the others, for example, mapping of data should require minimal cost, technical expertise, and upkeep.

The final asset mapping infrastructure includes a flexible and cloud-based data management system, an interactive web map, and community asset data stream. These are collectively shared and used by participating community organizations and health systems. The interactive map and data stream are distributed as publicly accessible web services and can be consumed by the public (see Fig. 3 for screenshot of web application).

A major component of the final product was how the community coalition wanted to explore the data and final mapping product. Following a focus group session at a monthly community health coalition meeting, 37 possible resource activities were defined and grouped into 11 subcategories; these were then aggregated into five major taxonomic groupings: Emergency Needs and Social Services (emergency social services, housing, shelter, personal essentials, food bank, food pantry, social justice, advocacy services, legal clinics, childcare resources, family

**Fig. 3** Screenshot of West Humboldt Park Community Resource Map

support services); Medical Providers and Health Services (primary health, community clinic, free clinic, hospital, health system navigation, health access support services); Wellness and Healthy Living (community gardens, open space, urban farming, cultural programs, art, dance, music, theater, fitness resources, gym, exercise classes, outdoor activity); Education and Job Resources (libraries, schools, job training, job placement, education); and Behavioral Support and Counseling (mental health, behavioral health, addiction services, meditation, spiritual services, counseling, coaching). These categories followed the priorities and sensitivities presented by multiple community organizations and their experiences with community members. For example, food pantries and community gardens both could be categorized as food resources; however, it was important to list the pantry as an emergency food resources and garden as a wellness resource. Furthermore, coalition members did not only want to click and view resources by address and/or buffer but also be able to generate a curated selection tailored to their clients. As such, we generated a "Resource Cart" selection tool that would facilitate that. Representatives noted that they may only have a few minutes to generate such a list, so the ease of product usability proved essential.

A comparison of data between the final data collective and a proprietary dataset showed variations in data quality and completeness, and highlighted smaller services only found within the community model. The proprietary data had considerably more resources over all for the community area of Humboldt Park (741 for this area alone versus 223 resources for the entire community model catchment area). However, it proved difficult to meaningfully compare between datasets because of varying categorization of data. For example, "food services" included multiple types of food stores (convenience stores, grocery stores, restaurants) in the proprietary model, and mainly emergency food services and farmers markets in the community model. "Health services" included all pharmacies and medical practices in the proprietary model, and predominantly community health clinics and major hospital systems in the community model (that likely had free or sliding scale services). Furthermore, each was missing data the other model contained; emergency food services like soup kitchens and smaller food pantries were absent using available queries in the proprietary model, and a thorough inventory of grocery stores was missing in the community model. In the proprietary model and search tool available on the Healthy Chicago Atlas, 14 topics were available as taxonomic categories tags for exploration (care, childcare, education, emergency, goods, health, housing, legal, mental health, money, transit, work, youth services) in comparison to the five major groups in the community model and web map. In the proprietary model, data was only available for querying using community area or zip code selection, or by selecting existing or known data categories. In contrast, the attribute search field in the community model scoured multiple columns to identify potential matches, rather than standardized categories. For the category or tag of "emergency services" that was common to both, there were zero resources available using the proprietary model.

## 4 Discussion

While VGI methods may not always generate the most complete datasets, here shown by the magnitude difference in number of resources between the West Humboldt pilot dataset and a proprietary model, the insight gained by participatory methodology cannot be overlooked. The generated data collective serves as a curated data source customized to the needs of its community members, and is not easily transferable to an organization outside of the group. This approach also found new data sources not found in other proprietary models. Furthermore, the West Humboldt pilot dataset highlights that the interface accessibility of digital information as potentially even more important than the data itself. Community coalition members sought to access data with flexible spatial and attribute queries, and were more interested in further customizing and saving query discoveries than simply viewing data. This finding was consistent with prior work that demonstrated difficulties of spatial data handling as a major challenge and opportunity in incorporating grassroots stakeholders in GIS-enabled research (Elwood 2008b). Finally, access to flexible and timely data sharing and mapping remained a priority throughout for organizations of varying technological abilities and time commitments. This underscores the need and interest to bridge digital divides and ensure that technological achievements remain accessible. Equitable access to geospatial systems and data, using well-designed and accessible software, remains crucial in challenging power relations for the empowerment of communities (Ghose and Welcenbach 2018).

Information diffusion is a spatial process; users tend to contribute to topics that are near them. The user-generated content of VGI, even in massive contributions like *Wikipedia,* tends to exhibit localized spatial behaviors (Hecht and Gergle 2010; Hardy et al. 2012). While this can prove beneficial for tech-savvy and tech-able populations, communities without are at best digitally underrepresented and, at worst, digitally misrepresented. Digital representation inequalities challenge the goals of community empowerment that drive much VGI work. In the bold work by Cochrane et al. (2017), a review of 10 years of literature from the top 10 GIScience journals found that only 128 of 1652 published articles referred to social justice, empowerment, or social change, perhaps caused by a "preoccupation on technical matters of mapping" (Pramono et al. 2006, p. 12) rather than the complex interactions impacting digital data and mapping representations. By blurring or even collapsing the differentiation between "VGI contributor" and "community member" and shifting the "expert" technical role to support an interface between, we may better work toward empowering communities.

By extending VGI with cloud-based systems and geospatial SaaS, stakeholders can build on the interactive and user-friendly principles of VGI (Goodchild 2007) and move beyond traditional concepts of how VGI have been created and shared (Elwood et al. 2012). While VGI concepts are not new to public health, the literature tends to focus on one-off data creations or visualizations, or person-level data contributions that include privacy dangers and pitfalls (data (Stensgaard et al. 2009;

Boulos et al. 2011; Goranson et al. 2013)). However, VGI concepts can extend to groups, rather than individuals, thus empowering community organizations in new means of data management, data sharing, mapping, and more. Such "community-engaged VGI" is essential in integrating previously siloed data systems and facilitating means of collaboration with health systems in urban health research and practice. Cloud-based mapping systems and spatially explicit SaaS leverage participatory GIS systems and VGI by building on the strengths of different stakeholders. The resulting user-friendly support system serves as a mash-up for both neo-geographic stakeholder (i.e., average web user) and tech-savvy programming expert, pushing past the data-divide in VGI systems (as referenced in Cinnamon and Schuurman 2013). By allowing for a dynamic, reproducible, adaptive, and participatory asset mapping system, health systems infrastructures can further support community health improvement frameworks by facilitating shared data and decision support implementations across health partners.

This approach has limitations, however, as a still emergent methodology with a process characterized by its continuous change. To minimize issues, we based the application on established technologies and templates that had been successfully tested over time. While these free, low-cost, and/or open-source technologies were implemented in the West Humboldt pilot to allow for affordability and longevity, not all components may remain free, low cost, or open over a lifetime. Google Map API components were used successfully in prototypes, for example, though portions may be phased out or transitioned to cost-based systems unexpectedly by Google. In the same manner, open-source technologies may become outdated if not maintained. For bulk updates, the system still requires access to a server to run scripts for automated processes. While the West Humboldt Park project necessarily requires these different processes, the components can be updated and transitioned over time as well. For example, alternate options can be used to facilitate data services instead of Google Fusion Tables, and could be interchanged with ease. If the basic data model remained similar, a new API service could be consumed within the web mapping application without interruption. As such, the inverted infrastructure can and should be adapted over time, serving as both a limitation and essential characteristic of flexibility.

Participatory Asset Mapping frameworks that incorporate VGI, GeoSaaS, and accessible interfaces for stakeholders will be crucial for future health planning and public policy research. As data is shared and explored across traditionally siloed environments, new insights are anticipated. For example, after learning about the project, a neighborhood federally qualified health center (Erie Humboldt Park Health Center) paired the piloted collective data stream (also using geospatial SaaS technology) with its electronic health records to prioritize health interventions. Areas of higher diet-related illness were found to have less emergency food services. Clinic members began to attend health coalition meetings, where their affordable health services and community connections were discovered as an overlooked asset for several community groups, despite a close proximity. Through the collaborative work of participatory mapping where data representation is central yet democratic, communities and health systems can identify shared and needed resources, (re)prioritize goals, and engage across differences.

## References

Ananthakrishnan, R., Chard, K., Foster, I., & Tuecke, S. (2015). Globus platform-as-a-service for collaborative science applications. *Concurrency and Computation: Practice and Experience, 27*(2), 290–305.

Bittner, C., Glasze, G., & Turk, C. (2013). Tracing contingencies: Analyzing the political in assemblages of web 2.0 cartographies. *GeoJournal, 78*(6), 935–948.

Bote-Lorenzo, M. L., Dimitriadis, Y. A., & G mez-Sánchez, E. (2004). Grid characteristics and uses: A grid definition. In *Grid computing* (pp. 291–298). Berlin/Heidelberg: Springer.

Boulos, M. N. K., Resch, B., Crowley, D., Breslin, J., Sohn, G., Burtner, R., et al. (2011). Crowdsourcing, citizen sensing and sensor web technologies for public and environmental health surveillance and crisis management: Trends, OGC standards and application examples. *International Journal of Health, 10*, 67.

Brandusescu, A., & Sieber, R. E. (2018). The spatial knowledge politics (SKP) of crisis mapping for community development. *GeoJournal, 1*, 1–16.

Brown, G., Rhodes, J., & Dade, M. (2018). An evaluation of participatory mapping methods to assess urban park benefits. *Landscape and Urban Planning, 178*, 18–31.

Burns, R. (2015). Rethinking big data in digital humanitarianism: Practices, epistemologies, and social relations. *GeoJournal, 80*(4), 477–490.

Center for Disease and Control (CDC). (2015). Community health improvement navigator.

Chicago Department of Public Health. (2017). Chicago health atlas resources. https://www.chicagohealthatlas.org/resources.

Cinnamon, J., & Schuurman, N. (2013). Confronting the data-divide in a time of spatial turns and volunteered geographic information. *GeoJournal, 78*(4), 657–674.

Cochrane, L., & Corbett, J. (2018). Participatory mapping. Handbook of communication for development and social change, 1–9.

Cochrane, L., Corbett, J., Evans, M., & Gill, M. (2017). Searching for social justice in GIScience publications. *Cartography and Geographic Information Science, 44*(6), 507–520.

Coetzee, S., & Wolff-Piggott, B. (2015). A review of sdi literature: Searching for signs of inverse infrastructures. In *Cartography-maps connecting the world* (pp. 113–127). Cham: Springer.

Eder, D. (2015). Searchable map template with Google fusion tables. https://github.com/derekeder/FusionTable-Map-Template.

Egyedi, T. M., & Mehos, D. C. (Eds.). (2012). *Inverse Infrastructures: Disrupting networks from below*. Edward Elgar Publishing.

Egyedi, T. M., Vrancken, J. L., & Ubacht, J. (2007). Inverse infrastructures: Coordination in self-organizing systems. In *Standardization and innovation in information technology, 2007. SIIT 2007. 5th international conference on* (pp. 23–36). IEEE.

Elwood, S. (2006). Critical issues in participatory GIS: Deconstructions, reconstructions, and new research directions. *Transactions in GIS, 10*(5), 693–708.

Elwood, S. (2008). Volunteered geographic information: Future research directions motivated by critical, participatory, and feminist GIS. *GeoJournal, 72*(3–4), 173–183.

Elwood, S. (2008b). Volunteered geographic information: key questions, concepts and methods to guide emerging research and practice. *GeoJournal, 72*(3-4), 133–135.

Elwood, S. (2009). Multiple representations, significations, and epistemologies in community-based GIS. In *Qualitative GIS: A mixed methods approach* (pp. 57–74).

Elwood, S., Goodchild, M. F., & Sui, D. Z. (2012). Researching volunteered geographic information: Spatial data, geographic research, and new social practice. *Annals of the Association of American Geographers, 102*(3), 571–590.

English, P. B., Richardson, M. J., & Garzón-Galvis, C. (2018). From crowdsourcing to extreme citizen science: Participatory research for environmental health. *Annual Review of Public Health, 39*, 335–350.

Fast, V., & Rinner, C. (2018). Toward a participatory VGI methodology: Crowdsourcing information on regional food assets. *International Journal of Geographical Information Science, 1*, 1–16.

Foster, I., & Kesselman, C. (1999). "The globus toolkit." The grid: blueprint for a new computing infrastructure: 259-278. Morgan Kaufmann Publishers Inc. San Francisco, CA, USA. ISBN:1-55860-475-8.

Foster, I., Kesselman, C., & Tuecke, S. (2001). The anatomy of the grid: Enabling scalable virtual organizations. *The International Journal of High Performance Computing Applications, 15*(3), 200–222.

Gao, S., Li, L., Li, W., Janowicz, K., & Zhang, Y. (2017). Constructing gazetteers from volunteered Big Geo-Data based on Hadoop. *Computers, Environment and Urban Systems, 61*, 172–186.

Ghose, R., & Welcenbach, T. (2018). "Power to the people": Contesting urban poverty and power inequities through open GIS. *The Canadian Geographer/Le Géographe canadien, 62*(1), 67–80.

Glasze, G., & Perkins, C. (2015). Social and political dimensions of the OpenStreetMap project: Towards a critical geographical research agenda. In *OpenStreetMap in GIScience* (pp. 143–166). Cham: Springer.

Goodchild, M. F. (2007). Citizens as sensors: The world of volunteered geography. *GeoJournal, 69*(4), 211–221.

Goranson, C., Thihalolipavan, S., & di Tada, N. (2013). VGI and public health: Possibilities and pitfalls. In *Crowdsourcing geographic knowledge* (pp. 329–340). Dordrecht: Springer.

Groves, P., Kayyali, B., Knott, D., & Van Kuiken, S. (2013). The 'big data' revolution in healthcare. *McKinsey Quarterly, 2*, 3.

Hardy, D., Frew, J., & Goodchild, M. F. (2012). Volunteered geographic information production as a spatial process. *International Journal of Geographical Information Science, 26*(7), 1191–1212.

Healthy People. Secretary's Advisory Committee on Health Promotion and Disease Prevention Objectives for 2020. Healthy People 2020: An opportunity to address the societal determinants of health in the United States. http://www.healthypeople.gov/2020/topicsobjectives2020/overview.aspx?topicid=39.

Hecht, B. J., & Gergle, D. (2010, February). On the localness of user-generated content. In *Proceedings of the 2010 ACM conference on Computer supported cooperative work* (pp. 229–232). ACM.

Hobona, G., Jackson, M., & Anand, S. (2012). Implementing Geospatial Web Services for Cloud Computing. In I. Management Association (Ed.), *Grid and cloud computing: Concepts, methodologies, tools and applications* (pp. 615–636). Hershey, PA: IGI Global. https://doi.org/10.4018/978-1-4666-0879-5.ch305

Kang, S. Y., & Lee, Y. H. (2014). *The implementation of geo-cloud SaaS system for supporting the civil engineering design using BRMS open software. 2014 fifth international conference on computing for geospatial research and application* (pp. 49–50).

Kerka, S. (2003). Community asset mapping. Trends and Issues Alert, (47). ERIC Clearinghouse on Adult, Career, and Vocational Education, Columbus, OH.

Kolak, M., & Stepanoe, M. (2016). "HumboldtResources: Alpha" Zenodo. https://doi.org/10.5281/zenodo.44691.2016.

Korpilo, S., Virtanen, T., Saukkonen, T., & Lehvävirta, S. (2018). More than A to B: Understanding and managing visitor spatial behaviour in urban forests using public participation GIS. *Journal of Environmental Management, 207*, 124–133.

Kramer, S., Amos, T., Lazarus, S., & Seedat, M. (2012). The philosophical assumptions, utility and challenges of asset mapping approaches to community engagement. *Journal of Psychology in Africa, 22*(4), 537–544.

Kretzmann, J. P., & McKnight, J. (1993). *Building communities from the inside out* (pp. 2–10). Evanston: Center for Urban Affairs and Policy Research, Neighborhood Innovations Network.

Lightfoot, E., McCleary, J. S., & Lum, T. (2014). Asset mapping as a research tool for community-based participatory research in social work. *Social Work Research, 38*(1), 59–64.

Mandarano, L., Meenar, M., & Steins, C. (2010). Building social capital in the digital age of civic engagement. *Journal of Planning Literature, 25*(2), 123–135.

Meier, P. (2011). Verifying crowdsourced social media reports for live crisis mapping: An introduction to information forensics. iRevolution blog.

Pramono, A. H., Natalia, I., & Janting, Y. (2006). Ten years after: Counter-mapping and the Dayak lands in West Kalimantan, Indonesia. Digital Library of the Commons.

Qazi, N., Smyth, D., & McCarthy, T. (2013). Towards a GIS-based decision support system on the Amazon cloud for the modelling of domestic wastewater treatment solutions in Wexford, Ireland. 2013 Uksim 15Th international conference on computer modelling and simulation, 236–240.

Raymond, C. M., Gottwald, S., Kuoppa, J., & Kyttae, M. (2016). Integrating multiple elements of environmental justice into urban blue space planning using public participation geographic information systems. *Landscape and Urban Planning, 153*, 198–208.

Sadler, R. C. (2016). Integrating expert knowledge in a GIS to optimize siting decisions for small-scale healthy food retail interventions. *International Journal of Health Geographics, 15*(1), 19.

Sieber, R. E., Robinson, P. J., Johnson, P. A., & Corbett, J. M. (2016). Doing public participation on the geospatial web. *Annals of the American Association of Geographers, 106*(5), 1030–1046.

Solís, P., McCusker, B., Menkiti, N., Cowan, N., & Blevins, C. (2018). Engaging global youth in participatory spatial data creation for the UN sustainable development goals: The case of open mapping for malaria prevention. *Applied Geography, 98*, 143–155.

Spinuzzi, C. (2005). The methodology of participatory design. *Technical Communication, 52*(2), 163–174.

Stensgaard, A. S., Saarnak, C. F. L., Utzinger, J., Vounatsou, P., Simoonga, C., Mushinge, G., et al. (2009). Virtual globes and geospatial health: The potential of new tools in the management and control of vector-borne diseases. *Geospatial Health, 3*(2), 127–114.

Tsou, M. H., & Buttenfield, B. P. (2002). A dynamic architecture for distributing geographic information services. *Transactions in GIS, 6*(4), 355–381.

Vaquero, L. M., Rodero-Merino, L., Caceres, J., & Lindner, M. (2008). A break in the clouds: Towards a cloud definition. *ACM SIGCOMM Computer Communication Review, 39*(1), 50–55.

Vree, W. G. (2003). Internet en Rijkswaterstaat: een ICT-infrastructuur langs water en wegen.

Wang, L., Tao, J., Kunze, M., Castellanos, A. C., Kramer, D., & Karl, W. (2008, September). Scientific cloud computing: Early definition and experience. In *High performance computing and communications, 2008. HPCC'08. 10th IEEE international conference on* (pp. 825–830). IEEE.

West Humboldt Park Development Council, 2013.

Yang, C., Li, W., Xie, J., & Zhou, B. (2008). Distributed geospatial information processing: Sharing distributed geospatial resources to support Digital Earth. *International Journal of Digital Earth, 1*(3), 259–278.

Yu, J., Wu, J., & Sarwat, M. (2015). Geospark: A cluster computing framework for processing large-scale spatial data. In *Proceedings of the 23rd SIGSPATIAL international conference on advances in geographic information systems* (p. 70).

Zaharia, M., Chowdhury, M., Das, T., Dave, A., Ma, J., McCauley, M., Franklin, M. J., Shenker, S., & Stoica, I. (2012). Resilient distributed datasets: A fault-tolerant abstraction for in-memory cluster computing. In *Proceedings of the 9th USENIX conference on networked systems design and implementation* (Vol. 2).

Zaharia, M., Franklin, M., Ghodsi, A., Gonzalez, J., Shenker, S., Stoica, I., et al. (2016). Apache spark. *Communications of the ACM, 59*(11), 56–65.

Zhan, J., Sha, Y., & Yan, J. (2012). *Design and implementation of logistics vehicle monitoring system based on the SaaS model. 2012 fifth international conference on business intelligence and financial engineering* (pp. 524–526).

**Marynia Kolak,** MS, MFA, PhD, is a Social Determinants of Health geographer using open science tools and an exploratory data analytic approach to investigate issues of equity across space and time. Her research centers on how "place" impacts health outcomes in different ways, for different people, from opioid risk environments to chronic disease clusters. She is the Assistant

Director of Health Informatics and Lecturer in GIScience at the Center for Spatial Data Science, University of Chicago, and serves as a Public Service Intern at the Chicago Department of Public Health. She received her PhD in Geography at ASU, M.F.A in Writing from Roosevelt University, M.S. in GIS from John Hopkins University, and B.S. in Geology from the University of Illinois at Urbana-Champaign.

**Michael Steptoe** is a PhD student in the School of Computing, Informatics and Decision Systems Engineering at Arizona State University. Steptoe obtained his B.S.E. and M.S. degrees from ASU in Computer Systems Engineer and Computer Science. His research interests include data visualization and mobile applications.

**Holly Manprisio,** MPH, serves as Program Manager of External Affairs for Northwestern Memorial HealthCare in Chicago, Illinois. Holly has 15 years of experience working in community engagement and health education, and she currently oversees the hospital's community health needs assessment and implementation process. She is dedicated to improving health equity and does so through innovative program implementation to address priority needs in collaboration with a wide range of community partners. Holly received her undergraduate degree in Community Health Education from Illinois State University in 2003 and completed her Master of Public Health from DePaul University in 2010.

**Lisa Azu-Popow,** MHA, is a Program Director in the Department of Community Services/ External Affairs for Northwestern Memorial HealthCare. Lisa has over 20 years of healthcare experience with the majority of her time in community health. In her current role, Lisa oversees programs that directly impact the lives of the underserved patient populations such as care coordination programs, access to health programs, and a community engagement initiative focusing on health and wellness education and programs in the West Humboldt Park Community. Lisa has Bachelors of Arts in Biology from Lake Forest College and a Master in Health Administration from Tulane University School of Public Health and Tropical Medicine.

**Megan Hinchy,** MPH, is a Program Coordinator who works with children and families in Chicago to decrease rates of childhood obesity and improve health outcomes. She is employed by Ann and Robert H. Lurie Children's Hospital with the Consortium to Lower Obesity in Chicago's Children (CLOCC). Megan understands the importance of improving access to healthy foods especially in neighborhoods lacking full scale grocery stores, "food deserts." Megan has partnered with many community-based organizations, parks, schools, and residents in Chicago to provide resources that promote health and wellness. Megan holds a Master's Degree in Public Health from Florida International University.

**Geraldine Malana,** MPH, DO graduated from medical school with a dual degree (DO/MPH) at A.T. Still University School of Osteopathic Medicine in Arizona. Her public health thesis focused on how primary care providers detect and notice social determinants of health during their visits. She went on to McGaw Northwestern's Family Medicine Residency in Humboldt Park, where her training was rooted in providing primary care to underserved populations. Her residency research focused on determining the distribution of common diseases seen in her clinic within the greater Chicago area to find "hot-spot" areas of disease that could be used as high priority outreach areas. She now practices family medicine at Cambridge Health Alliance, the Malden Family Health Center, and helps with clinical teaching of medical students and family medicine residents.

**Ross Maciejewski,** PhD, is an Associate Professor at Arizona State University in the School of Computing, Informatics, and Decision Systems Engineering and Director of the Center for Accelerating Operational Efficiency (CAOE) – a Department of Homeland Security Center of Excellence. His primary research interests are in the areas of geographical visualization and visual analytics focusing on homeland security, public health, dietary analysis, social media, criminal incident reports, and the food-energy-water nexus. Professor Maciejewski is a recipient of an NSF CAREER Award (2014) and was named a Fulton Faculty Exemplar (2017) and Global Security Fellow at Arizona State. His work has been recognized through a variety of awards at the IEEE Visual Analytics Contest (2010, 2013, 2015), a best paper award in EuroVis 2017, and a CHI Honorable Mention Award in 2018.

# Improving Urban and Peri-urban Health Outcomes Through Early Detection and Aid Planning

**Kathryn Grace, Alan T. Murray, and Ran Wei**

**Abstract** Chronic food insecurity significantly constrains short- and long-term health, as well as the development of individuals and households, ultimately impacting economic progress in some of the poorest and fastest growing communities on the planet. One of the strategies used to combat household- and individual-level food insecurity is food aid. Ensuring that food aid reaches the neediest people, however, is an ongoing challenge. In this chapter, we explore the use of geospatial technologies as part of a framework for improving food aid targeting in Bamako, Mali. We develop and apply quantitative models that rely on remotely sensed data and health survey data to highlight the importance of different aspects of demand for food aid in urban spaces. The results highlight the usefulness of this approach for food aid planning in urban areas where food need is unevenly distributed over a densely populated area.

## 1 Introduction

Urban food aid systems in sub-Saharan Africa are essential because estimates suggest that among the 472 million people living in these areas, high proportions are chronically and persistently undernourished (Lall et al. 2017; Van de Poel et al. 2007;

K. Grace (✉)
Department of Geography, Environment and Society, University of Minnesota, Twin Cities, MN, USA
e-mail: klgrace@umn.edu

A. T. Murray
Department of Geography, University of California, Santa Barbara, CA, USA
e-mail: amurray@ucsb.edu

R. Wei
School of Public Policy and Center for Geospatial Sciences, University of California, Riverside, CA, USA
e-mail: ranwei@ucr.edu

 
Y. Lu, E. Delmelle (eds.), *Geospatial Technologies for Urban Health*, Global Perspectives on Health Geography, https://doi.org/10.1007/978-3-030-19573-1_12

Abuya et al. 2012). Such undernutrition is responsible for around three million child deaths annually and contributes to poor health and well-being, slows recovery times from infections or illness, and adversely impacts cognitive development (Gillespie et al. 2013; Black et al. 2013; Bhutta et al. 2014). Furthermore, undernutrition limits adult labor force participation and is associated with reduced educational attainment among children, thereby impacting earnings potential later on in life (FAO, IFAD, and WFP 2015). Undernutrition in urban sub-Saharan Africa significantly constrains short- and long-term health and development of individuals and households, hindering economic progress in some of the poorest and fastest growing communities on the planet.

Reducing household- and individual-level undernutrition in urban and peri-urban[1] sub-Saharan Africa requires addressing deficiencies in some combination of the four pillars of food insecurity: availability, access, utilization and stability. In an urban context, availability refers to the presence of food in a particular place, like a grocery store, a market or a small household plot or garden. Access to food relates to affordability and proximity of food or food resources. Utilization includes the nutritional value of food and the body's ability to obtain nourishment from food. Stability is the reliability of each of the other pillars and can be impacted by broad-scale political, economic, and environmental factors (Brown et al. 2015). Large-scale urban food insecurity is most often associated with food price increases, conflict/political failures, and/or global market fluctuations (FAO 1996; Smith et al. 2000; Sen 1990, 1997; Pinstrup-Andersen 2009; Misselhorn et al. 2012; Brown et al. 2017). However, research has highlighted the presence and potential importance of urban agriculture[2] for meeting the needs of urban dwellers, especially the poorest among them (Castillo 2003; Zezza and Tasciotti 2010; FAO 2012; Lerner and Eakin 2011).

Food aid is another potentially significant avenue for bringing sustenance to households facing undernutrition issues. International food aid is one of the primary sources of assistance provided by wealthy countries to sub-Saharan African countries. Interestingly, such aid represents the largest component of US aid expenditures, as an example. A major challenge facing the distribution of food aid is developing effective and efficient targeting systems. In other words, mechanisms are needed for food aid distribution that ensure that the people with the greatest need receive the aid. Because poverty, food insecurity, and urban agricultural practices may vary within a city and over time, a high level of spatial and temporal detail to capture this heterogeneity is required. Accordingly, geographic-level detail associated with food outlets in relation to anticipated need is crucial for the delivery of essential food aid resources.

[1] Peri-urban areas are defined here as neighborhoods on the outskirts of dense urban centers where city infrastructure like electricity or piped water is limited. These can either be informal/formal settlements that developed as a result of need for affordable housing near to the urban center, or they may have been rural areas that have been incorporated into the city boundaries as the city expands.

[2] Includes crops, gardens, and livestock goods.

While there is a substantial amount of research on the effectiveness of international food aid once it reaches the individual or the household in sub-Saharan Africa (e.g., Violette et al. 2013; Hampshire et al. 2009; Lentz and Barrett 2013; Gautam and Andersen 2017), considerably less research exists on the effectiveness of food aid distribution programs (see Clay et al. 1999; Maxwell et al. 2013; Rancourt et al. 2015; Grace et al. 2017). In this chapter, we focus on a major urban area in one of the poorest countries in the world, Bamako, Mali, and investigate the use of geospatial technologies to support improved food aid targeting. Bamako is one of the fastest growing urban areas in sub-Saharan Africa and is characterized by high (and potentially growing) levels of malnutrition (Castle et al. 2014). Our study area considers both urban and peri-urban areas of Bamako and uses spatial optimization models combined with GIS and remote sensing to inform food aid distribution. Data from the Demographic and Health Survey (DHS) as well as remotely sensed imagery of vegetation in Bamako provide critical information on potential food aid demand areas.

## 2 Background

The goal of food aid targeting and distribution is simple – to ensure that individuals in need are able to gain access to free or low-cost food so that they may live active and healthy lives (FAO 1996). Because aid resources are limited, effectively identifying (or targeting) individuals who most need food and then getting it to them is vital (Jaspars and Young 1996; Clay et al. 1999). There are two types of food aid. One type involves emergency or crisis situations, like droughts or earthquakes. The other is non-emergency food aid designed to meet chronic issues with production, distribution, access, etc. resulting in malnutrition and undernourishment. In this chapter, we focus on targeting and distribution characteristics of non-emergency food aid.[3] Traditional methods employed by agencies such as the World Food Programme (WFP), a division of the United Nations, underlie our understanding of food aid targeting. The WFP is the world's largest humanitarian agency and has been instrumental in international food aid and distribution since the middle of the twentieth century. Because WFP is so foundational and influential in developing and maintaining approaches to international food aid, their approach serves as a primary strategy that most other agencies adhere to.

WFP, often in combination with country governments, identifies communities vulnerable to food insecurity. After a community has been targeted, a range of

[3] Note that the WFP is using the term food aid as part of a broader concept of "food assistance". Rather than focusing only on feeding hungry people, the food assistance approach aims to consider long-term needs and diverse approaches to meeting these needs. Some aspects of food assistance are included in the type of food "aid" we mention in this chapter, namely cash transfers. We use the term food aid throughout the chapter, however, as reflects its most common usage in academic research.

different strategies are used to identify individuals and families in greatest need. Food aid is then distributed using a variety of approaches. Among the most common ways that food aid is targeted to meet the needs of the most deprived individuals is through cash transfers, by focusing on children in school meal programs, via clinic or hospital nutrition education programs, or through voucher programs for low-income families which provide access to free or reduced cost culturally relevant food staples (USAID 2013, 2014; Maxwell et al. 2013; Lentz and Barrett 2013).

The geographic aspects of food aid distribution – where should access sites be located to most effectively reach targeted populations with the most intense need – vary by country and community (Clay et al. 1999). Notably, the food aid distribution system as it currently functions is somewhat dependent on existing infrastructure. For example, markets where vouchers can be distributed and used or hospitals and schools of adequate size may be required to support the necessary components of the distribution system. Furthermore, there is evidence that while distribution for non-emergency food aid does reflect local environmental conditions that may exacerbate food insecurity (a poor growing season in a given area, for example), there is also an indication that food aid distribution may be based on historical practice (Jayne et al. 2001; Clay et al. 1999). In other words, communities that at one point demonstrated notable need for food aid, continue to receive food aid regardless of their current needs.

Research has investigated the different ways that food aid is used to benefit individuals and households (e.g., Hidrobo et al. 2014; Gentilini 2014; Gelli et al. 2007; Leroy et al. 2009). This research has helped to identify the effectiveness of different types of aid (i.e., nutrition education during prenatal appointments versus education of influential community members). Further, it has also highlighted the potential for certain groups, usually the very poor, to face major barriers in accessing aid intended for them, while other groups, those that are slightly better off economically or those with certain household characteristics, benefit more from certain types of food aid (Hidrobo et al. 2014). And while this research has provided insight into the micro conditions impacting the effectiveness of food aid, broader macro-level and spatial questions of how to identify vulnerable communities (or neighborhoods) in the first place, remain largely ad hoc (see Maxwell et al. 2013; Lentz and Barrett 2013).

In application, delivery of food aid resources is heavily dependent on the location of distribution outlets. Culture, land use, history, topography, accessibility and a range of other quantitative and qualitative factors influence where food aid outlets are located. In this chapter, we aim to demonstrate the use of an explicit framework that incorporates dynamic and varied factors into quantitative approaches for locating distribution outlets that provide an additional perspective on food aid targeting. The application to urban West Africa is particularly relevant to contemporary issues facing many developing countries. Urban areas represent a heterogeneous mix of intense and entrenched poverty. Such areas increasingly concentrate poverty and food insecurity in slum communities, especially among newly arrived immigrants with limited access to resources (FAO 2012). Peri-urban areas are reliant on local rainfed agriculture as well as depend on low-paying and temporary employment in the urban centers (see Grace et al. 2017b). In both urban and peri-urban settings of West Africa, children and families face high levels of poverty and food insecurity.

Spatial optimization combined with GIS and remote sensing technologies offers an important path forward in better targeting individuals and neighbors in need of food aid. An overview of spatial optimization can be found in Tong and Murray (2012), highlighting that optimization involves decisions to be made (using variables), and objective(s) and constraining conditions that are geographically explicit in some manner. Grace et al. (2017) demonstrated the utility of spatial optimization for strategic-level food aid provision across a region. However, the issues unique to urban areas and the lack of data that directly measures these factors – specifically income and the presence or absence of urban agriculture – present significant research challenges for which geospatial technologies have much to offer.

## 3 Data and Methods

This research focuses on food aid delivery in Bamako, Mali, an urban area in sub-Saharan Africa. In order to support programs like the WFP that wish to provide aid, it is necessary to identify food distribution outlets in this urban area. Challenges include neighborhood-level detail about the nature of food availability through formal and informal mechanisms. Help and utilization is highly dependent on access, and a poorly configured aid distribution system will mean that food is not getting to those most in need.

In our analysis, we combine existing survey-based measures of food insecurity (using child health outcomes related to chronic undernutrition) and vegetation characteristics to derive estimates of demand for food aid. We construct different measures of demand for food aid using environmental data on local vegetation and two different sources of population/health data. We describe the data below.

The Normalized Difference Vegetation Index (NDVI) is derived from the Moderate Resolution Imaging Spectroradiometer (MODIS) on board NASA's Terra satellite (Carroll et al. 2004). NDVI can be considered a measure of vegetation and is particularly useful for drought and famine early warning systems. We use 250 m NDVI data and calculate the seasonal maximum NDVI value (for 2011, an average year) for each demand area within Bamako. NDVI has been widely used to determine food availability and agricultural production in communities without detailed agricultural data. While Bamako is largely urban, many households have small gardens and peri-urban areas of Bamako contain rainfed agricultural plots that produce food used to meet household nutritional demands or generate income. Therefore, vegetation measures, like NDVI, contribute to better understanding and describing local demand for food aid. The use of NDVI to measure urban agriculture as it relates to capturing food availability has been explored in a number of settings (see Brown and McCarty 2017). Figure 1 depicts this situation and demonstrates an example of urban agriculture in Bamako that would likely be captured by vegetation measures.

**Fig. 1** Urban agricultural plot in Bamako, Mali. (Photo by: Ibrahim TRAORE)

In addition to NDVI as a component of estimating food need or supply, also considered are demographic and health aspects of the population. The population count within a given area using 100 m data is available from the World Population Organization (http://www.worldpop.org.uk/). While urbanization offers opportunities for wage earning employment, education, and health care, many urban areas in sub-Saharan Africa are characterized by intense poverty and food insecurity. The number of people in an area serves as a useful measure of "mouths to feed." But is only part of the picture as some areas within the city will be characterized by higher levels of deprivation. Children's anthropometric information (height-for-age) provides a commonly used indicator of chronic undernutrition and can be used as a measure of the prevalence of food insecurity in a community. We use data on children's height-for-age scores from the 2012–2013 DHS survey for Mali (ICF 2013). This data is available for approximately 60 spatially referenced clusters representing nearly 2000 households in urban and peri-urban Bamako. Estimates of undernutrition prevalence are calculated for each cluster (number of undernourished children under 5 years of age among those sampled in the cluster).

The Bamako region is represented as regular grid/raster/tessellation surface with individual areas 500 × 500 m size. Each area is considered a neighborhood of potential demand for food aid. The DHS clusters are represented as points but the actual cluster locations could be anywhere within 2 km of the reported point to ensure confidentiality and mask individual identities.

The demand for food aid can be estimated in two different ways. Grace et al. (2017) presented a general approach to estimate food aid demand as follows:

$$w_i^v = f\left(\gamma_i, \delta_j \mid j \in \Omega_i^\delta\right) \tag{1}$$

where $w_i^v$ is the demand anticipated area $i$ based on vegetation (denoted with the superscript $v$) and total population, $f()$ is a function, $\gamma_i$ is the population in area $i$, $\Omega_i^\delta$ is the set of neighbors of area $i$ likely impact its food insecurity, and $\delta_j$ is the vegetation index for neighboring area $j$. The specification of function $f()$ might vary across space and time, but will generally reflect that a higher vegetation index suggests more potential local food resources and less demand for food aid, whereas a larger population indicates more demand for food aid. In this study, we estimate $w_i^v$ as follows:

$$w_i^v = \gamma_i * \left(1 - \frac{\sum_{j \in \Omega_i^\delta} \delta_j}{\left|\Omega_i^\delta\right|}\right) \tag{2}$$

where $\Omega_i^\delta = \left\{ j \mid d_{ij} \leq 500\,\text{m} \right\}$ and $d_{ij}$ is the Euclidean distance between areas $i$ and $j$. Distance between areas is measured as the distance between the centroids of areas. There are 2,445,615 individuals making up the population in Bamako, resulting in a total of 1,466,701 food aid demand after being weighted by the vegetation index.

The spatial distribution of undernourished children is an important factor for determining areas of poverty and food insecurity within a city (FAO 2012). As a result, we also estimate food aid demand by children as follows:

$$w_i^c = f\left(\theta_i, \eta_k \mid k \in \Omega_i^\eta\right) \tag{3}$$

where $w_i^c$ is the demand estimate for the children population (denoted with the superscript $c$) in area $i$, $\theta_i$ is the children population in area $i$, $\Omega_i^\eta$ is the set of DHS clusters influential for estimating food insecurity in area $i$, and $\eta_k$ is the percentage of food-insecure children in DHS cluster $k$. Again, while the specification of function $f()$ might vary across space and time, higher percentage of food-insecure children and a larger population of children generally suggest more demand for food aid. In this study, we estimate $w_i^c$ as follows:

$$w_i^c = \theta_i * \left(\frac{\sum_{k \in \Omega_i^\eta} \eta_k}{\left|\Omega_i^\eta\right|}\right) \tag{4}$$

where $\Omega_i^\eta = \left\{ k \mid d_{ik} \leq 2\,\text{km} \right\}$ and $d_{ik}$ is the Euclidean distance between area $i$ and cluster $k$. Distance between areas and clusters is measured as the distance between the centroids of areas and clusters. Figure 2 demonstrates the process of deriving the set $\Omega_i^\eta$. A 2 km buffer is generated for each DHS cluster and then we identify the set $\Omega_i^\eta$ by overlaying the DHS buffer layer with the demand area. After determining

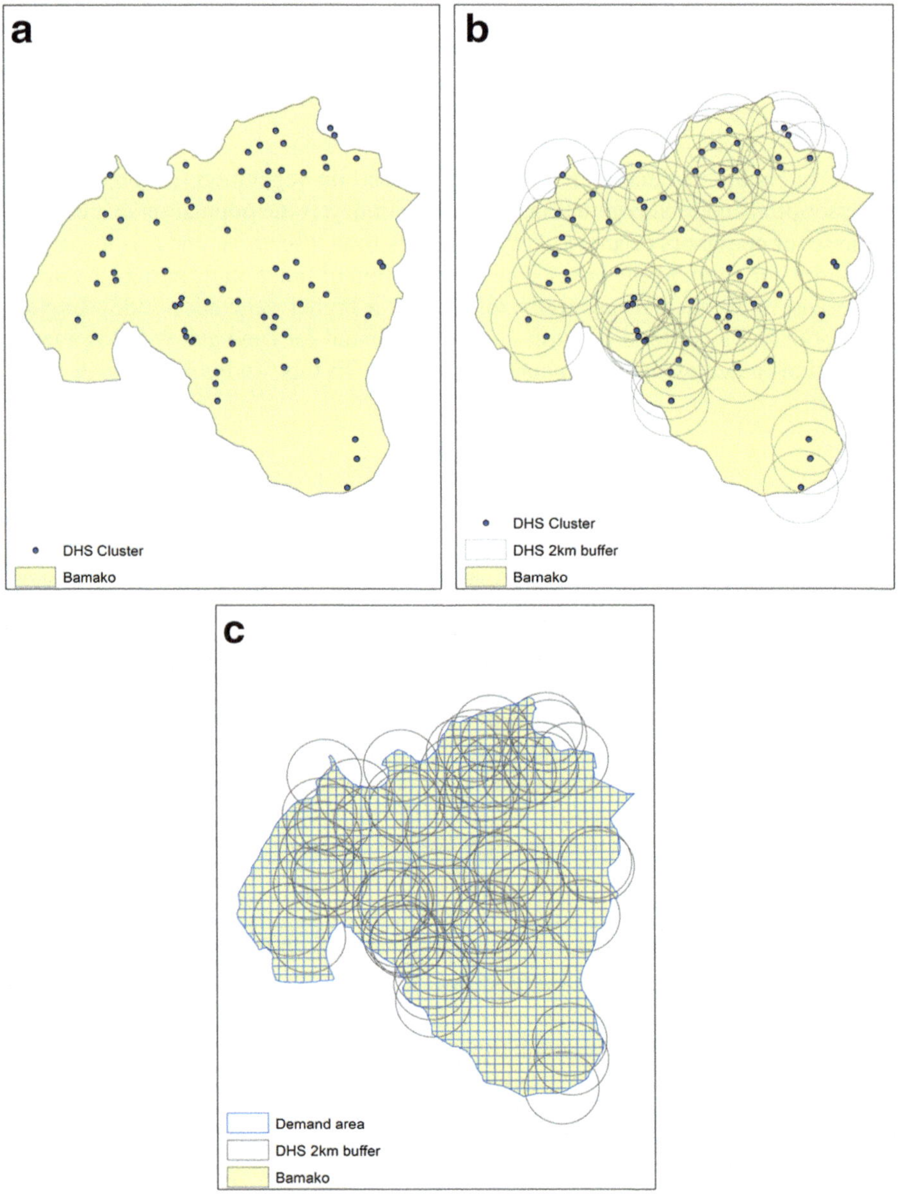


**Fig. 2** Demand estimate for children population

the set $\Omega_i^\eta$, the average food-insecure children percentage of the DHS clusters is estimated for each demand area. There are 289,762 children population in Bamako area, resulting in a total of 12,275 food aid demand after being weighted by the percentage of food-insecure children.

In addition to DHS, population and vegetation data, we relied upon road infrastructure data to determine potential food aid outlets. As discussed in Grace et al.

(2017), food outlets are often most accessible when they are sited near or along the road network. This facilitates delivery of food aid resources. Consistent with this work, the road network was used in the identification of potential outlet locations. Road network data can be obtained through online GIS databases, such as OpenStreetMap, World Street Map (Esri), etc. In this research, road network data is obtained from DIVA-GIS (http://diva-gis.org/). Locations along the major/primary road network components were identified as potential food distribution outlets. This was done approximately every 100 m along the road network. Figure 3 shows the study area delineated by demand area, along with the road network and potential outlet locations.

With demand for food aid and potential outlet locations specified, a spatial optimization model is used to identify the optimal locations for food distribution outlets so that average distance to demand areas from their closest outlet is a minimum. Consider the following notation:

$i$ = index of demand areas;
$n$ = index of potential food aid distribution outlets;
$\Psi$ = total budget limitation;
$\beta_n$ = cost associated with siting outlet $n$;
$d_{in}$ = travel distance from demand area $i$ to outlet $n$;

$$X_n = \begin{cases} 1 & \text{if an outlet is sited at potential location } n; \\ 0 & \text{otherwise;} \end{cases}$$

$$Z_{in} = \begin{cases} 1 & \text{if demand at area } i \text{ served by outlet at } n; \\ 0 & \text{otherwise;} \end{cases}$$

As indicated previously, the index $i$ represents demand areas and potential outlet locations are denoted using the index $n$. Potential outlets are assumed more accessible if they are along major roads. As both demand areas and outlet locations are finite and identified prior to the application of the model, the travel distance between them, $d_{in}$, can therefore be derived in advance as well. The binary decision variables $X_n$ represent whether potential location $n$ is selected for a food aid distribution outlet. The variables $Z_{in}$ are used to track the closest sited outlet for each demand area.

Given this notation, a bi-objective spatial optimization problem is used to support food aid distribution, and can be structured as follows:

$$\text{Minimize} \frac{\sum_i \sum_n w_i^v d_{in} Z_{in}}{\sum_i w_i^v} \tag{5}$$

$$\text{Minimize} \frac{\sum_i \sum_n w_i^c d_{in} Z_{in}}{\sum_i w_i^c} \tag{6}$$

$$\text{Subject to} \sum_j Z_{in} = 1, \quad \forall i \tag{7}$$

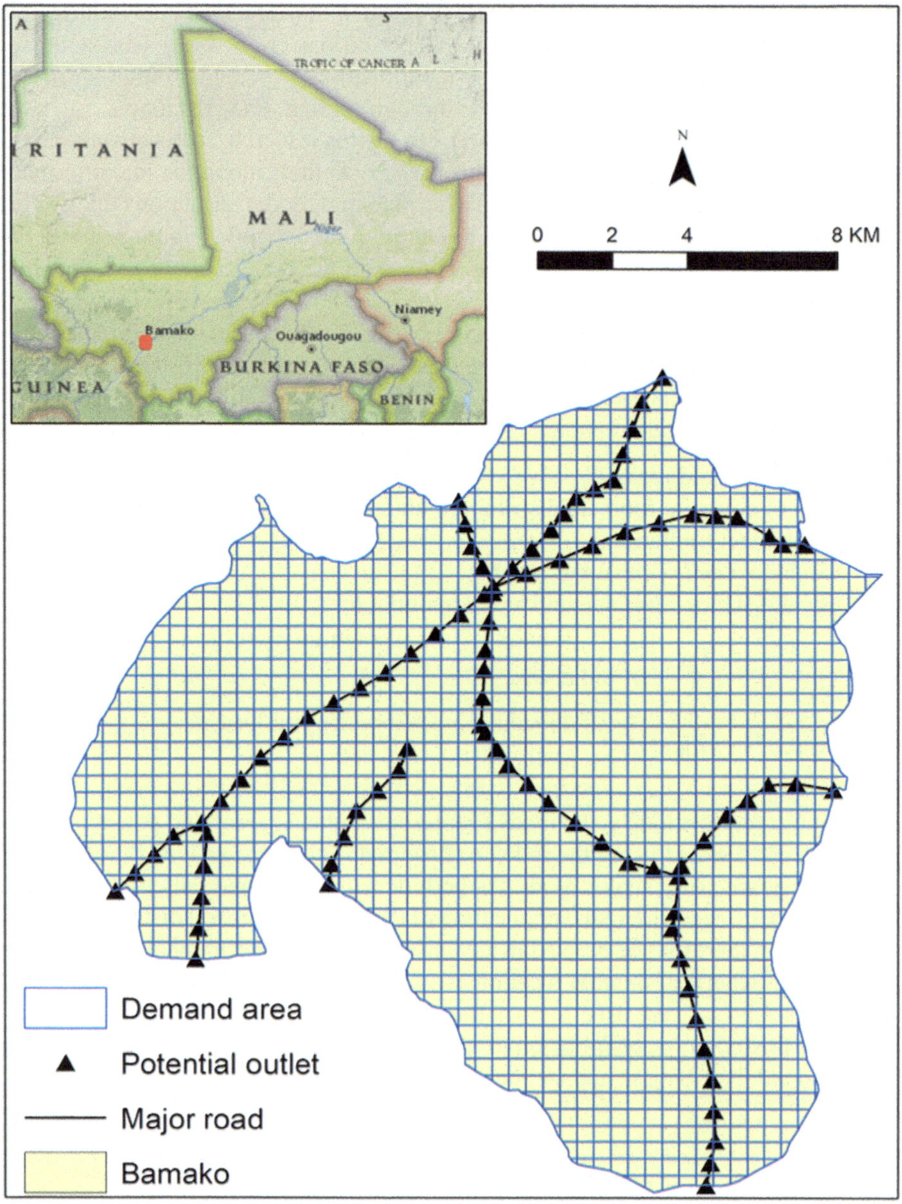


**Fig. 3** Study area

$$Z_{in} \le X_n, \quad \forall i, n \tag{8}$$

$$\sum_n \beta_n X_n \le \Psi \tag{9}$$

$$X_n = \{0,1\}, \quad \forall j \tag{10}$$

$$Z_{in} = \{0,1\}, \quad \forall i$$

The first objective of the model, (5), is to minimize the average travel distance of expected food aid demand based on vegetation index and general population, Eq. (2). The second objective, (6), is to minimize the average travel distance of expected food aid demand based on the number of food-insecure children, Eq. (4). Constraints (7) ensure that each demand cell is served by one outlet. Constraints (8) require that demand at cell *i* can be served by an outlet at *j* only if an outlet is sited at *j*. Constraint (9) sets a budget limitation on total food aid investment. Constraints (10) impose binary integer restrictions on decision variables.

This formulation can be thought of as an extension of the p-median problem detailed in ReVelle and Swain (1970) and Church and Murray (2009), where budget constraint (9) is used to impose limits on the number of outlets to be sited. A complication in solving this model is the existence of multiple objectives, but also problem size and other associated structural characteristics. A number of options are possible for generating the associated Pareto trade-off curve. One is the weighting method (see Cohon 1978). The two objectives in the model can be combined through the use of a weight, $\varepsilon$. Specifically, objectives (4) and (5) can be integrated as follows:

$$\text{Minimize}\,\varepsilon \frac{\sum_i \sum_n w_i^v d_{in} Z_{in}}{\sum_i w_i^v} + (1-\varepsilon)\frac{\sum_i \sum_n w_i^c d_{in} Z_{in}}{\sum_i w_i^c} \tag{11}$$

Such an approach can convert this bi-objective problem to a single objective, and can then be solved using commercial or open-source MIP solvers, such as Gurobi, CPlex, and GLPK. As the p-median and related models are NP-hard (Garey and Johnson 1979), heuristic methods are often needed when the problem size and structure exceeds the computational limits of exact mixed-integer programming solvers. A review of solution techniques for p-median problem can be found in Murray and Church (1996), Mladenović et al. (2007), Church (2008), and Li et al. (2011). When $\varepsilon = 1$, this model focuses solely on minimizing the average demand weighted travel distance, where demands are based on the vegetation index and total population. When $\varepsilon = 0$, the emphasis is on minimizing the average demand-weighted travel distance using the number of food-insecure children. By varying the weight between 0 and 1, trade-offs likely exist, representing Pareto solutions. Identifying and examining these trade-offs are essential for informed planning and decision-making for best serving those in need of food aid.

# 4 Results

ArcGIS along with ERDAS IMAGINE are used for spatial data acquisition, processing and manipulation. Additionally, Shapely (a Python geometry library) is used to support GIS operation in spatial optimization model specification for deriving proximity and other spatial relationships. Gurobi (a commercial optimization package) was used to identify optimal solutions for each problem instance. All processing and computation were done on a desktop personal computer (Intel Xeon E5 CPU, 2.30 GHz with 96 GB RAM).

Figure 4 shows the spatial distribution of estimated demand based on vegetation composition and total population, along with the number of food-insecure children. Significant demand is observed in central and northeast Bamako in both scenarios shown in Fig. 4. However, northwest Bamako also shows large need for aid to food-insecure children.

The spatial optimization model is applied to identify an optimal configuration of food distribution outlets under various investment scenarios. Here, we assume the cost of siting food outlets is the same across all potential locations, so $\beta_j$ are equal. For convenience, cost is assigned a value of one. As a result, $\Psi$ represents the total number of aid distribution outlets to be sited, consistent with the p-median problem formulated in ReVelle and Swain (1970). The number of outlets considered ranged from $p$ equal to 1 to 91.

Figure 5 shows the efficiency trade-off of aid outlets when $\varepsilon = 0$, 0.5 and 1. The $x$-axis shows the number of food aid distribution outlets to be sited, $\Psi$, and the

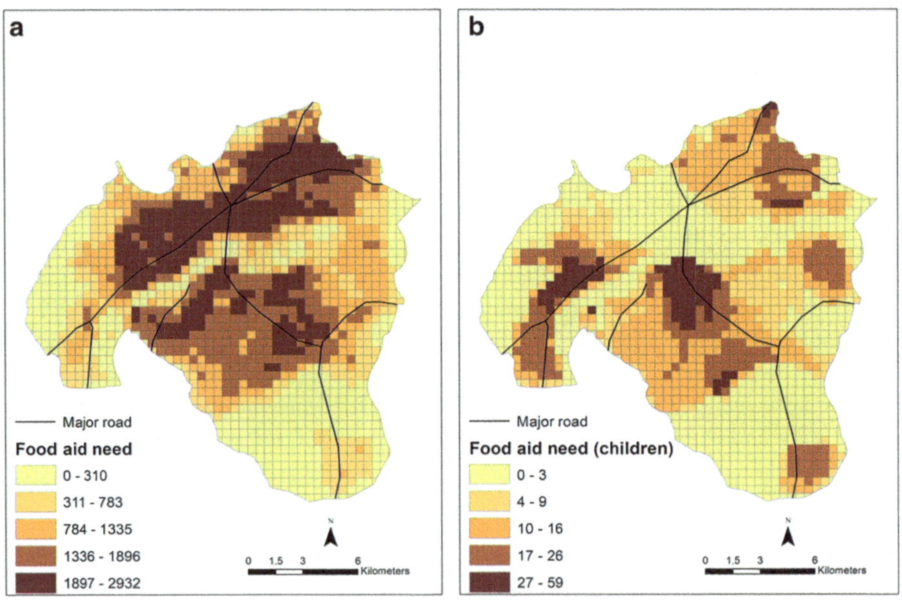


**Fig. 4** Spatial distribution of estimated demand (**a**) based on NDVI and general population (**b**) based on the number of food-insecure children

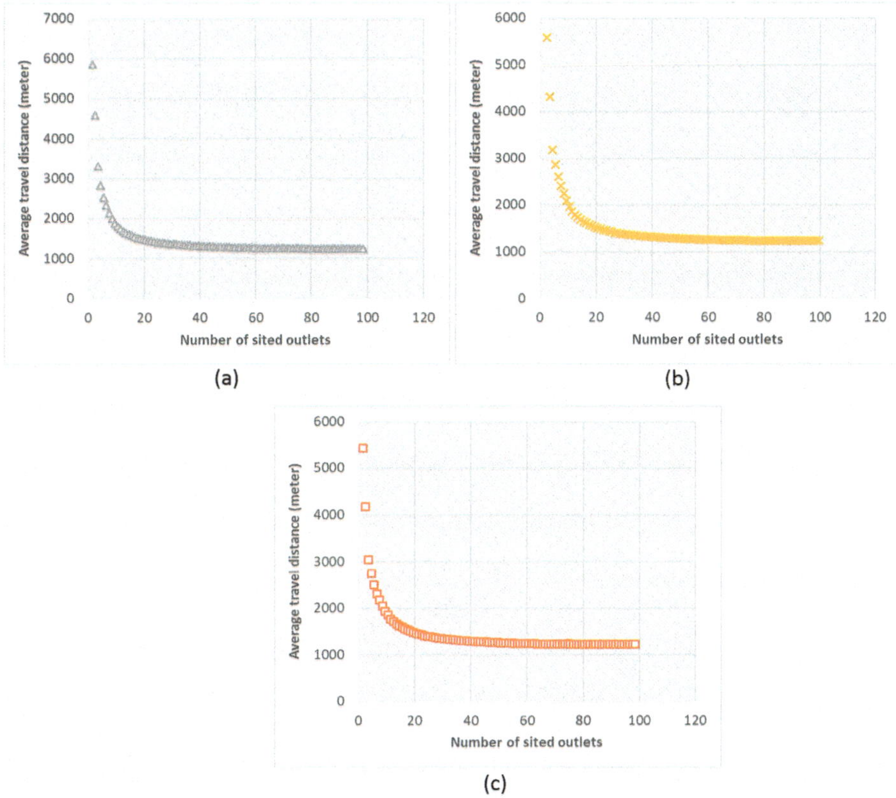


**Fig. 5** Locational efficiency trade-off of food aid distribution outlets (**a**) $\varepsilon = 0$ (**b**) $\varepsilon = 0.5$ (**c**) $\varepsilon = 1$

*y*-axis indicates the average travel distance from demand to the closest sited outlet, objective (10). When demand is solely estimated based on vegetation composition and total population ($\varepsilon = 1$), the average travel distance decreases from 5485 m where only one outlet is sited to 1267 m where 91 outlets are sited. Alternatively, when demand is solely estimated based on the number of food-insecure children ($\varepsilon = 0$), the average travel distance decreases from 5871 m for one outlet to 1277 m when 90 outlets are sited. When total population and children demand are equally weighted ($\varepsilon = 0.5$), the average travel distance decreases from 5631 m for one outlet to 1270 m when 91 outlets are sited. It is also interesting to note that access improves only marginally after 20 outlets in all three scenarios. For instance, 71 outlets are needed to reduce average travel distance by 249 m when $\varepsilon = 1$, and 70 outlets are needed to reduce average travel distance by 213 m when $\varepsilon = 0$. Again, the average distance measure is significant when total demand is considered, 1,466,701 in the case of vegetation weighted population, objective (5), and 12, 275 in the case of food-insecure children, objective (6). Average distance is therefore per person, so any difference found is significant considering the entire region.

Figure 6a shows the trade-offs between the two objectives when $\Psi = 1, 5, 10$, and 20. The *x*-axis represents the average travel distance for children demand,

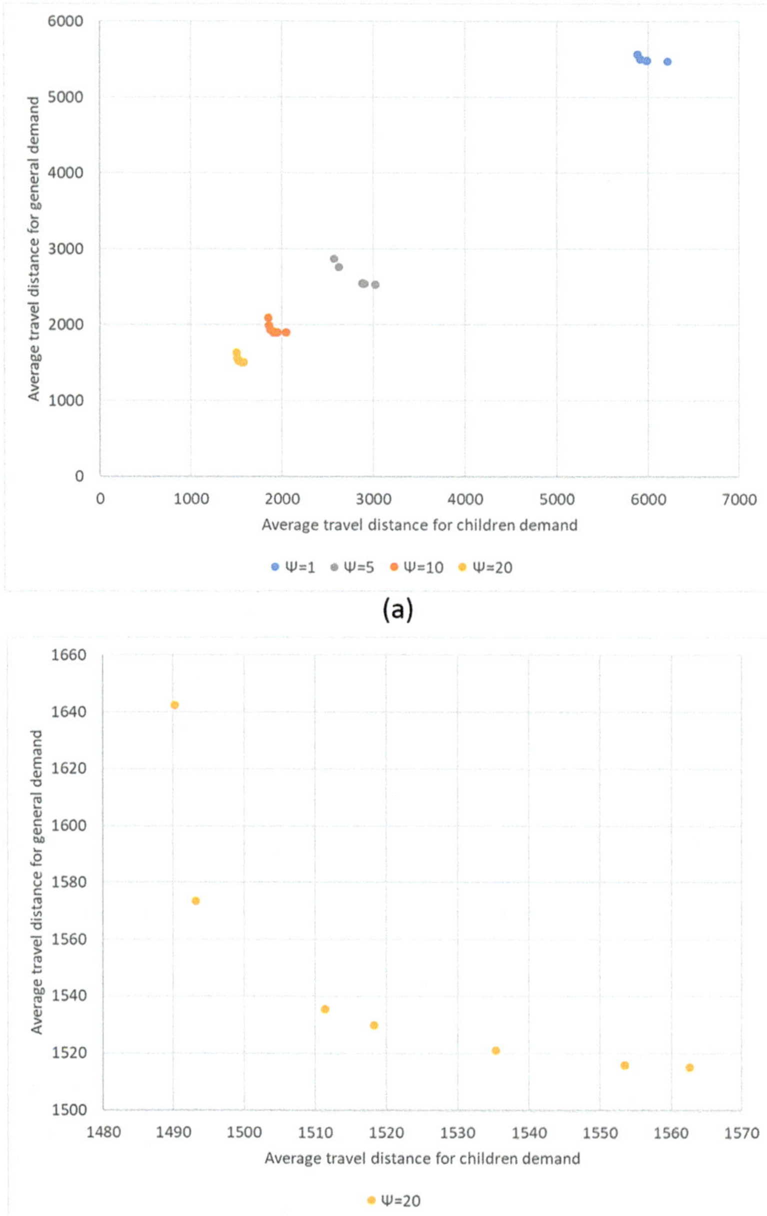


**Fig. 6** Locational efficiency trade-off between general and children demand

objective (5), and the *y*-axis is the average travel distance for total demand, objective (4). There are clear trade-offs between resulting average travel distance. A closer look at the trade-offs for $\Psi = 20$ is provided in Fig. 6b, where the average travel distance for total demand increases from 1490 to 1562 m as average travel distance for food-insecure children decreases from 1643 to 1516 m. The identified locations for 20 food distribution outlets is shown in Fig. 7. The optimal configuration varies across scenarios. For example, when $\varepsilon = 1$, six outlets are sited along a major road in north Bamako because the greatest amount of demand/need is distributed along

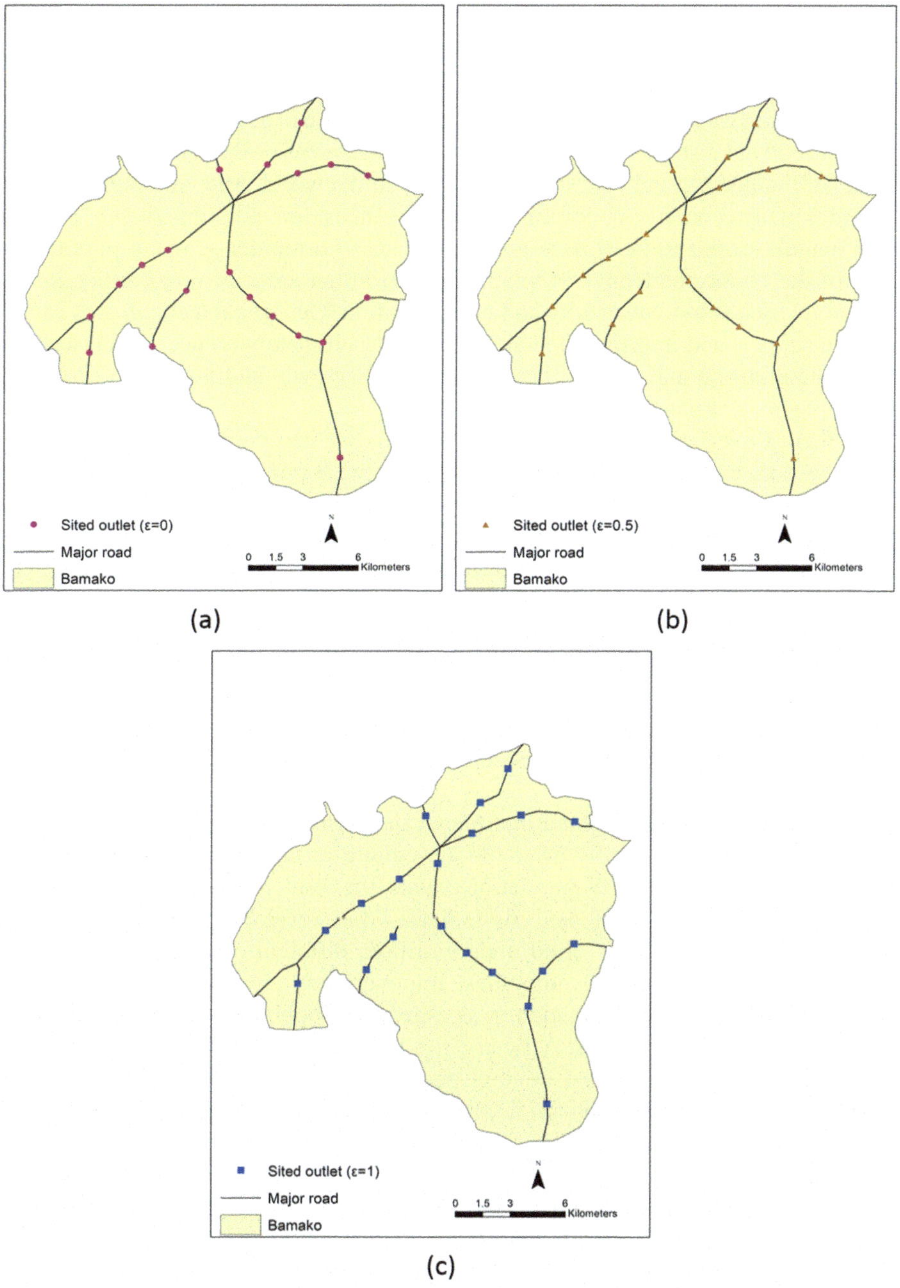


**Fig. 7** Spatial configuration of 20 sited outlets (**a**) $\varepsilon = 0$ (**b**) $\varepsilon = 0.5$ (**c**) $\varepsilon = 1$

this road. However, when $\varepsilon = 0$, the sited outlets along the major road are mainly distributed on the west and east. No outlet is sited in the middle portion. The reason is that food-insecure children demand is lower in this area.

## 5 Discussion

In this chapter, we investigated and demonstrated the use of geospatial technologies to support food aid distribution in urban and peri-urban areas of Bamako, Mali. Bamako is one of the poorest and fastest growing cities in the world. And while urbanization provides many economic opportunities for individuals, limited infrastructure in cities like Bamako strain to keep pace with the needs of a rapidly growing population. One of the major challenges in urban and peri-urban areas of major cities in sub Saharan Africa is food insecurity and undernutrition. Geospatial technologies can be used to support and improve food aid targeting to ultimately reduce food insecurity and undernutrition and support healthy individual growth and long-term economic development of urban centers.

In this chapter, we investigated the application of geospatial technologies using remotely sensed data of vegetation combined with population data to estimate demand for food aid based on the potential for urban agriculture to support nutritional demands. We also developed an estimate of demand using child anthropometric measures that reflect chronic undernutrition and household food insecurity. The results highlight the unique perspective provided by different demand estimates and the importance of considering that food insecurity and food aid need can arise from different processes, depending on who you are and where a household or individual is in an urban area. In other words, even within a relatively small geographic area (like Bamako), considering the neighborhood context and the ways that people gain access to food is a vitally important component of urban policies aimed at reducing food insecurity.

Past research has highlighted that fine-scale estimates of vegetation provide an important measure of locally available agriculture in urban, peri-urban, and rural West Africa. Combined with population counts, the results highlight areas of potential need for food aid. Our analysis indicates that developing demand estimates based on the prevalence of child undernutrition, different outcomes are likely to emerge. Using child health outcomes, impoverished areas may not always be characterized as lacking vegetation or as especially population dense. For example, peri-urban areas of large West African cities are often characterized by high levels of poverty (these areas often contain people or households who have recently migrated from rural areas and are in need of low-cost housing) and, in some cases, higher than central-urban levels of vegetation. However, because of poverty, distance to the urban center, and the complex challenges facing peri-urban areas with limited city infrastructure (often no access to piped water, no electricity, and costly commutes to jobs in the urban-center), the presence of local agriculture does not meet the nutritional needs of households.

Considering these different model results together allows analysts and policy-makers to reflect on different priorities for food aid distribution with respect to specific needs of a given population. For example, where there is a high prevalence of child undernutrition and where vegetation is low and the population count is high (northwestern area in Fig. 2), these areas might benefit most from food aid that is able to meet the nutrition needs of a variety of people at different life spots – children as well as adults. Areas that show child undernutrition only (southern area in Fig. 2) would likely benefit from locating food aid in specific areas to facilitate easy access for people with small children but also would ensure that the food aid is customized to meet the needs of young children.

As conditions change – population grows and becomes denser and the city expands into new areas – the approach we have developed here can be easily modified to accommodate new data or to incorporate different dimensions of food insecurity. Additionally, if different indicators of food insecurity or poverty or related factors are of interest for aid targeting, the models can also accommodate different demand specifications.

## 6 Conclusion

Urban and peri-urban communities in much of sub-Saharan Africa are rapidly growing and straining limited infrastructure along with expanding geographic boundaries. Partly as a natural response to these changes, many urban and peri-urban dwellers are often dependent on local agriculture and urban gardens to meet some of their nutrition and income needs. At the same time, food insecurity, often measured by child malnutrition outcomes, remains a persistent challenge to urban dwellers and the economic development of the city. Food aid provides one important means of reducing food insecurity in urban areas. Geospatial technologies can be used to improve food aid targeting and planning in urban areas. This has the potential to ultimately reduce geographic barriers to accessing food aid. In applying these technologies with readily available health survey data, timely and spatially detailed models of food aid can be developed to help guide interventions that will bring more food into people's lives.

This research represents an integration of different types of data to provide a quantitative perspective on food aid distribution. Importantly, our application engages with data that are already used to explore food insecurity and estimate food aid demand. However, there are many important limitations to the approach that we have proposed. Among the most important limitations are the lack of data on the type of food aid – different aid types likely require different logistical support, the lack of data on community characteristics and safety, and the lack of information about factors that may change seasonally, like road networks. Given these important limitations and constraints, we note that this quantitative approach is intended to give additional insight into food aid distribution and should be used as part of a multi-method decision-making process.

## References


Abuya, B. A., Ciera, J., & Kimani-Murage, E. (2012). Effect of mother's education on child's nutritional status in the slums of Nairobi. *BMC Pediatrics, 12*(1), 80.

Bhutta, Z., Das, J., Bahl, R., Lawn, J., Salam, R., Paul, V., Sankar, M., et al. (2014). Can available interventions end preventable deaths in mothers, newborn babies, and stillbirths, and at what cost? *The Lancet, 384*(9940), 347–370.

Black, R., Alderman, H., Bhutta, Z., Gillespie, S., Haddad, L., Horton, S., Lartey, A., et al. (2013). Maternal and child nutrition: Building momentum for impact. *The Lancet, 382*(9890), 6–375.

Brown, M. E., & McCarty, J. L. (2017). Is remote sensing useful for finding and monitoring urban farms? *Applied Geography, 80*, 23–33.

Brown, M.E., Antle, J.M., Backlund, P., Carr, E.R., Easterling, W.E., Walsh, M.K., Ammann, C., Attavanich, W., Barrett, C.B., Bellemare, M.F., Dancheck, V., Funk, C., Grace, K., Ingram, J.S.I., Jiang, H., Maletta, H., Mata, T., Murray, A., Ngugi, M., Ojima, D., O'Neill, B., & Tebaldi, C. (2015). Climate Change, Global Food Security, and the U.S. Food System. USDA Technical Document, Washington DC. https://doi.org/10.7930/J0862DC7.

Brown, M. E., Carr, E. R., Grace, K. L., Wiebe, K., Funk, C. C., Attavanich, W., et al. (2017). Do markets and trade help or hurt the global food system adapt to climate change? *Food Policy, 68*, 154–159.

Carroll, M.L., DiMiceli, R.A., Sohlberg, R.A., & Townshend, J.R.G. (2004). 250m MODIS Normalized Difference Vegetation Index, University of Maryland, College Park, Maryland, Day 289, 2003.

Castillo, G. E. (2003). Livelihoods and the city: An overview of the emergence of agriculture in urban spaces. *Progress in Development Studies, 3*(4), 339–344.

Castle, S., Scott, R., & Mariko, S. (2014). Child health and nutrition in Mali: Further analysis of the 2012–13 demographic and health survey. DHS Further Analysis Reports No. 92. Rockville, Maryland, USA: ICF International.

Church, R. L. (2008). BEAMR: An exact and approximate model for the p-median problem. *Computers & Operations Research, 35*(2), 417–426.

Church, R. L., & Murray, A. T. (2009). *Business site selection, location analysis, and GIS*. Hoboken: Wiley.

Clay, D. C., Molla, D., & Habtewold, D. (1999). Food aid targeting in Ethiopia: A study of who needs it and who gets it. *Food Policy, 24*(4), 391–409.

Cohon, J. L. (1978). *Multiobjective programming and planning*. New York: Academic Press.

FAO. (1996). *World Food Summit: Rome Declaration on World Food Security*. Rome: United Nations Food and Agriculture Organization.

FAO. (2012). *Food, agriculture and cities: The challenges of food and nutrition security, agriculture and ecosystem management in an urbanizing world*. In (p. 48). Rome, Italy: United Nations Food and Agriculture Organization.

FAO, IFAD & WFP. (2015). The State of Food Insecurity in the World 2015. Meeting the 2015 international hunger targets: Taking stock of uneven progress. Rome, Italy, FAO.

Garey, M. R., & Johnson, D. S. (1979). Computers and Intractability: A Guide to the Theory of NP-Completeness. New York: W. H. Freeman.

Gautam, Y., & Andersen, P. (2017). Aid or abyss? Food assistance programs (FAPs), food security and livelihoods in Humla, Nepal. *Food Security, 9*(2), 227–238.

Gelli, A., Meir, U., & Espejo, F. (2007). Does provision of food in school increase girls' enrollment? Evidence from schools in sub-Saharan Africa. *Food and Nutrition Bulletin, 28*(2), 149–155.

Gentilini, U. (2014). *Our daily bread: What is the evidence on comparing cash versus food transfers*. Washington, DC: The World Bank Group.

Gillespie, S., Haddad, L., Mannar, V., Menon, P., Nisbett, N., & Maternal and Child Nutrition Study Group. (2013). The politics of reducing malnutrition: Building commitment and accelerating progress. *The Lancet, 382*(9891), 552–569.

Grace, K., Wei, R., & Murray, A. T. (2017). A spatial analytic framework for assessing and improving food aid distribution in developing countries. *Food Security, 9*(4), 867–880.

Grace, K., Lerner, A. M., Mikal, J., & Sangli, G. (2017b). A qualitative investigation of child-bearing and seasonal hunger in peri-urban Ouagadougou, Burkina Faso. *Population and Environment, 38*(4), 369–380.
Hampshire, K., Panter-Brick, C., Kilpatrick, K., & Casiday, R. (2009). Saving lives, preserving livelihoods: Understanding risk, decision-making and child health in a food crisis. *Social Science and Medicine, 68*(4), 758–765.
Hidrobo, M., Hoddinott, J., Peterman, A., Margolies, A., & Moreira, V. (2014). Cash, food, or vouchers? Evidence from a randomized experiment in northern Ecuador. *Journal of Development Economics, 107*, 144–156.
ICF International. (2013). *Demographic and health surveys Mali*. Rockville: ICF International.
Jaspars, S., & Young, H. (1996). *General food distribution in emergencies: From nutritional needs to political priorities*. UK: Overseas Development Institute (ODI).
Jayne, T. S., Strauss, J., Yamano, T., & Molla, D. (2001). Giving to the poor? Targeting of food aid in rural Ethiopia. *World Development, 29*(5), 887–910.
Lall, S. V., Henderson, J. V., & Venables, A. J. (2017). *Africa's cities: Opening doors to the world*. Washington, DC: World Bank. © World Bank. https://openknowledge.worldbank.org/handle/10986/25896 License: CC BY 3.0 IGO.
Lentz, E., & Barrett, C. (2013). The economics and nutritional impacts of food assistance policies and programs. *Food Policy, 42*, 151–163.
Lerner, A. M., & Eakin, H. (2011). An obsolete dichotomy? Rethinking the rural–urban interface in terms of food security and production in the global south. *The Geographical Journal, 177*(4), 311–320.
Leroy, J. L., Ruel, M., & Verhofstadt, E. (2009). The impact of conditional cash transfer programmes on child nutrition: A review of evidence using a programme theory framework. *Journal of Development Effectiveness, 1*(2), 103–129.
Li, X., Xiao, N., Claramunt, C., & Lin, H. (2011). Initialization strategies to enhancing the performance of genetic algorithms for the p-median problem. *Computers & Industrial Engineering, 61*(4), 1024–1034.
Maxwell, D., Parker, J., & Stobaugh, H. (2013). What drives program choice in food security crises? Examining the "response analysis". *question. World Development, 49*, 68–79.
Misselhorn, A., Aggarwal, P., Ericksen, P., Gregory, P., Horn-Phathanothai, L., Ingram, J., & Wiebe, K. (2012). A vision for attaining food security. *Current Opinion in Environmental Sustainability, 4*(1), 7–17.
Mladenović, N., Brimberg, J., Hansen, P., & Moreno-Pérez, J. A. (2007). The p-median problem: A survey of metaheuristic approaches. *European Journal of Operational Research, 179*(3), 927–939.
Murray, A. T., & Church, R. L. (1996). Applying simulated annealing to location-planning models. *Journal of Heuristics, 2*(1), 31–53.
Pinstrup-Andersen, P. (2009). Food security: Definition and measurement. *Food Security, 1*(1), 5–7.
Rancourt, M.-È., Cordeau, J., Laporte, G., & Watkins, B. (2015). Tactical network planning for food aid distribution in Kenya. *Computers and Operations Research, 56*, 68–83.
ReVelle, C. S., & Swain, R. W. (1970). Central facilities location. *Geographical Analysis, 2*(1), 30–42.
Sen, A. (1990). Food, economics and entitlements. In J. Dreze & A. Sen (Eds.), *The political economy of hunger* (pp. 10–45). New York: Clarendon Press.
Sen, A. (1997). Entitlement perspectives on hunger. In *Ending the inheritance of hunger*. Rome: World Food Programme.
Smith, L. C., El Obeid, A., & Jensen, H. (2000). The geography and causes of food insecurity in developing countries. *Agricultural Economics, 22*(2), 199–215.
Tong, D., & Murray, A. T. (2012). Spatial optimization in geography. *Annals of the Association of American Geographers, 102*(6), 1290–1309.
USAID. (2013). US Agency for International Development. New approaches to food assistance fact sheet. Downloaded July 11, 2016.

USAID. (2014). US Agency for International Development. How title II food aid works. Downloaded July 11, 2016.
Van de Poel, E., O'Donnell, O., & Van Doorslaer, E. (2007). Are urban children really healthier? Evidence from 47 developing countries. *Social Science & Medicine, 65*(10), 1986–2003.
Violette, W. J., Harou, A., Upton, J., Bell, S., Barrett, C., Gómez, M., & Lentz, E. (2013). Recipients' satisfaction with locally procured food aid rations: Comparative evidence from a three country matched survey. *World Development, 49*, 30–43.
Zezza, A., & Tasciotti, L. (2010). Urban agriculture, poverty, and food security: Empirical evidence from a sample of developing countries. *Food Policy, 35*(4), 265–273.

**Kathryn Grace** is an Associate Professor in the Department of Geography, Environment and Society and a faculty affiliate at the Minnesota Population Center, both at the University of Minnesota, Twin Cities, USA. Dr. Grace's research focuses on the way that contextual and environmental factors impact the lives of women and children. She relies on a diverse set of statistical and mathematical models to investigate complex relationships between individuals, their communities, and the local environmental context.

**Alan Murray** is a Professor in the Department of Geography at University of California, Santa Barbara, USA. Dr. Murray's research addresses public service systems, emergency response, public health, transportation, natural resource management, and urban growth and development through the development and application of spatial analytics, including GIS, spatial optimization, and spatial statistics, among other methods.

**Ran Wei** is an Assistant Professor in the School of Public Policy and a founding faculty of the Center for Geospatial Sciences at the University of California, Riverside, USA. Dr. Wei's areas of emphasis include GIScience, urban and regional analysis, spatial analysis, optimization, geovisualization, high-performance computing, and location analysis. Substantively, she has focused on a range of national and international issues, including urban/regional growth, transportation, public health, crime, housing mobility, energy infrastructure, and environmental sustainability.

# Index




Y. Lu, E. Delmelle (eds.), *Geospatial Technologies for Urban Health*,
Global Perspectives on Health Geography, https://doi.org/10.1007/978-3-030-19573-1